Lecture Notes in Artificial Intelligence 10654

Subseries of Lecture Notes in Computer Science

LNAI Series Editors

Randy Goebel
  University of Alberta, Edmonton, Canada
Yuzuru Tanaka
  Hokkaido University, Sapporo, Japan
Wolfgang Wahlster
  DFKI and Saarland University, Saarbrücken, Germany

LNAI Founding Series Editor

Joerg Siekmann
  DFKI and Saarland University, Saarbrücken, Germany

More information about this series at http://www.springer.com/series/1244

Yi Zeng · Yong He · Jeanette Hellgren Kotaleski
Maryann Martone · Bo Xu · Hanchuan Peng
Qingming Luo (Eds.)

Brain Informatics

International Conference, BI 2017
Beijing, China, November 16–18, 2017
Proceedings

 Springer

Editors
Yi Zeng
Chinese Academy of Sciences
Beijing
China

Yong He
Beijing Normal University
Beijing
China

Jeanette Hellgren Kotaleski
KTH Royal Institute of Technology and
  Karolinska Institute
Stockholm
Sweden

Maryann Martone
University of California, San Diego
San Diego, CA
USA

Bo Xu
Chinese Academy of Sciences
Beijing
China

Hanchuan Peng
Allen Institute for Brain Science
Seattle, WA
USA

Qingming Luo
Wuhan National Lab Optoelectronics
Huazhong University of Science and
  Technology
Wuhan
China

ISSN 0302-9743   ISSN 1611-3349   (electronic)
Lecture Notes in Artificial Intelligence
ISBN 978-3-319-70771-6   ISBN 978-3-319-70772-3   (eBook)
https://doi.org/10.1007/978-3-319-70772-3

Library of Congress Control Number: 2017957855

LNCS Sublibrary: SL7 – Artificial Intelligence

Printed on acid-free paper

This Springer imprint is published by Springer Nature
The registered company is Springer International Publishing AG
The registered company address is: Gewerbestrasse 11, 6330 Cham, Switzerland

Preface

This volume contains papers selected for presentation at the technical and invited special sessions of the 2017 International Conference on Brain Informatics (BI 2017), which was held at the Grand Gongda Jianguo Hotel, Beijing, China, during November 16–18, 2017. The conference was co-organized by Beijing University of Technology, Institute of Automation, Chinese Academy of Sciences, Web Intelligence Consortium (WIC), and IEEE Computational Intelligence Society Task Force on Brain Informatics (IEEE-CIS TF-BI).

Brain informatics (BI) started the exploration as a research field with the vision of investigating the brain from the perspective of informatics. First, brain informatics combines aspects of cognitive science, neuroscience, machine learning, big data analytics, AI, and ICT to study the brain as a general information processing system. Second, new measurements and informatics equipment, tools, and platforms are revolutionizing the way we understand the brain. Third, staring from its proposal as a field, brain informatics has the goal of inspiring future artificial intelligence, especially Web intelligence.

The series of Brain Informatics conferences started with the WICI International Workshop on Web Intelligence Meets Brain Informatics, held in Beijing, China in 2006. The next four conferences of Brain Informatics in 2009, 2010, 2011, and 2012 were held in Beijing, China, Toronto, Canada, Lanzhou, China, and Macau, China, respectively. Since 2013, "health" was added to the conference title with an emphasis on real-world applications of brain research in human health and well-being and BHI 2013, BIH 2014, BIH 2015, and BIH 2016 were held in Maebashi, Japan, Warsaw, Poland, London, UK, and Omaha, USA, respectively. This year, we celebrated the tenth anniversary of Brain Informatics. This grand event in Beijing was co-hosted by the Beijing University of Technology, and the Institute of Automation, Chinese Academy of Sciences.

BI addresses the computational, cognitive, physiological, biological, physical, ecological, and social perspectives of brain informatics, as well as topics related to mental health and well-being. It also welcomes emerging information technologies, including but not limited to Internet/Web of Things (IoT/WoT), cloud computing, big data analytics, and interactive knowledge discovery related to brain research. BI also encourages research exploring how advanced computing technologies are applied to and make a difference in various large-scale brain studies and their applications.

Informatics-enabled studies are transforming brain science. New methodologies enhance human interpretive powers when dealing with big data sets increasingly derived from advanced neuro-imaging technologies, including fMRI, PET, MEG, EEG, and fNIRS, as well as from other sources like eye-tracking and wearable, portable, micro and nano devices. New experimental methods, such as in toto imaging, deep tissue imaging, opto-genetics and dense-electrode recording, are generating massive amounts of brain data at very fine spatial and temporal resolutions. These technologies allow for the measuring, modeling, managing, and mining of multiple forms of big brain data. Brain informatics techniques for the analysis of the data will help achieve a better understanding of human thought, memory, learning, decision-making, emotion, consciousness, and social behaviors. These methods and related studies will also assist in building brain-inspired intelligence, brain-inspired computing, and human-level wisdom-computing paradigms and technologies, thereby improving the treatment efficacy of mental health and brain disorders.

BI 2017 provided a broad forum for academics, professionals, and industry representatives to exchange their ideas, findings, and strategies in utilizing the powers of the mammalian brain, especially human brains and man-made networks to create a better world. In the 2017 version of the Brain Informatics conference, we argued that brain research is a grand challenge to all countries and various scientific communities. Its potential impact not only will help us to understand who we are, and provide us with a better life, but will also enable a paradigm shift in the new intelligence era. Hence, future brain research needs joint efforts and collaboration from different countries worldwide. It also calls for interdisciplinary studies to investigate the brain from various perspectives.

Therefore, the theme of BI 2017 was "Informatics Perspective of Investigation on the Brain and Mind."

BI 2017 involved an inspiring cadre of world leaders in brain informatics research, including keynote speakers Alan Evans, James McGill Professor at McGill University, Canada, Tom Mitchell, E. Fredkin University Professor at Carnegie Mellon University, USA, and feature speakers Yanchao Bi, Professor at Beijing Normal University, China, Adam R. Ferguson, Associate Professor at University of California, San Francisco, USA, Bin Hu, Professor at Lanzhou University, China, Mike Hawrylycz, Investigator at Allen institute for Brain Science, USA, and Dinggang Shen, Professor at University of North Carolina at Chapel Hill, USA. BI 2017 also included a panel discussion among the leaders of brain informatics research worldwide.

Here we would like to express our gratitude to all members of the Conference Committee for their instrumental and unwavering support. BI 2017 had a very exciting program with a number of features, ranging from keynote talks to technical sessions, workshops/special sessions, and panel discussion. This would not have been possible without the generous dedication of the Program Committee members in reviewing the conference papers and abstracts, the BI 2017 workshop and special session chairs and organizers, and our keynote and feature speakers in giving outstanding talks at the conference. BI 2017 could not have taken place without the great team effort of the local Organizing Committee and generous support from our sponsors. We would especially like to express our sincere appreciation to our kind sponsors, including the Beijing Advanced Innovation Center for Future Internet Technology/BJUT (http://bjfnc.bjut.edu.cn), the Faculty of Information Technology/BJUT (http://xxxb.bjut.edu.cn), the Research Center for Brain-inspired Intelligence, Institute of Automation, Chinese Academy of Sciences (http://bii.ia.ac.cn), the Web Intelligence Consortium (http://wi-consortium.org), the Chinese Society for Cognitive Science (http://www.cogsci.org.cn), the Chinese Association for Artificial Intelligence (http://caai.cn), the International Neural Network Society (https://www.inns.org), the IEEE Computational

Intelligence Society (http://cis.ieee.org), the Allen Institute for Brain Science (https://alleninstitute.org), Springer LNCS/LNAI (http://www.springer.com/gp/computer-science/lncs), PsyTech Electronic Technology Co., Ltd. (http://www.psytech.com.cn), Beijing 7Invensun Technology Co., Ltd. (https://www.7invensun.com), John Wiley & Sons, Inc. (http://www.wiley.com), and Synced Technology Inc. (https://syncedreview.com).

Special thanks go to the Steering Committee co-chairs, Ning Zhong and Hanchuan Peng, for their help in organizing and promoting BI 2017. We also thank Juzhen Dong, Yang Yang, Jiajin Huang, Tielin Zhang, and Liyuan Zhang for their assistance with the CyberChair submission system and local affairs. We are grateful to Springer's *Lecture Notes in Computer Science* (LNCS/LNAI) team for their support. We thank Springer for their help in coordinating the publication of this special volume in an emerging and interdisciplinary research field.

September 2017

Yi Zeng
Yong He
Jeanette Hellgren Kotaleski
Maryann Martone
Bo Xu
Hanchuan Peng
Qingming Luo

Organization

General Chairs

Bo Xu   Chinese Academy of Sciences, China
Hanchuan Peng   Allen Institute for Brain Sciences, USA
Qingming Luo   Huazhong University of Science and Technology, China

Program Committee Chairs

Yi Zeng   Chinese Academy of Sciences, China
Yong He   Beijing Normal University, China
Jeanette H. Kotaleski   KTH Royal Institute of Technology, Karolinska Institute,
                         Sweden
Maryann Martone   University of California, San Diego, USA

Organizing Chairs

Ning Zhong   Maebashi Institute of Technology, Japan
Jianzhou Yan   Beijing University of Technology, China
Shengfu Lu   Beijing University of Technology, China

Workshop/Special Session Chairs

An'an Li   Huazhong University of Science and Technology, China
Sen Song   Tsinghua University, China

Tutorial Chair

Wenming Zheng   South East University, China

Publicity Chairs

Tielin Zhang   Chinese Academy of Sciences, China
Yang Yang   Maebashi Institute of Technology, Japan, and Beijing
                     University of Technology, China
Shouyi Wang   University of Texas at Arlington, USA

Steering Committee Co-chairs

Ning Zhong   Maebashi Institute of Technology, Japan
Hanchuan Peng   Allen Institute for Brain Science, USA

WIC Co-chairs/Directors

Ning Zhong   Maebashi Institute of Technology, Japan
Jiming Liu   Hong Kong Baptist University, SAR China

IEEE-CIS TF-BI Chair

Ning Zhong   Maebashi Institute of Technology, Japan

Program Committee

Samina Abidi   Dalhousie University, Canada
Asan Agibetov   Italian National Research Council, University of Genova,
                       Italy
Jun Bai   Institute of Automation, Chinese Academy of Sciences,
                 China
Olivier Bodenreider   US National Library of Medicine, USA
Weidong Cai   The University of Sydney, Australia
Mirko Cesarin   University Milano-Bicocca, Italy
Phoebe Chen   La Trobe University, Australia
Netta Cohen   University of Leeds, UK
Hong-Wei Dong   University of Southern California, USA
Yuhong Feng   Shenzhen University, China
Huiguang He   Institute of Automation, Chinese Academy of Sciences,
                      China
Zhisheng Huang   Vrije University of Amsterdam, The Netherlands
Qingqun Kong   Institute of Automation, Chinese Academy of Sciences,
                       China
Peipeng Liang   Xuanwu Hospital, Capital Medical University, China
Sidong Liu   School of Information Technologies, University of
                     Sydney, Australia
Xin Liu   Institute of Automation, Chinese Academy of Sciences,
                 China
Roussanka Loukanova   Stockholm University, Sweden
Antonio Moreno-Ribas   Rovira i Virigili University, Spain
David M.W. Powers   Flinders University, Australia
Shouyi Wang   University of Texas at Arlington, USA
Abdel-Badeeh Salem   Ain Shams University, Egypt
Dinggang Shen   UNC School of Medicine, USA
Bailu Si   Shenyang Institute of Automation, Chinese Academy
                   of Sciences, China
Andrzej Skowron   Warsaw University, Poland
Dominik Slezak   University of Warsaw, Poland
Neil Smalheiser   University of Illinois, USA
Diego Sona   Instituto Italiano di Tecnologia, Italy
Marcin Szczuka   University of Warsaw, Poland

Predrag Tosic	Washington State University, USA
Egon L. Van den Broek	Utrecht University, The Netherlands
ZhiJiang Wan	Maebashi Institute of Technology, Japan
Guoyin Wang	Chongqing University of Posts and Telecommunications, China
Yuwei Wang	Institute of Automation, Chinese Academy of Sciences, China
Zhijiang Wang	Peking University Sixth Hospital, China
Mingrui Xia	Beijing Normal University, China
Yong Xia	Northwestern Polytechnical University, China
Tianming Yang	Institute of Neuroscience, Chinese Academy of Sciences, China
Yang Yang	Maebashi Institute of Technology, Japan
Yiyu Yao	University of Regina, Canada
Kaori Yoshida	Kyushu Institute of Technology, Japan
Fabio Massimo Zanzotto	University of Rome Tor Vergata, Italy
Qian Zhang	Institute of Automation, Chinese Academy of Sciences, China
Tielin Zhang	Institute of Automation, Chinese Academy of Sciences, China
Yanqing Zhang	Georgia State University, USA
Ning Zhong	Maebashi Institute of Technology, Japan
Haiyan Zhou	Beijing University of Technology, China
Hongyin Zhu	Institute of Automation, Chinese Academy of Sciences, China

Contents

Cognitive and Computational Foundations of Brain Science

Human Information Processing Systems

Special Session on Brain Informatics in Neurogenetics (BIN 2017)

Cognitive and Computational Foundations of Brain Science

Speech Emotion Recognition Using Local and Global Features

Yuanbo Gao[1], Baobin Li[1(✉)], Ning Wang[2], and Tingshao Zhu[3]

[1] School of Computer and Control, University of Chinese Academy of Sciences,
Beijing 100190, China
libb@ucas.ac.cn
[2] Beijing Institue of Electronics Technology and Application, Beijing 100091, China
[3] Institute of Psychology Chinese Academy of Sciences, Beijing 100101, China
tszhu@psych.ac.cn

Abstract. Speech is an easy and useful way to detect speakers' mental and psychological health, and automatic emotion recognition in speech has been investigated widely in the fields of human-machine interaction, psychology, psychiatry, etc. In this paper, we extract prosodic and spectral features including pitch, MFCC, intensity, ZCR and LSP to establish the emotion recognition model with SVM classifier. In particular, we find different frame duration and overlap have different influences on final results. So, Depth-First-Search method is applied to find the best parameters. Experimental results on two known databases, EMODB and RAVDESS, show that this model works well, and our speech features are enough effectively in characterizing and recognizing emotions.

Keywords: Speech · Emotion · SVM · EMODB · RAVDESS

1 Introduction

As a natural and effective way of human communication, speech is an important biometric feature classifying speakers into categories ranging from age, identity, idiolect and sociolect, truthfulness, cognitive health. In particular, it is also one of the most expressive modalities for human emotions, and how to recognize emotion automatically from human speech has been widely discussed in the fields of the human-machine interaction, psychology, psychiatry, behavioral science, etc. [1,2].

Finding and extracting good and suitable features is one of challenging and important tasks in the study of speech emotion recognition. In 1989, Cummings et al. reported that the shape of the glottal pulse varies with different stressed conditions in a speech signal [3]. Seppänen et al. extracted pitch, formant and energy features on MediaTeam emotional speech corpus, and got the 60% accuracy [4]. In 2005, Luengo et al. used prosodic features to recognize speech emotion based on Basque Corpus with the three different classifiers, and the best result

© Springer International Publishing AG 2017
Y. Zeng et al. (Eds.): BI 2017, LNAI 10654, pp. 3–13, 2017.
https://doi.org/10.1007/978-3-319-70772-3_1

was up to 98.4% by using GMM classifier. Origlia et al. extracted 31 dimensional pitch and energy features, and got almost 60% accuracy for a multilingual emotional database including four European languages in 2010 [5].

Moreover, prosodic and spectral features including the linear predictor cepstral coefficients (LPCC) [6] and mel-frequency cepstral coefficients (MFCC) [7] have been also widely used in speech emotion recognition [8,9]. In 2000, Bou-Ghazale and Hansen [10] found that features based on cepstral analysis, such as LPCC and MFCC, outperformed ones extracted by linear predictor coefficients (LPC) [11] in detecting speech emotions. In 2003, New et al. proved that the performance of log-frequency power coefficient was better than LPCC and MFCC when using hidden Markov model as a classifier [12]. Wu et al. proposed modulation spectral features for the automatic recognition of human affective information from speech, and an overall recognizing rate of 91.6% was obtained for classifying seven emotions [13].

In addition, voice quality features are also related to speech emotion including format frequency and bandwidth, jitter and shimmer, glottal parameter, etc. [14,15]. Li et al. extracted jitter, shimmer features mixed with MFCC features as voice quality parameters to identify emotions on SUSAS database, and compared to MFCC, the accuracy increased by 4% [14]. Lugger et al. combined prosodic features with voice quality parameters for speech emotion recognition and the best accuracy was up to 70.1% with two ways of combining classifiers [15].

Recently, the combination of many kinds of features has been widely used for automatic speech emotion recognition. In 2012, Pan et al. found the combination of MFCC, mel-energy spectrum dynamic coefficients and Energy, obtained high accuracies on both Chinese emotional database (91.3%) and Berlin emotional database (95.1%) [16]. Chen et al. extracted the energy, zero crossing rate (ZCR), pitch, the first to third formants, spectrum centroid, spectrum cut-off frequency, correlation density, fractal dimension, and five Mel-frequency bands energy for every frame. Average accuracies were 86.5%, 68.5% and 50.2%, respectively [17]. In 2013, Deng et al. merged ZCR, MFCC and harmonics-to-noise features, and used the autoencoder method to classify emotions [18]. In 2014, Han et al. proposed to utilize deep neutral networks to extract MFCC and pitch-based features from raw data on Emotional Dyadic Motion Capture to distinguish excitement, frustration, happiness, neutral and surprise emotions, and the best accuracy obtained by DNN-KELM was 55% [19].

In this paper, we extract different features including MFCC, intensity, pitch, Line spectral pairs (LSP) and ZCR features, which have been proved to be effective, and then, a smooth and normalization operation is needed to reduce the noise. The output with delta regression coefficients are regarded as local features. In the end, we use 15 statistics methods to extract global features based on local features. In particular, we find the features of samples not only can effect the results of classification, but the frame duration and overlap are also two key factors, so the Depth First Search (DFS) is used to select frame duration and overlap and find the most appropriate combination between features and frame duration and overlap, which works well on two public databases: EMODB and RAVDESS.

This paper is organized as follows. Section 2 introduces and describes the information of two databases: EMODB and RAVDESS. Section 3 presents the process of extracting features including local and global features. Section 4 shows the classified results, and the average accuracy of cross-validation is 87.3% for EMODB and 79.4% for RAVDESS respectively. Finally, the conclusion and future work are outlined in Sect. 5.

2   Materials and Methods

2.1   Database

In this paper, two databases are considered: Berlin emotional corpus (EMODB) [20] and Ryerson Audio-Visual Database of Emotional Speech and Song (RAVDESS) [21]. EMODB has been built by Institute of Communication Science at the Technical University of Berlin, which contains 535 utterances of 10 actors (5 males, 5 females) with 7 different emotions—angry, anxiety/fear, happiness, sadness, disgust, boredom and neutral. The distributions of each class are shown in Table 1.

Table 1. The distributions of EMODB Database

Emotion	Angry	Boredom	Disgust	Fear	Happy	Neutral	Sad
Number	127	81	46	69	71	79	62

RAVDESS includes the audio-visual of 24 performers (12 male, 12 female) speaking and singing the same two sentences with various emotions, where the speech recordings include angry, calm, disgust, fearful, happy, neutral, sad, surprise eight emotions, and the song recording have six emotions except for digust and surprise emotions. RAVDESS contains over 7000 files (audio-only, video-only, full audio-video). We only use audio utterances with normal intensity, and the number of each class is 92.

2.2   Features for Speech Emotion Recognition

We choose frame duration and overlap by DFS. The range of frame durations is from 20 ms to 100 ms, and every frame durations acts as a father node. Every overlaps is a child node and start at 10 ms, increasing by 1 ms each time until equal to the half of its father node.

Every speech sample is divided into a sequence of frames. A hamming window is applied to remove the signal discontinuities at the ends of each frame. Within each window, we extract pitch, LSP, MFCC, intensity and ZCR features, and combine these features with a smooth operation. The output with their delta regression coefficients are our speech features. The following subsections will describe these features in details.

Local Features. The local features are comprised of pitch, MFCC, LSP, Intensity and ZCR features.

Pitch Features. Each frame $x^i(t)$ pass a hamming window, and a short-time Fourier transform function is applied on it,

$$H(w) = \sum_n^N x^i(t) * Ham(len(x^i(t))) * e^{-jwn}, \tag{1}$$

where x^i means the ith frame's sample points, N means the length of ith frame. Then, we use above results to compute the autocorrelation function (ACF) and Cepstrum coefficients,

$$C^i = \sum_w^N log^{|H(w)|} * e^{jwn}, R^i = \sum_w^N |H(w)|^2 * e^{jwn}, \tag{2}$$

where C^i denotes the ith frame's Cepstrum coefficient and R^i is ith frame's ACF. Pitches are computed as follows,

$$
\begin{aligned}
C_p^i &= \frac{f}{idx(max(C^i[f * 0.02 : f * 0.2])) + f * 0.02 - 1}, \\
R_p^i &= \frac{f}{idx(max(R^i[f * 0.02 : f * 0.2])) + f * 0.02 - 1},
\end{aligned}
\tag{3}
$$

where C_p^i and R_p^i denote the ith frame pitch computed by Cepstrum coefficient and ACF, respectively, f stands for the sampling rate, and the idx means the index of values.

MFCC Features. MFCC features are commonly used for human speech analysis. MFCCs use frequency bands based on the Mel-spaced triangular frequency bins, then a Discrete Cosine Transform (DCT) is applied to calculate the desired number of cepstral coefficients. The first twelve cepstral coefficients are extracted as our MFCC features.

LSP Features. LSP is a way of uniquely representing the LPC-coefficients. It decomposes the p order linear predictor $A(z)$ into a symmetrical and anti-symmetrical part denoted by the polynomial $P(z)$ and $Q(z)$, separately.

The p order linear prediction system can be written as the Eq. (4) in z-domain, and a prediction error $A(z)$ is produced by (5),

$$x(n) = \sum_{k=1}^p a_k x(n-k) \longrightarrow x(z) = \sum_{k=1}^p a_k x(z), \tag{4}$$

$$A(z) = 1 - \sum_{k=1}^p a_k * z^{-k}. \tag{5}$$

Two polynomials $P(z)$ and $Q(z)$ are given by

$$P(z) = A(z) + z^{-(p+1)}A(z^{-1}), \quad Q(z) = A(z) - z^{-(p+1)}A(z^{-1}). \tag{6}$$

LSP parameters are expressed as the zeros of P(z) and Q(z), and we choose the first eight parameters as our LSP features.

Intensity and ZCR features. Intensity, one of the major vocal attribute, can be computed by

$$I_i = \frac{1}{N} \sum_n^N Ham^2(x^i(n)). \tag{7}$$

ZCR is the rate at which the signal changes from positive to negative or back. It has been used in both speech recognition, being a key feature to classify percussive sounds.

Smoothing and Normalization. Smoothing is an useful tool to reduce the noise and slick data contours. In smoothing, the data points of a signal are modified so that individual points that are higher than the immediately adjacent points, are reduced, and points that are lower than the adjacent points are increased. In this paper, we adopt a moving average filter of 3-length to smooth data,

$$s_i = \frac{x_{i-1} + x_i + x_{i+1}}{3}. \tag{8}$$

It is essential to normalize data for a strong emotion recognition system. The goal of normalization is to eliminate speaker and recording variability while keeping the emotional discrimination. In this paper, the min-max method that limits the value of each feature ranging between 0 and 1 is adopted for feature scaling. For every sample, the normalized feature is estimated by the following

$$s_i = \frac{s_i - min(s_i)}{max(s_i) - min(s_i)}. \tag{9}$$

Global Features. Global features are calculated as statistics of all speech features extracted from an utterance. The majority of researchers have agreed that global features are superior to local ones in terms of classification accuracy and classification time. Therefore, the maximum, minimum, mean, maximum position, minimum position, range (maximum-minimum), standard deviation, skewness, kurtosis, linear regression coefficient (slope and offset), linear regression error (linear error and quadratic error), quartile and inter-quartile range are computed based on local features, and combined into the global features.

Every frame consists of pitch, MFCC, LSP, intensity and ZCR features. After smoothing these features, the delta regression coefficients are merged, then, the global features are computed based on local features. Figure 1 shows the block diagram for the processing of feature extraction. We extract these features with the help of OpenSMILE [22].

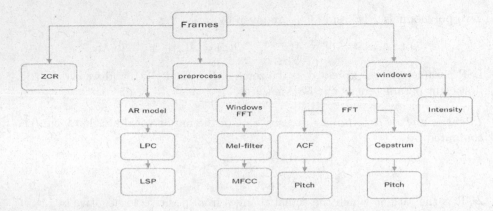

Fig. 1. Block diagram of processing of features extraction

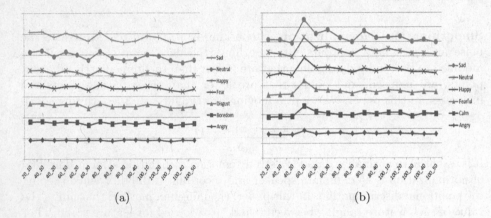

Fig. 2. Frame Duration

3   Results/Discussion

In the speech emotion recognition, many classifiers including GMM, artificial neutral networks (ANN), and support vector machine (SVM), has been used. Specially, SVM classifier is widely used in many pattern recognition applications, and outperforms other well-known classifiers. This study use a linear kernel SVM with sequential minimal optimization (SMO) to build model.

In particular, by DFS, we find the best durations between two database are different. Every broken line in Fig. 2 means the change of accuracy in different duration and overlap for corresponding emotion. For German Database, Fig. 2(a) shows that the best frame duration is 40 ms and overlap is 10 ms while Fig. 2(b) tells us that the best frame duration is 60 ms and overlap is 10 ms in terms of RAVDESS Database.

3.1 Classification Results for EMODB

For EMODB database, the number of samples is 535. We adopt the ten-fold cross validation to avoid overfitting, and the results are shown in Table 2. It can be seen the average accuracy of cross-validation is 87.3%. The previous results on EMODB database such as in [23], the average accuracy was 83.3%. In [24], they aggregated features from MPEG-7 descriptors, MFCCs and Tonality. The best results was 83.39% by using SVM classifier. In [25], they extracted 1800 dimensional features, and average accuracy was 73.3% for classifying six categories ignoring the disgust emotion.

Table 2. Confusion Matrix on the EMODB (%)

	Angry	Boredom	Disgust	Fear	Happy	Neutral	Sad
Angry	92.9	0	0	2.3	4.6	0	0
Boredom	0	87.5	0	0	0	4.9	7.4
Disgust	2.1	2.1	86.9	4.2	4.2	0	0
Fear	5.7	0	1.4	85.5	4.3	0	2.8
Happy	14.1	0	1.4	9.8	74.6	0	0
Neutral	0	6.4	0	0	0	93.6	0
Sad	0	9.6	0	0	0	1.7	90.3

Table 3 shows the results of each feature (e.g. pitch, MFCC, etc.) and MFCC+LSP to compare with our original features. The reason why we choose MFCC+LSP feature is that MFCC and LSP are the best two types of features in term of classifying accuracy.

Table 3. Comparisons the accuracy of each feature and MFCC+LSP with fusion features (%)

	Angry	Boredom	Disgust	Fear	Happy	Neutral	Sad
ZCR	79.5	52.5	36.9	51.4	30.9	51.2	56.4
Intensity	81.1	61.2	26.9	42.6	11.2	48.7	53.2
Pitch	83.4	72.5	47.8	55.8	36.6	57.5	80.6
LSP	81.8	83.7	69.5	70.5	59.1	81.2	66.1
MFCC	86.6	85.0	80.4	79.4	64.7	83.7	83.8
MFCC+LSP	92.1	88.7	84.7	82.3	74.6	87.5	79.0
Original	92.9	87.5	86.9	85.5	74.6	93.6	90.3

3.2   Classification Results for RAVDESS

RAVDESS database composed by six emotions which are angry, calm, fearful, happy, neutral and sad. Each emotions has 92 samples, and we adopt the same processes as EMODB. The results are shown in Table 4. The average accuracy of angry is 94.5%, calm is 84.7%, fearful is 86.9%, happy is 79.3%, neutral is 69.5% and sad is 60.8%. In [26], it has the accuracy 82% for angry, 96% for happy, 70% for neutral, 58% for sad, 84% for calm and 88% for fearful.

Table 4. Confusion Matrix on the RAVDESS (%)

	Angry	Calm	Fearful	Happy	Neutral	Sad
Angry	94.5	5.5	0	0	0	0
Calm	2.5	84.7	1.0	0	3.2	8.6
Fearful	0	1.0	86.9	3.2	1.0	7.6
Happy	0	0	7.6	79.3	9.7	3.2
Neutral	0	4.2	1.0	6.5	69.5	18.4
Sad	0	3.2	11.9	3.2	20.6	60.8

Table 5 shows the results of each feature (e.g. pitch feature, MFCC features, etc.) and MFCC+LSP features to compare with our original features.

Table 5. Comparison the accuracies between different screening methods on RAVDESS (%)

ZCR	39.1	18.4	25	5.4	39.1	20.6
Intensity	60.8	41.3	25.0	40.2	31.5	17.3
Pitch	52.1	44.5	44.5	31.5	16.3	10.8
LSP	70.6	51.1	59.7	59.7	53.2	28.2
MFCC	71	59.7	64.1	64.1	63.2	30
MFCC+LSP	94.5	82.6	88.1	80.6	69.5	59.6
Original	94.5	84.7	86.9	79.3	69.5	60.8

3.3   SFFS

Moreover, we use the sequential floating forward search (SFFS) which is simple, fast, effective and widely accepted technique to select features. The central idea of SFFS is that we choose the best feature firstly by forward tracking, then exclude a number of features by backtracking, this process will be repeated until the number of features which have been selected is unchange. For EMODB corpus, the number of features which are selected by using SFFS is 35 and the accuracy of Happy improved by 1.4%. For RAVDESS corpus, the dimension reduce to 27. The accuracy of Happy improved by 1.3% and Fearful improved by 1.2%. The results for EMODB corpus and RAVDESS corpus are described in Tables 6 and 7 respectively.

Table 6. Comparison the accuracies between different screening methods on EMODB (%)

	Angry	Boredom	Disgust	Fear	Happy	Neutral	Sad
SFFS	92.9	87.5	86.9	85.5	76	93.6	90.3
Original	92.9	87.5	86.9	85.5	74.6	93.6	90.3

Table 7. Comparison the accuracies between different screening methods on EMODB (%)

	Angry	Calm	Fearful	Happy	Neutral	Sad
SFFS	94.5	84.7	88.1	80.6	69.5	60.8
Original	94.5	84.7	86.9	79.3	69.5	60.8

4   Conclusions

In this paper, we extract two types of features for emotion recognition based on two well-known databases: EMODB and RAVDASS. Then we carry out the fusion, smooth and normalization operation on these features, and the accuracies obtained on the two databases are very promising using SVM classifier, but we still not find a useful reduce dimensional method to get better results. In addition, the limitation of this paper is that we only used two speech emotional corpora. In the future, we will use more databases to identify the effect of our method and build more persuasive model using some methods of deep learning like Bidirectional Long ShortTerm Memory (BLSTM) networks and Gaussian RBM, and we also improve our reducing methods to get better results so that we can further throw invalid features, and use less features to get better results.

Acknowledgments. The research was supported in part by NSFC under Grants 11301504 and U1536104, in part by National Basic Research Program of China (973 Program2014CB744600).

References

1. Minker, W., Pittermann, J., Pittermann, A., Strauß, P.M., Bühler, D.: Challenges in speech-based human-computer interfaces. Int. J. Speech Technol. **10**(2–3), 109–119 (2007)

2. Ntalampiras, S., Potamitis, I., Fakotakis, N.: An adaptive framework for acoustic monitoring of potential hazards. EURASIP J. Audio Speech Music Process. **2009**, 13 (2009)
3. Cummings, K.E., Clements, M.A., Hansen, J.H.: Estimation and comparison of the glottal source waveform across stress styles using glottal inverse filtering. In: Proceedings of the IEEE Energy and Information Technologies in the Southeast. Southeastcon 1989, pp. 776–781. IEEE (1989)
4. Seppänen, T., Väyrynen, E., Toivanen, J.: Prosody-based classification of emotions in spoken finnish. In: INTERSPEECH (2003)
5. Origlia, A., Galatà, V., Ludusan, B.: Automatic classification of emotions via global and local prosodic features on a multilingual emotional database. In: Proceeding of the 2010 Speech Prosody. Chicago (2010)
6. Atal, B.S.: Effectiveness of linear prediction characteristics of the speech wave for automatic speaker identification and verification. J. Acoust. Soc. Am. **55**(6), 1304–1312 (1974)
7. Davis, S.B., Mermelstein, P.: Comparison of parametric representations for monosyllabic word recognition in continuously spoken sentences. IEEE Trans. Acoust. Speech Signal Process. **28**(4), 357–366 (1980)
8. Ververidis, D., Kotropoulos, C., Pitas, I.: Automatic emotional speech classification. In: Proceedings of the IEEE International Conference on Acoustics, Speech, and Signal Processing, (ICASSP 2004), vol. 1, IEEE I-593 (2004)
9. Fernandez, R., Picard, R.W.: Classical and novel discriminant features for affect recognition from speech. In: Interspeech, pp. 473–476 (2005)
10. Bou-Ghazale, S.E., Hansen, J.H.: A comparative study of traditional and newly proposed features for recognition of speech under stress. IEEE Trans. Speech Audio Process. **8**(4), 429–442 (2000)
11. Rabiner, L.R., Schafer, R.W.: Digital processing of speech signals (prentice-hall series in signal processing) (1978)
12. Nwe, T.L., Foo, S.W., De Silva, L.C.: Speech emotion recognition using hidden markov models. Speech Commun. **41**(4), 603–623 (2003)
13. Wu, S., Falk, T.H., Chan, W.Y.: Automatic speech emotion recognition using modulation spectral features. Speech Commun. **53**(5), 768–785 (2011)
14. Li, X., Tao, J., Johnson, M.T., Soltis, J., Savage, A., Leong, K.M., Newman, J.D.: Stress and emotion classification using jitter and shimmer features. In: IEEE International Conference on Acoustics, Speech and Signal Processing. ICASSP 2007, vol. 4, IEEE IV-1081 (2007)
15. Lugger, M., Janoir, M.E., Yang, B.: Combining classifiers with diverse feature sets for robust speaker independent emotion recognition. In: 2009 17th European Signal Processing Conference, pp. 1225–1229. IEEE (2009)
16. Pan, Y., Shen, P., Shen, L.: Speech emotion recognition using support vector machine. Int. J. Smart Home **6**(2), 101–108 (2012)
17. Chen, L., Mao, X., Xue, Y., Cheng, L.L.: Speech emotion recognition: features and classification models. Digit. Signal Process. **22**(6), 1154–1160 (2012)
18. Deng, J., Zhang, Z., Marchi, E., Schuller, B.: Sparse autoencoder-based feature transfer learning for speech emotion recognition **7971**, 511–516 (2013)
19. Han, K., Yu, D., Tashev, I.: Speech emotion recognition using deep neural network and extreme learning machine. In: Interspeech, pp. 223–227 (2014)
20. Burkhardt, F., Paeschke, A., Rolfes, M., Sendlmeier, W.F., Weiss, B.: A database of german emotional speech. Interspeech **5**, 1517–1520 (2005)

21. Livingstone, S., Peck, K., Russo, F.: Ravdess: the ryerson audio-visual database of emotional speech and song. In: 22nd Annual Meeting of the Canadian Society for Brain, Behaviour and Cognitive Science (CSBBCS) (2012)
22. Eyben, F., Wöllmer, M., Schuller, B.: Opensmile: the munich versatile and fast open-source audio feature extractor. In: Proceedings of the 18th ACM international conference on Multimedia, pp. 1459–1462. ACM (2010)
23. Kotti, M., Paternò, F.: Speaker-independent emotion recognition exploiting a psychologically-inspired binary cascade classification schema. Int. J. Speech Technol. **15**(2), 131–150 (2012)
24. Lampropoulos, A.S., Tsihrintzis, G.A.: Evaluation of MPEG-7 Descriptors for Speech Emotional Recognition (2012)
25. Wang, K., An, N., Li, B.N., Zhang, Y., Li, L.: Speech emotion recognition using fourier parameters. IEEE Trans. Affect. Comput. **6**(1), 69–75 (2015)
26. Zhang, B., Essl, G., Provost, E.M.: Recognizing emotion from singing and speaking using shared models. In: 2015 International Conference on Affective Computing and Intelligent Interaction (ACII), pp. 139–145. IEEE (2015)

Advertisement and Expectation in Lifestyle Changes: A Computational Model

Seyed Amin Tabatabaei[✉] and Jan Treur

VU University Amsterdam, Amsterdam, The Netherlands
{s.tabatabaei,j.treur}@vu.nl

Abstract. Inspired by elements from neuroscience and psychological literature, a computational model of forming and changing of behaviours is presented which can be used as the basis of a human-aware assistance system. The presented computational model simulates the dynamics of mental states of a human during formation and change of behaviour. The application domain focuses on sustainable behavior.

Keywords: Computational modeling · Temporal-casual network · Cognitive states · Behaviour change · Decision making

1 Introduction

Human-aware assisting computing systems have been proposed as a promising tool to support behaviour and habit changes toward a more healthy [1], or a more sustainable [2] lifestyle. Such a system is supposed to collect many different types of data from the environment and user about a specific context. It can use the ubiquitous power of mobile systems in combination with cloud computing to collect data all the time, make computational processes on them and deliver interventions at proper time. Doing this, it is possible to equip such a system with understanding of behavioural and mental processes of users. In this way, psychological theories (like [3]) are translated into formal and dynamic models implemented in the system, which can be used for understanding, prediction and anticipation of mental processes and behaviour [4].

There is a growing interest in understanding different ways to change the behaviour of people in different contexts (e.g. [5–7]). Performing a specific action or behaviour in general is due to a series of attitudes and related goals. So it can be argued that changing attitudes and goals can change the behaviours. In addition, behaviour change can sometimes be affected through regulation, or economic instruments [8]. This strategy is applied in the energy market of many countries. In these countries, a dynamic price is offered for the electricity consumers by defining time-based prices for electricity.

Some works also drawn attention to the importance of cultural elements, norms, routines, habits, social networks, fashion and advertising. It should be noticed that strongest degree of behaviour change occurs when different strategies are combined [9]. In this work, the focus is mostly on the combination of two

© Springer International Publishing AG 2017
Y. Zeng et al. (Eds.): BI 2017, LNAI 10654, pp. 14–25, 2017.
https://doi.org/10.1007/978-3-319-70772-3_2

strategies: the economical motivations in combination with the advertisements that promise people benefits of these motivations.

The expectancy theory of motivation, introduced by Victor H. Vroom, says that a person will behave in a specific way because he intends to select that behaviour over others due to what he expects the result of that selected behaviour will be [10]. In the other words, the intention or motivation of the behaviour selection is determined by the desirability of the expected outcomes. According to [11], Vroom asserts that "intensity of work effort depends on the perception that an individual's effort will result in a desired outcome".

Expectation confirmation theory [3] is a cognitive theory which seeks to explain post-adaptation satisfaction as a function of expectation and disconfirmation of beliefs. Although this theory originally appeared in psychology and marketing literature, it has been adopted in several other fields. This theory involves four primary constructs: expectations, perceived performance, disconfirmation of beliefs, and satisfaction. In our model, we took inspiration from such theories. However, there are also some small differences. For example, our model also contains expectation but the pre-consumption expectation is replaced by post-consumption expectations (the same is done in [12]).

In this work, a computational model is proposed for understanding the role of economical motivations and advertisements on changing behaviours toward a more sustainable lifestyle. The model is at the cognitive level, which abstracts from too specific neurological details, but still reflects the underlying neurological concepts.

The rest of this paper is organized as follows: the next section is a review on the main concepts of temporal-causal modeling approach. In Sect. 3, the proposed model is explained. In Sect. 4, the results of some experimental simulations are depicted and discussed. Section 5 is a mathematical analysis on the results of the simulations, which validate their accuracy. Finally, the conclusion will come.

2   Temporal-Causal Modeling

The proposed model in this work was designed as a temporal-causal network model based on the Network-Oriented Modeling approach described in [13]. The dynamic perspective takes the form of an added continuous time dimension. This time dimension enables causal relations, where timing and combination of causal effects can be modeled in detail.

This approach is used in many different domains, for example: habit learning [14], contagion and network change in social networks [15], emotion regulation [16] and the role of social support on mood [17]. For more information about this Network-Oriented Modeling approach, see [18].

Based on this approach, a model can be designed at the conceptual level. Figure 1 shows the graphical representation of a part of model. In a graphical conceptual representation, states are represented by nodes, and their connections are represented by arrows which connect the nodes.

A complete model has some labels in addition to its nodes and arrows, representing some more detailed information:

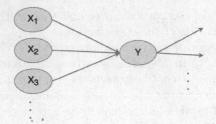

Fig. 1. The graphical representation of a part of a model

- The weight of a connection from a state X to a state Y, in the range of $[-1, 1]$, denoted by $\omega_{X,Y}$
- A speed factor ηY for each state, in the range of $[0, 1]$
- The type of combination function $cY(...)$ for each state Y shows how to aggregate the multiple causal impacts on that state.

The following rules show how such a model works from a numerical perspective:

- The variable t indicates the time and varies over the non-negative real numbers
- At any time point, any state of the model has a real value, $X(t)$, in the range of $[0, 1]$. (0 means that state is deactivated and 1 means it is fully activated)
- If there is a connection from state X to state y, the impact of X to Y at time t is equal to: $impact_{X,Y} = \omega_{X,Y} X(t)$
- The aggregated impact of multiple states (X1, X2, X3,..) connected to state Y at time t is calculated using the combination function of node Y:

$$\mathbf{aggimpact}_Y(\mathbf{t}) = \mathbf{c}_Y(\mathbf{impact}_{X1,Y}, \mathbf{impact}_{X2,Y}, \mathbf{impact}_{X3,Y},)$$
$$= \mathbf{c}_Y(\omega_{X1,Y} X_1(t), \omega_{X2,Y} X_2(t), \omega_{X3,Y} X_3(t),) \tag{1}$$

- The effect of aggimpactY(t) on state Y is exerted on this state gradually; its speed of change is dependent on the speed factor η_Y of Y:

$$Y(t + \Delta t) = Y(t) + \eta_Y[\mathbf{aggimpact}_Y(t)Y(t)]t \tag{2}$$

For each state the speed factor represents how fast its value is changing in response to the casual impacts. States with fleeting behaviour (like emotional states) have a high value for speed factor. Thus the following equations show the difference equation for state Y:

$$Y(t + \Delta t) = Y(t) + \eta_Y[\mathbf{c}_Y(\omega_{X1,Y} X1(t), \omega_{X2,Y} X2(t), \omega_{X3,Y} X3(t),)Y(t)]\Delta t \tag{3}$$

The above numerical representations can be used for mathematical and computational analysis and simulations.

3   The Computational Model

The proposed computational model is based on the literature and the concepts introduced in Introduction. This model was designed at the conceptual cognitive level, which is based on the neurological theories but abstracts from low level details. It uses temporal relations between different cognitive states to describe the mechanisms for action selection and behaviour changes.

3.1   Graphical Representation of the Model

Figure 2 shows the conceptual representation of the causal model of cognitive states related to two long-term goals. For the sake of presentation, it is assumed that there are just one short-term goal and one behaviour paired to each long-term goal. In this figure, the states related to one long-term goal are in the same color. As depicted, long-term goals lead to short term goals. In general, there is a many-to-many relation between long-term goals and short-term goals.

In the next step, a short-term goal affects the intention for a specific behaviour, which in turn leads to that behaviour. In this work, it is assumed that there is a competition between (assumed at least partly mutually excluding) behaviours: the intentions of different behaviours have negative relations to each other. Thus, activation of one behaviour has a negative effect on the others, which models a winner-takes-it-all principle. The other state which affects the intention is the expectation about the results of a behaviour. If the person expects good results from performing a behaviour, then he or she has a stronger intention to do it. On the other hand, if he or she is dissatisfied about that behavior, the intention will become down.

The expectation about the results of a behaviour is affected both by the advertisements about that behaviour and by the observation of real results. In the proposed model, in the cases that a person does not have any experience in performing a behaviour, this expectation is mostly based on the advertisements. By performing the behaviour for a while and increasing the confidence about the observed real results, the expectations would be mostly based on these observations. It should be noticed that unrealistic expectations (a big difference between promised results in the advertisements and real observed results) leads to dissatisfaction. A short definition of each state and its role is explained in Table 1.

In this paper, some assumptions are made for the sake of simplicity in presentation. These are some assumption that applied here but do not limit the model and the model can work beyond them:

Assumption 1: There are just two long-term goals available, which are not fully antithetical. For instance, one goal can be living environmental friendly; and another one living with more comfort.

Assumption 2: There are just two short-term goals, which are aligned with long-term goals. Instances for short term goals can be: saving money by not

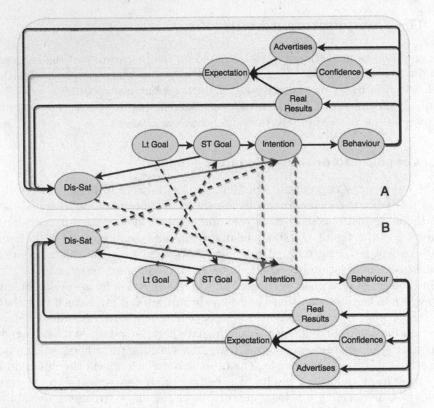

Fig. 2. The graphical conceptual representation of proposed model. This representation contains just two goal-behaviour pairs and their relevant states. In this figure, ovals represent the nodes and arrows show the connections. The blue arrows have positive weight and red ones a negative one. Dashed arrows are the connections between the states of two goals. For a brief description about each node, please look at Table 1. (Color figure online)

using high power electrical devices during peak hours, and being completely comfortable by using devices when they are needed.

Assumption 3: There are just two behaviours, behaviour 1 and 2 that are respectively more related long-term goal 1 and 2. As instances for possible behaviours: using high power electrical devices during of peak hours, and using electrical devices when they are needed.

As it can be seen, goals and behaviours are not fully antithetical, while there are obvious differences. These conditions aim at simplifying the presentation of the model and the corresponding simulation results. And as explained, the model is not limited to these conditions.

Table 1. Definition of states of conceptual model

State name	Definition
Lt Goal	This stated indicates a long-term goal of the person, which can be related to a few short-term goals. Examples: living environmentally friendly; living comfortable
St Goal	A short-term goal of the person which can be aligned with a few long term goals and can lead to a few behaviours in different contexts. Examples: saving money through using high power devices during off-peak hours
Intention	This state shows the intention of performing a specific behaviour
Behaviour	This state indicates a behaviour in a specific context. It is assumed that in each time point, just one behaviour can be activated. Examples: Do not using high power devices during peak hours
Real Result	The observed real results of a behaviour after doing it. Example: the amount of money saving on the electricity bill due to the using high power devices during off-peak hours
Confidence	The confidence of the person about his expectation. If he or she performs a behaviour for a longer time, he or she would be more confident about his/her expectation
Advertisement	The promised results for doing a behaviour via advertisements
Expectation	The expectations about the results of performing a behaviour in a specific context. This expectation can be form due to the personal observations of the real results, or due to the advertisements
Dissatisfaction	Dissatisfaction in performing an action can happen due to a big difference between the observed result and expectations

3.2 Numerical Representations and Parameters

As explained in Sect. 3.1, to be able to use a temporal-causal network model in simulations, some elements need to be known: the weight of connections, speed factors and the type of combination function for each state, and the parameters (if any) of these combination functions. However, reading this subsection is not necessary for understanding the model and simulation results.

Connection Weights. The following table shows the connection weights for the connections of one goal and its related states.

It should be noticed that the connection weights between the states for the other goal are the same. The only exceptions are the weights of the connections from st-goal to intention, and from behaviour to real result which both are 0.2 in the second pair of goal-behaviour.

Table 2. Weight of connections in the proposed model. The graphical representation of model is presented in Fig. 2

	Behaviour	Intention	St goal	Dissat	Real result	Confidence	Expectation
Behaviour				1	0.5	1	
Intention	1						
St goal		0.35		0.05			
Lt goal			0.45				
Dissat		−1					
Real result				−0.5			1
Confidence						0.9	1
Ads							1
Expectation		0.3		0.4			

In addition to the weights shown in Table 2, there are some connections which connect the states of two parts of the model (from one goal-behaviour pair to the other one). The weights of these connections are these: $\omega_{ltGoal,st-goal} = 0.05, \omega_{intention,intention} = -0.35, \omega_{Dissatisfaction,intention} = 0.05$

Combination Functions and Parameters. In this work, states which are not affected by other states, are external states (advertisement and long term goal). So, the value of these states are not dependent on the model and are used to define different scenarios (Sect. 4.2). As explained, before performing a behaviour, the expectation of the person about the results is mostly based on the advertisements, and by performing the behaviour, the expectation would gradually become based on the real observations. To have such dynamics, for the expectation, the following combination function is used, expressing that the aggregated impact is a weighted average of the impacts by advertising and real result, where the weights are Conf and 1-Conf:

$$CF(Ads, Real-result, Conf) = Conf * Real-result + (1-Conf) * Ads \quad (4)$$

For the Real result state, an identity combination function id(.) is used. The output of this function is equal to its input. For the other states of the model, a simple logistic function is used:

$$\textbf{alogistic}_{\sigma,\tau}(V1, V2, V3 \cdots) = 1/(1 + exp(-\sigma(V1 + V2 + V3 \cdots - \tau))) \quad (5)$$

As it is clear, this function has two parameters (steepness σ, and threshold τ) which can be defined for each states. The following Table 3 shows the values of σ and τ for the states with logistic function, in addition to the speed factor of all states.

4   Simulation Experiments

The human-aware network model presented above was used to make a comparison between what the model predicts and what actually holds in the real world (based on the literature [3,10]). In the first subsection, some of hypotheses are explained. In Subsect. 4.2 three different scenarios and their corresponding results are discussed. Finally, the defined hypotheses are analysed using the results of simulations.

4.1   Hypotheses

Some of the expected behaviours of the model can be formulated as follows:

- **Hypothesis 1.** Due to conflicts between the different behaviours, at any time point, just one behaviour can be activated.
- **Hypothesis 2.** In the case that a behaviour is activated and the person is not dissatisfied with its result and there is no external motivation to stop it, its activation should continue.
- **Hypothesis 3.** External motivations (like advertisements via television or social networks) can motivate a person to change the behaviour via increasing the expectation about its corresponding results.
- **Hypothesis 4.** After performing an action (due to advertisements) for a while, if the observed results are much lower than the promises, the person will become dissatisfied about it, and stop that behaviour.

4.2   Scenarios and Results

To have more realistic scenarios, "domestic energy usage" was selected as the application domain of simulated scenarios. By the growth of electricity generation by non-schedulable sources, matching supply and demand becomes a more important challenge in this context. An often used way to encourage the costumers to shift their loads to off-peak hours is time-based pricing. In this strategy, the price of electricity during the peak hours (when demand is high and supply is low) is higher than off-peak hours. In addition to such an economical motivations, it is important to inform the costumers about these economical advantages. In the defined scenarios, the role of advertisements about these commercial advantages is analysed.

To study the matching of simulation results and hypothesis, the model was used with two long-term goals:

- Long term goal 1: living environmentally friendly
- Long term goal 2: living with comfort

Table 3. Parameters of the model. Speed factors and parameters of logistic functions

	Behave 1	Behave 2	Intention 1& 2	St-goal 1& 2	Dissat 1	Dissat 2	Real result 1 & 2	Confidence 1& 2	Expect 1& 2
Speed factor	0.5	0.5	0.3	0.2	0.5	0.5	0.5	0.025	0.2
Steepness σ	50	50	50	30	50	50	The combination function of these states is not Logistic		
Threshold τ	0.4	0.6	0.1	0.3	1.05	1.15			

Each of them has one short-term goal:

- Short term goal 1: saving money by using the high power electrical devices in off peak hours (when the price is lower)
- Short term goal 2: having comfort by using the devices whenever they are needed

And each short time goal leads to a different behaviour:

- Behaviour 1: using the high power electrical devices in off peak hours
- Behaviour 2: using the electrical devices whenever they are needed

In the defined scenarios both goals are available in parallel. Only the availability of advertisements and the promising results in the advertisements affect change in the scenarios. Moreover, the result of first behaviour is higher (0.5) than the second one (0.2). In all scenarios, it is assumed that the person in the past just always lived based on behaviour2 (using the electrical devices whenever they are needed) and there was no dissatisfaction.

In the **first scenario**, a very low advertisement (0.1) is assumed for living environmentally friendly. Figure 2 shows the results:

As it can be seen in Fig. 3, in the first scenario the person will continue doing behaviour 2. And due to the lack of external motivations, the person will not switch to the behaviour1.

In the **second scenario**, some advertisements inform the person about the advantages of behaviour1. However, the promised result about the behaviour 2 is much higher than actual result (1.0 vs. 0.5) of this behaviour. Figure 4 shows the results; the changed behaviour does not persist:

In the **third scenario**, some advertisements inform the person about the advantages of behaviour1, and the promised results are not too different the real

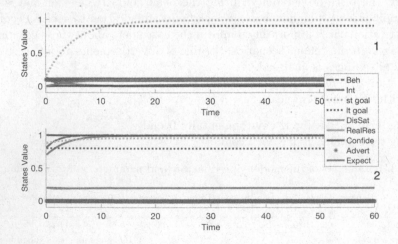

Fig. 3. The results of the 1st scenario. Upper graph is related to the Behaviour 1. Lower graph is related to Behaviour 2.

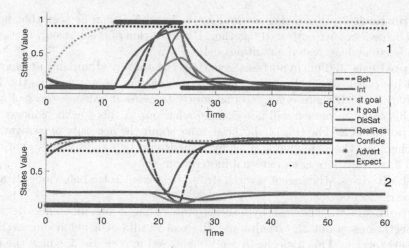

Fig. 4. Results of the 2nd scenario. Upper graph is related to the Behaviour 1. Lower graph is related to Behaviour 2.

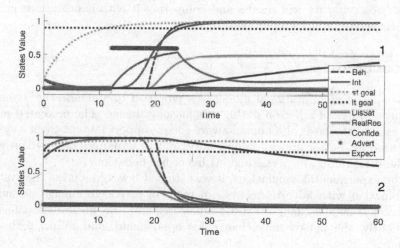

Fig. 5. Results of the 2nd scenario. Upper graph is related to the Behaviour 1. Lower graph is related to Behaviour 2.

results (0.6 vs. 0.5). Figure 5 shows the results of this simulation; the changed behavior persists.

4.3   Explanation

In Sect. 4.1, four hypothesis are defined based on the literature. In this subsection, these hypotheses are analyzed based on the results of the scenarios in Sect. 4.2:

- **Hypothesis 1.** By looking at the Figs. 3, 4 and 5, it can be seen that never two behaviours are activated together. By activation of a behaviour, the other one become deactivated simultaneously.
- **Hypothesis 2.** This hypothesis says that without any stimulant, no changes will happen in the behaviours. This hypothesis can be analysed in the first scenario, which there is no external motivation. As it can be seen in Fig. 2, in this case the person will not change behaviour in this specific context.
- **Hypothesis 3.** The third hypothesis talks about the necessity of motivations in changing the behaviours. This hypothesis can be analysed in scenarios 2 and 3, where there is an external motivation.
  In both cases, the agent switch from behavior 2 to behaviour 1 after advertisements.
- **Hypothesis 4.** The last hypothesis states that a big difference between expectations about the results and the real results of a behaviour leads to dissatisfaction. This hypothesis can be analysed in scenario 3, where the person changes behaviour due to the very high expectation about the results of behaviour1. This high expectation is built due to the unrealistic advertisements. As it can be seen in Fig. 4, after performing behaviour 1 for a while, observing its real results and comparing it with expectations creates dissatisfaction, which stops this behaviour.

5 Conclusion

In this paper, a computational model was proposed to simulate the dynamics of cognitive states of a person during a behaviour change. The presented model can be used as the basis of a human-aware smart support system. Such a system can be used for better understanding the cognitive dynamics of a person while he or she is changing (or persisting on) his or her behaviour.

In the experimental simulations, it was studied how economical motivations in combination with advertisements can tempt a person to change behaviour. However, it is shown that unrealistic advertisements can make dissatisfaction. Subsequently, the person may change his mind again, and return to his old behaviour.

The results of the simulations match to the expectations. The model was verified via a mathematical analysis which had a positive outcome.

As mentioned, the probability of behaviour change toward a better lifestyle is higher, when different strategies are combined together. Therefore, in future work, the role of other elements (like social networks) will also be considered in the model.

References

1. Riley, W.T., Rivera, D.E., Atienza, A.A., Nilsen, W., Allison, S.M., Mermelstein, R.: Health behavior models in the age of mobile interventions: are our theories up to the task? Transl. Behav. Med. **1**(1), 53–71 (2011)

2. Kjeldskov, J., Skov, M.B., Paay, J., Pathmanathan, R.: Using mobile phones to support sustainability: a field study of residential electricity consumption. In: Proceedings of the SIGCHI Conference on Human Factors in Computing Systems, pp. 2347–2356. ACM (2012)
3. Oliver, R.L.: A cognitive model of the antecedents and consequences of satisfaction decisions. J. Mark. Res. **17**, 460–469 (1980)
4. Zhang, C., van Wissen, A., Lakens, D., Vanschoren, J., De Ruyter, B., IJsselsteijn, W.A.: Anticipating habit formation: a psychological computing approach to behavior change support. In: Proceedings of the 2016 ACM International Joint Conference on Pervasive and Ubiquitous Computing: Adjunct, pp. 1247–1254. ACM (2016)
5. Michie, S., van Stralen, M.M., West, R.: The behaviour change wheel: a new method for characterising and designing behaviour change interventions. Implementation Sci. **6**(1), 42 (2011)
6. Hargreaves, T.: Practice-ing behaviour change: applying social practice theory to pro-environmental behaviour change. J. Consum. Cult. **11**(1), 79–99 (2011)
7. Free, C., Phillips, G., Galli, L., Watson, L., Felix, L., Edwards, P., Patel, V., Haines, A.: The effectiveness of mobile-health technology-based health behaviour change or disease management interventions for health care consumers: a systematic review. PLoS Med. **10**(1), e1001,362 (2013)
8. Owens, S., Driffill, L.: How to change attitudes and behaviours in the context of energy. Energy Policy **36**(12), 4412–4418 (2008)
9. Gardner, G.T., Stern, P.C.: Environmental Problems and Human Behavior. Allyn & Bacon, Boston (1996)
10. Oliver, R.L.: Expectancy theory predictions of salesmen's performance. J. Mark. Res. **11**, 243–253 (1974)
11. Holdford, D., Lovelace-Elmore, B.: Applying the principles of human motivation to pharmaceutical education. J. Pharm. Teach. **8**(4), 1–18 (2001)
12. Hossain, M.A., Quaddus, M.: Expectation-confirmation theory in information systems research: a review and analysis. In: Dwivedi, Y., Wade, M., Schneberger, S. (eds.) Information Systems Theory. Integrated Series in Information Systems, vol. 28, pp. 441–469. Springer, New York (2012). doi:10.1007/978-1-4419-6108-2_21
13. Treur, J.: Dynamic modeling based on a temporal-causal network modeling approach. Biol. Inspired Cogn. Archit. **16**, 131–168 (2016)
14. Klein, M.C.A., Mogles, N., Treur, J., van Wissen, A.: A computational model of habit learning to enable ambient support for lifestyle change. In: Mehrotra, K.G., Mohan, C.K., Oh, J.C., Varshney, P.K., Ali, M. (eds.) IEA/AIE 2011. LNCS, vol. 6704, pp. 130–142. Springer, Heidelberg (2011). doi:10.1007/978-3-642-21827-9_14
15. Blankendaal, R., Parinussa, S., Treur, J.: A temporal-causal modelling approach to integrated contagion and network change in social networks. In: ECAI, pp. 1388–1396 (2016)
16. Abro, A.H., Klein, M.C., Manzoor, A.R., Tabatabaei, S.A., Treur, J.: Modeling the effect of regulation of negative emotions on mood. Biol. Inspired Cogn. Archit. **13**, 35–47 (2015)
17. Abro, A.H., Klein, M.C.A., Tabatabaei, S.A.: An agent-based model for the role of social support in mood regulation. In: Bajo, J., Hallenborg, K., Pawlewski, P., Botti, V., Sánchez-Pi, N., Duque Méndez, N.D., Lopes, F., Julian, V. (eds.) PAAMS 2015. CCIS, vol. 524, pp. 15–27. Springer, Cham (2015). doi:10.1007/978-3-319-19033-4_2
18. Treur, J.: Network-Oriented Modeling. UCS. Springer, Cham (2016). doi:10.1007/978-3-319-45213-5

A Computational Cognitive Model
of Self-monitoring and Decision Making
for Desire Regulation

Altaf Hussain Abro$^{(\boxtimes)}$ and Jan Treur

Behavioural Informatics Group, Vrije Universiteit Amsterdam,
De Boelelaan 1081, 1081 HV Amsterdam, The Netherlands
{a.h.abro, j.treur}@vu.nl

Abstract. Desire regulation can make use of different regulation strategies; this implies an underlying decision making process, which makes use of some form of self-monitoring. The aim of this work is to develop a neurologically inspired computational cognitive model of desire regulation and these underlying self-monitoring and decision making processes. In this model four desire regulation strategies have been incorporated. Simulation experiments have been performed based for the domain of food choice.

Keywords: Cognitive modelling · Computational modelling · Desire regulation · Regulation strategies

1 Introduction

The study of desire and desire regulation has become an important focus for researchers addressing human behaviour. In particular, the desire for eating and its regulation has received much attention in recent years from various fields, including psychology, social science, cognitive science and neuroscience; e.g., [1]. What to eat, when and how much to eat all are influenced or controlled by a brain reward mechanism. Persons who cannot control their eating because of poor regulation of their desires for food often suffer from overweight and obesity. Various studies show that overweight and obesity has become a major problem that affects health in many ways, for example, concerning cardiovascular issues, diabetes, high blood pressure, and more [2]. There has been a strongly increasing trend in the occurrence of obesity globally and in western societies in particular [3]. Often individuals become habitual in taking certain types of foods, and changing such habits is a difficult process. On the other hand, the role of environmental factors is also crucial, such as an environment which is full of appealing food. Much literature is available on the human neurological basis of the regulatory mechanism, for example, dealing with certain desires, e.g., for food, drugs, or alcohol. In general the regulation mechanism for such aspects has a common neurological underpinning; evidence is provided that the neural basis of desire regulation has some characteristics that are similar to those underlying the regulation of negative emotion. This includes the involvement of areas such as the dorsolateral prefrontal

© Springer International Publishing AG 2017
Y. Zeng et al. (Eds.): BI 2017, LNAI 10654, pp. 26–38, 2017.
https://doi.org/10.1007/978-3-319-70772-3_3

cortex, inferior frontal gyrus, and dorsal anterior cingulate cortex [4, 5] This suggest that regulatory mechanisms can be considered for any mental state.

Desire is considered here as a mental state that can lead to certain actions for its fulfilment: a desire state has forward causal connections to action (preparation) states. Regulatory mechanisms that can be used to regulate a desire (and a food desire in particular) will indirectly affect these action states. There is a body of research available that provides evidence and suggests that desires can be down-regulated using different strategies; e.g., [6, 7]. For the case of emotion regulation, [8, 9] have put forward a process model that describes how emotion regulation takes place in different stages. Some of the elements of this process model provide useful inspiration for a generalisation and in particular to develop a model for desire regulation.

The design of human-aware or socially aware intelligent systems that support people in their daily routine can benefit from more attention from a computational perspective. A dynamical cognitive model can be used to obtain some awareness for such a system of a person. For that purpose it is important to use a modelling approach that addresses dynamical aspects and interaction of various mental states underlying behaviour. In this paper the processes of self-monitoring and decision making concerning desire regulation strategies are explored computationally. This work is an extension of the model of desire generation and regulation presented in [10]. Four regulation strategies have been considered: (1) reinterpretation (modulating beliefs), (2) attention deployment (changing the own sensing and focusing of the world), (3) suppression (suppressing a desire and the actions triggered by it), and (4) situation modification (modifying the world itself, for example, by avoidance). The rest of the paper contains in Sect. 2 neurological background literature, Sect. 3 presents and provides the details of the computational model in detail. Next, Sect. 4 presents the results of simulation experiments. Section 5 concludes the paper.

2 Background

The computational cognitive model of self-monitoring and decision making for desire regulation presented in this paper is inspired by cognitive neuroscience literature; in this section some neurological background is discussed. Generally a desire is considered as a mental state of wanting to have or achieve something; it could be for food, smoking, alcohol or any physiological or other aspect. It often implies strong intentions towards the desired aspect, such as to fulfil the physiological need, and sometimes it leads to physically or mentally unhealthy behaviours. In this work the considered application domain is food desire and its regulation. Food desire is considered as the subjective sense of wanting a certain type of food; it has motivational characteristics such as wanting to eat that particular food or being motivated towards that food, it also involve cognitions as well, for example, intrusive beliefs or thoughts about the food [11]. Taking the neuroscience perspective into account, numerous theoretical and empirical studies are available, including fMRI studies about food-related behaviour that show how persons become food addicted; for example, it explains that how appealing food environment affect the brain reward pathways such as how greatly appetizing foods activates reward pathways leading to obesity [12, 13]. Other studies pointing towards the dysfunction or dysregulation of brain reward pathways may also contribute to more consumption of highly appetizing foods, which ultimately leads to more weight gain and obesity [14].

Literature shows that food reward mechanisms and regulation involves various parts of brain, and while regulating a desire various other factors are also involved [1], such as homeostatic requirements, cognition, and the pleasure from food eating [15]. For example, the pleasure factor of food intake is mediated by reward-related cortical and sub-cortical systems of the brain, including the hypothalamus, ventral striatum and the orbitofrontal cortex (OFC) [16, 17].

The monitoring and executive control of various functions in the brain involves the frontal lobes, in particular the prefrontal cortex, orbitoprefrontal cortex also referred as ventromedial PFC (vmPFC); the vmPFC is involved in predicting actions and their outcomes and this is associated with the selection of actions in response to a stimulus. So, the monitoring mechanism of desire regulation can perform selection from an available pool of regulation strategies based on the assessment of the desire as well as based on the preference of that person to make use of certain regulation strategies. In this way decisions about strategies will select any particular regulation strategy or multiple strategies to regulate the desire.

Often regulation takes place when a desire starts to compete with certain important goals. For instance, if a person wants to control his or her diet to lose weight and is in a palatable food environment he or she is trying to avoid certain types of food with high calories. Desire regulation can make use of a variety of specific strategies to modulate the level of desire. Some inspiration can be found in models addressing emotion regulation. Reinterpretation (cognitive change) is a form of regulation strategy often used to change the meaning of stimuli by changing beliefs about these stimuli. The neurological underlying mechanism of such regulation processes [18] shows that the ventrolateral PFC and posterior parietal cortex seem more active for the duration of both cognitive upregulation and downregulation, and were functionally coupled with vmPFC and dlPFC. Moreover for the reinterpretation strategy it is plausible that the control from the PFC affects the interpretation change to change the thoughts about appetizing food. To control different pathways in order to perform regulation according to different strategies the PFC seems involved together with different areas within the brain [19, 20]. For example, during a desire suppression strategy there was more activity in the prefrontal cortex [21]. So, particularly the dlPFC, plays prominent roles in the suppression of a desire to eat, and increases activation in a right prefronto-parietal regulation network [22].

Various empirical studies on the attention deployment strategy [21], explain that this regulation strategy works by moving the attention away from the particular food or deploy attention to something else; e.g. to any other type of food. Besides this, other studies [23, 24] show that the interaction between prefrontal PFC and sensory areas occurs in the sense that PFC has a prominent role for controlling and directing attention to a location, or to an object.

3   Conceptual Representation of the Model

This section shows how the different states in the model affect each other. The proposed model is based on the literature described in the introduction and background sections; an overview is depicted in Fig. 1. A short description of each state is provided in Table 1. The model shows how desire regulation strategies work to regulate a desire using an explicit self-monitoring and selection mechanism of regulation strategies. The model is based on a temporal-causal network [25] reflecting the underlying neurological mechanisms in an abstracted manner. The conceptual cognitive model has been applied to the food desires domain as illustration for this paper.

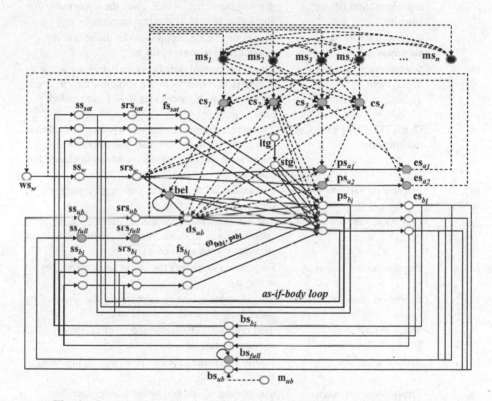

Fig. 1. Conceptual model of self-monitoring and desire regulation strategies.

3.1   Desire Generation and Choosing Actions

The cognitive model adopts some elements from [10], but adds the self-monitoring and regulation selection process. It is assumed that desire generation involves both physiological and environmental factors, so both aspects have been integrated. It shows how such bodily unbalance (e.g., hungriness) and environmental factors (e.g., appetizing food) can generate the desire. Physiological states are based on metabolic activity

Table 1. Overview of the states of the model (see also Fig. 1)

Name	Informal name	Description
ws_w	World state w	The world situation the person is facing, the stimulus, in the example w is a food stimulus
ss_w	Sensor state for w	The person senses the world through the sensor state
srs_w	Sensory representation state for w	Internal representation of sensory world information on w
bel_i	Beliefs	These represent the interpretation of the world information
cs_1	Control state for reinterpretation of the stimulus	By becoming activated this control state suppresses the positive belief, which gives the opportunity for alternative beliefs to become dominant
cs_2	Control state for desire suppression	This control state suppresses the desire and the associated preparations for b_i
cs_3	Control state for attention deployment	When this control state becomes active, by action a_1 it diverts attention from the stimulus, e.g., to divert attention from high calories food to low calories food
cs_4	Control state for situation modification	This state monitors determines whether a situation is unwanted. If so, the control state activates the preparation and execution of action a_2 to change this situation
ps_{a1}	Preparation for action a_1	Preparation to deploy attention (e.g., change gaze direction) by an action a_1
es_{a1}	Execution state for action a_1	The action a_1 regulates the desire by focusing the attention away from the stimulus, so the intensity of stimulus can be reduced
ps_{a2}	Preparation for action a_2	Preparation to modify the world situation ws_w by an action a_2
es_{a2}	Execution state for action a_2	This state for action a_2 is changing the situation by decreasing the level of world state w
m_{ub}	Metabolism for ub	This represents the metabolic level affecting unbalance state ub
bs_{ub}	Body state of ub	Bodily unbalance, for example, underlying being hungry
ss_{ub}	Sensor state for body unbalance ub	The person senses bodily unbalance state ub, providing sensory input
srs_{ub}	Sensory representation of ub	Internal sensory representation of body unbalance ub
bs_{full}	Fullness body state of $full$	A bodily state that represents fullness (in particular for the food scenario)
ss_{full}	Sensor state for bodily state $full$	The person senses body state $full$, providing sensory input
srs_{full}	Sensory representation of $full$	Internal sensory representation of bodily state $full$

(*continued*)

Table 1. (*continued*)

Name	Informal name	Description
ds_{ub}	Desire for unbalance ub	Desire to compensate for body unbalance ub (e.g., desire to eat to get rid of being hungry)
ps_{bi}	Preparation for an action b_i	Preparation for an action b_i to fulfil the desire (e.g., b_i represent the available food choices)
fs_{bi}	Feeling b_i	Feeling state fs_{bi} for the effect of action b_i
srs_{bi}	Sensory representation of b_i	Internal sensory representation of body state for b_i in the brain
ss_{bi}	Sensor state for b_i	The person senses the body states through the sensor state
ss_{sat}	Sensor state for satisfaction sat	The person senses body states providing sensory input to the feelings of satisfaction
srs_{sat}	Sensory representation of sat	Internal representation of the body aspects of feelings of satisfaction sat
fs_{sat}	Feeling for satisfaction sat	These are the feelings about the considered action choice, how much satisfactory it is
es_{bi}	Execution state for action b_i	An action choice to reduce unbalance ub (in the food scenarios b_i is to eat a chosen food)
bs_{bi}	Body state for b_i	This is the body state resulting from action b_i (e.g., to eating that particular food)
ltg	Long term goal	This represents the long term goal (to lose weight, for example)
stg	Short term goal	This refers to an incremental way of achieving long term goals (e.g., avoid fast food)
$\omega_{fs_{bi}, ps_{bi}}$	Learnt connections	This models how the generated feeling affect the preparation for response b_i
mon_i	Self-monitoring states	These states assess the intensity of the desire, e.g., from mon_1 for low to mon_4 for very high intensity. This assessment is input for the decision on which regulation strategies to apply

(e.g., concerning energy level) and are modelled by using connections from a metabolic state m_{ub} that leads to the bodily unbalance represented by body state bs_{ub}.

A person senses the bodily unbalance via the internal sensor state ss_{ub} that enables the person to form the sensory body representation state srs_{ub} representing body unbalance in the brain and this in turn affects the desire. Environmental factors also can lead to a desire (e.g., palatable food) generation. They start with the environment situation that is represented by *world state* ws_w and the person senses it through *sensor state* ss_w and represents it by *sensory representation state* srs_w. Based on these representations the person forms beliefs to interpret the world state (e.g., the food). If positive beliefs are strong, then the desire will be stronger. Alternatively, if a person has more negative beliefs (e.g., this food is not healthy) then the desire may be less strong. The desire in turn affects preparations for actions to fulfil the desire (e.g., get the body in balance). This is modelled by the connection from the desire state ds_{ub} to preparation states ps_{bi}. In the considered scenario, these prepared responses ps_{bi} lead to a (predicted) sensory body representation srs_{bi} and to feeling states fs_{bi}. Subsequently, the states fs_{bi} have strengthening impacts on the preparation state ps_{bi}, which in turn has an impact on feeling state, fs_{bi}, through srs_{bi} which makes the process recursive: an as-if body loop (e.g., [26]). The weights $\omega_{fs_{bi},ps_{bi}}$, of the connections from feeling states fs_{bi} to preparation states ps_{bi} are learnt over time based on a Hebbian learning principle [27].

Making food choices for actions also involves long term goals *ltg* and short term goals *stg*; for example, if a person has the long term goal to reduce body weight, he or she may start to achieve that by generating a more specific short term goal to eat low calories food instead of high calories food.

3.2 Self-monitoring and Regulation Strategies

The self-monitoring mechanism continuously performs monitoring and assessment of the desire intensity. It provides input for the decision making on whether to activate any or multiple desire regulation strategies. In the example scenario four monitoring states were used, but it could be any number. This combined self-monitoring and decision model provides some freedom to persons in selection of desire regulation strategies. The self-monitoring process is modelled through links from the desire to a number of (self-)monitoring states. The decision making process involves links from and to the four control states for the four regulation strategies covered. The role of the self-monitoring states is to assess the intensity of a desire. As discussed in Sect. 2, in general the upward connections to the PFC are used for monitoring and assessment purposes. In this model four different desire regulation strategies have been incorporated; depending on the situation and personality of an individual, one, two, three or all of these regulation strategies are selected. Note that all self-monitoring states have connections to every control state for a regulation strategy, so any monitoring state can contribute to activation of any regulation strategy. This can involve personal characteristics of the person such as preferences regarding the regulation strategy and sensitivity of the person for the stimulus.

Regulation strategies can be used in different phases of the causal pathways in which the desire occurs. The causal pathways considered are those *leading to* a desire, from a world state and body state via sensing and representing it and beliefs interpreting it, and those pathways *following* the desire, in particular the action preparation options triggered by the desire. Desire regulation strategies addressing the first type of causal pathways include cognitive reinterpretation (modifying the beliefs on the world situation, controlled by cs_1), attention deployment (modifying the sensing of the situation, controlled by cs_3), and situation modification (modifying the situation itself, controlled by cs_4). Desire regulation strategies addressing the second type of causal pathway include desire suppression and suppression of the preparation for actions triggered by the desire (controlled by cs_2).

Reinterpretation works by altering the interpretation of a stimulus (e.g., by thinking about the negative consequences of such food), which can in turn reduce the desire. To model reinterpretation two types of belief states bel_i (positive bel_1 and negative bel_2) have been taken into account, they represent alternative interpretations and suppress each other via inhibition connections. A person may interpret food in a different way (e.g., this food is not healthy) by the control state cs_1 becoming active that suppresses the positive belief, so the alternative negative belief gets more strength and the desire becomes less.

The second regulation strategy used in the model is suppression of the desire and the triggered action option preparation states $prep_{bi}$. For example, if a person is just sitting in front of appetizing food and a desire to eat develops, to regulate this desire he/she may start to suppress the desire and the preparations for eating actions. To model this strategy control state cs_2 is used with suppressing connections to the desire ds_{ub} and preparation states $prep_{bi}$.

The third regulation strategy considered in this model is attention deployment. It is used to focus attention away from the sensed cue, for example palatable food, so that the person may get a lower desire level. To model the attention deployment strategy control state cs_3 is used. By using this control mechanism a person can prepare and undertake action a_1 through ps_{a1} and es_{a1} to turn the gaze away to focus attention on something less craving thus decreasing the influence of stimulus ws_w on the sensory state ss_w, so then the environmental influence becomes low which may lead to less desire or even no desire anymore.

The fourth desire regulation strategy is situation modification. It refers to changing or leaving the craving environment which leads to the desire. In this model the control state cs_4 is used for this; this state receives impact from the desire ds_{ub} and the stimulus through the sensory representation of the world srs_w and activates preparation state ps_{a_2} and execution state es_{a_2} so that action a_2 can change the environment state ws_w (reducing its level).

3.3 Numerical Representation of the Model

The conceptual representation of the cognitive model presented above (Fig. 1 and Table 1) shows a collection of mental states with their connections. In a numerical format the states Y have activation values $Y(t)$ (real numbers ranging between 0 and 1) over time t where also the time variable is modelled by real numbers. The update mechanism of a state from time point t to time point $t + \Delta t$ depends on all incoming connections to that state. The following elements are considered to be given as part of a conceptual representation (which makes the graph shown in Fig. 1 a labelled graph):

- For each connection from state X to state Y a *weight* $\omega_{X,Y}$ (a number between -1 and 1), for strength of impact; a negative weight is used for suppression
- For each state Y a *speed factor* η_Y (a positive value) for timing of impact
- For each state Y a *combination function* $c_Y(...)$ used to aggregate multiple impacts from different states on Y

The conceptual representation of this cognitive model can be transformed in a systematic or even automated manner into a numerical representation as follows [25] obtaining the following *difference* and *differential equation*:

$$Y(t + \Delta t) = Y(t) + \eta_Y \left[c_Y \left(\omega_{X_1, Y} X_1(t), \ldots, \omega_{X_k, Y} X_k(t) \right) - Y(t) \right] \Delta t$$
$$dY(t)/dt = \eta_Y \left[c_Y \left(\omega_{X_1, Y} X_1(t), \ldots, \omega_{X_k, Y} X_k(t) \right) - Y(t) \right]$$

In the model considered here, for all states the *advanced logistic sum combination function* **alogistic**$_{\sigma,\tau}$(...) is used:

$$c_Y(V_1, \ldots, V_k) = \textbf{alogistic}_{\sigma,\tau}(V_1, \ldots, V_k)$$
$$= \left(\frac{1}{1 + e^{-\sigma(V_1 + \ldots + V_{K-\tau})}} - \frac{1}{1 + e^{\sigma\tau}} \right) (1 + e^{-\sigma\tau})$$

Here σ is a *steepness* parameter and τ a *threshold* parameter. The advanced logistic sum combination function has the property that activation levels 0 are mapped to 0 and it keeps values below 1. For example, for the preparation state ps_{bi} the model is numerically represented as

$$\textbf{aggimpact}_{psbi}(t) = \textbf{alogistic}_{\sigma,\tau}(\omega_{srs_{ub},\, psbi}\, srs_{ub}(t),$$
$$\omega_{fs_{bi},\, ps_{bi}} fs_{bi}(t), \omega_{fs_{sat},\, ps_{bi}} fs_{sat}(t), \omega_{stg,\, ps_{bi}} stg_{,}(t))$$
$$ps_{bi}(t + \Delta t) = ps_{bi}(t) + \eta_{ps_{bi}} [\textbf{aggimpact}_{ps_{bi}}(t) - ps_{bi}(t)] \Delta t$$

The numerical representation of the Hebbian learning connections is:

$$\omega_{fs_{bi},\, ps_{bi}}(t + \Delta t) = \omega_{fs_{bi},\, ps_{bi}}(t) + [\eta\, fs_{bi}(t) ps_{bi}(t)(1 - \omega_{fs_{bi},\, ps_{bi}}(t)) - \zeta \omega_{fs_{bi},\, ps_{bi}}(t)] \Delta t$$

Simulations are based on these numerical representations. All difference equations for states and adaptive connections have been converted in a form which can be computed in a program (in Matlab).

4   Simulation Results

This section provides the details of simulation results in relation with one of the various scenarios that have been explored. These scenarios address food desire regulation strategies. The selection of regulation strategies varies and depends on the personal characteristics and the sensitivity of the person towards the type of food.

The simulations have been executed for 180 time points; the time step was $\Delta t = 0.1$ and all update speeds were $\eta = 0.1$. The rest of the parameter values for connections weights are presented in Table 2 and the values for parameters threshold τ and steepness σ are given in Table 3 (due to less space both table have been uploaded as separate file see [28]. These parameters values have been obtained by taking into consideration the patterns that are known from literature and searching for the ranges of parameter values that provide such patterns. The initial values for all states were set to 0. Note that connections from self-monitoring states to control states can be chosen according to the scenario. The following real world scenario is considered.

John is suffering from overweight and obesity and wants to loose weight. He wants to control his diet by stopping or reducing eating high calories food. John is present in an appetizing food environment with high calories food. He may apply various desire regulation strategies. He may start reinterpretation of the food by thinking about the health related consequences. He may also move his attention away from that appetizing food, or he may try to suppress his desire (e.g. he may even sit in front of that food but not eat) and he can change the situation by leaving.

The presented scenario represents one of the experiments performed. The simulation results shows how monitoring and regulation took place. Figure 2 shows that if the person initially has more positive beliefs about that type of food or stimuli it leads to a desire.

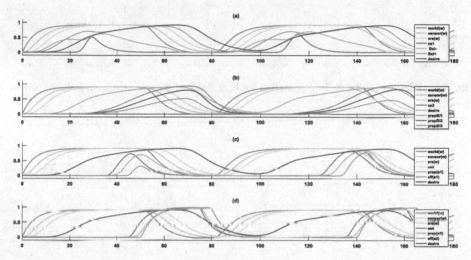

Fig. 2. (a) reinterpretation (b) suppression (c) attention deployment and (d) situation modification.

It is shown that initially this person wants to regulate the desire by applying the reinterpretation strategy see Fig. 2(a), as control state cs_1 becomes active to change the initial positive beliefs into negative beliefs (thinking about the consequences of that food type), but as this strategy is not successful, later the person tries to change his attention (e.g., changing gaze direction); this can be seen in Fig. 2(c) that shows how control state cs_3 becomes active for that purpose.

But still it didn't work well to reduce the desire, so as a last resort the person decides to leave the situation by activating control state cs_4 to perform the situation modification action which changes world state ws_w; so then there is no more stimulus and no more desire; see Fig. 2(d). Also Fig. 3 shows that there is no eating action due to the situation modification. Although the person was prepared (all preparation states become high, see Fig. 3(b)) for an eating action, there was no eating due to situation modification action (see Fig. 3(d)); the behaviour of the self-monitoring states and learnt connections can be seen in Fig. 4.

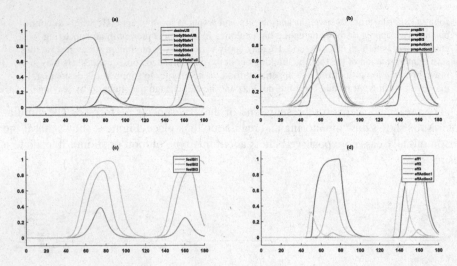

Fig. 3. (a) desire & fullness (b) preparation for b_i & actions (c) associated feelings b_i (d) actions options

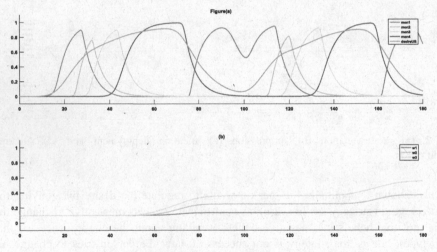

Fig. 4. (a) self-monitoring states. (b) learnt connections

5   Conclusion

In this paper, a neurologically inspired computational model of self-monitoring and decision making for desire regulation was presented. A number of simulation experiments have been performed according to different scenarios, that show how people can decide to apply different desire regulation strategies. The model incorporates a self-monitoring mechanism that is responsible for assessing the desire. These assessments are input for the decision making process on selection of desire regulation strategies. This process is based on various factors including the assessed desire level, preferences and personality characteristics such as sensitivity level towards the type of food. In this model four desire regulation strategies have been taken into account: reinterpretation, attention deployment, desire suppression and situation modification. Similar strategies are also known from approaches to emotion regulation such as [9, 29]. However, the current paper contributes a way in which these regulation strategies can be generalised to address any mental state, affective or not, and this generalisation is applied in particular to desire states.

References

1. Berridge, K.C., Ho, C.-Y., Richard, J.M., DiFeliceantonio, A.G.: The tempted brain eats: pleasure and desire circuits in obesity and eating disorders. Brain Res. **1350**, 43–64 (2010)
2. WHO | Obesity and overweight, WHO: World Health Organization (2016)
3. Vos, T., et al.: Global, regional, and national incidence, prevalence, and years lived with disability for 301 acute and chronic diseases and injuries in 188 countries, 1990–2013: a systematic analysis for the Global Burden of Disease Study 2013. Lancet **386**(9995), 743–800 (2015)
4. Buhle, J.T., Silvers, J.A., Wager, T.D., Lopez, R., Onyemekwu, C., Kober, H., Weber, J., Ochsner, K.N.: Cognitive reappraisal of emotion: a meta-analysis of human neuroimaging studies. Cereb. Cortex 1–10 (2013)
5. Giuliani, N.R., Mann, T., Tomiyama, A.J., Berkman, E.T.: Neural systems underlying the reappraisal of personally craved foods. J. Cogn. Neurosci. **26**(7), 1390–1402 (2014)
6. Hollmann, M., Hellrung, L., Pleger, B., Schlögl, H., Kabisch, S., Stumvoll, M., Villringer, A., Horstmann, A.: Neural correlates of the volitional regulation of the desire for food. Int. J. Obes. **36**(5), 648–655 (2012). (Lond)
7. Giuliani, N.R., Calcott, R.D., Berkman, E.T.: Piece of cake. Cognitive reappraisal of food craving. Appetite **64**, 56–61 (2013)
8. Gross, J.J.: The emerging field of emotion regulation: an integrative review. Rev. Gen. Psychol. **2**(3), 271–299 (1998)
9. Gross, J.J., Thompson, R.A.: Emotion regulation: conceptual foundations (2007)
10. Abro, A.H., Treur, J.: A cognitive agent model for desire regulation applied to food desires. In: Criado Pacheco, N., Carrascosa, C., Osman, N., Julián Inglada, V. (eds.) EUMAS/AT - 2016. LNCS, vol. 10207, pp. 251–260. Springer, Cham (2017). doi:10.1007/978-3-319-59294-7_20
11. Berry, L.-M., Andrade, J., May, J.: Hunger-related intrusive thoughts reflect increased accessibility of food items. Cogn. Emot. **21**(4), 865–878 (2007)
12. Caroline, D.: From passive overeating to "food addiction": a spectrum of compulsion and severity. ISRN Obes. **2013** (2013). Article ID 435027
13. Finlayson, G., King, N., Blundell, J.E.: Liking vs. wanting food: importance for human appetite control and weight regulation. Neurosci. Biobehav. Rev. **31**(7), 987–1002 (2007)
14. Berthoud, H.-R., Lenard, N.R., Shin, A.C.: Food reward, hyperphagia, and obesity. Am. J. Physiol. Regul. Integr. Comp. Physiol. **300**(6), R1266–R1277 (2011)
15. Rolls, E.T.: Understanding the mechanisms of food intake and obesity. Obes. Rev. **8**(s1), 67–72 (2007)
16. Berthoud, H.-R.: Mind versus metabolism in the control of food intake and energy balance. Physiol. Behav. **81**(5), 781–793 (2004)
17. Grill, H.J., Kaplan, J.M.: The neuroanatomical axis for control of energy balance. Front. Neuroendocrinol. **23**(1), 2–40 (2002)

18. Kober, H., Mende-Siedlecki, P., Kross, E.F., Weber, J., Mischel, W., Hart, C.L., Ochsner, K. N.: Prefrontal-striatal pathway underlies cognitive regulation of craving. Proc. Natl. Acad. Sci. **107**(33), 14811–14816 (2010)

19. Dörfel, D., Lamke, J.-P., Hummel, F., Wagner, U., Erk, S., Walter, H.: Common and differential neural networks of emotion regulation by detachment, reinterpretation, distraction, and expressive suppression: a comparative fMRI investigation. Neuroimage **101**, 298–309 (2014)

20. Ochsner, K.N., Gross, J.J.: The neural bases of emotion and emotion regulation: a valuation perspective. In: The Handbook Emotion Regulation, 2nd edn., pp. 23–41. Guilford, New York (2014)

21. Siep, N., Roefs, A., Roebroeck, A., Havermans, R., Bonte, M., Jansen, A.: Fighting food temptations: the modulating effects of short-term cognitive reappraisal, suppression and up-regulation on mesocorticolimbic activity related to appetitive motivation. Neuroimage **60**(1), 213–220 (2012)

22. Yoshikawa, T., Tanaka, M., Ishii, A., Fujimoto, S., Watanabe, Y.: Neural regulatory mechanism of desire for food: revealed by magnetoencephalography. Brain Res. **1543**, 120–127 (2014)

23. Esghaei, M., Xue, C.: Does correlated firing underlie attention deployment in frontal cortex? J. Neurosci. **36**(6), 1791–1793 (2016)

24. Ferri, J., Hajcak, G.: Neural mechanisms associated with reappraisal and attentional deployment. Curr. Opin. Psychol. **3**, 17–21 (2015)

25. Treur, J.: Dynamic modeling based on a temporal–causal network modeling approach. Biol. Inspir. Cogn. Archit. **16**(16), 131–168 (2016)

26. Damasio, A.R.: The Feeling of What Happens: Body and Emotion in the Making of Consciousness. Harcourt, New York (2000)

27. Hebb, D.O.: The Organization of Behavior. A Neuropsychological Theory (2002)

28. Abro, A.H., Treur, J.: Apendix: A Computational Cognitive Model of Self-monitoring and Decision Making for Desire Regulation, pp. 1–2. Researchgate (2017)

29. Gross, J.J.: Emotion regulation: affective, cognitive, and social consequences. Psychophysiology **39**(3), 281–291 (2002)

Video Category Classification
Using Wireless EEG

Aunnoy K Mutasim[✉], Rayhan Sardar Tipu, M. Raihanul Bashar,
and M. Ashraful Amin

Computer Vision and Cybernetics Group, Department of Computer Science
and Engineering, Independent University, Bangladesh, Dhaka, Bangladesh
{aunnoy,1330418,1320454,aminmdashraful}@iub.edu.bd

Abstract. In this paper, we present a novel idea where we analyzed EEG
signals to classify what type of video a person is watching which we believe is
the first step of a BCI based video recommender system. For this, we setup an
experiment where 13 subjects were shown three different types of videos. To be
able to classify each of these videos from the EEG data of the subjects with a
very good classification accuracy, we carried out experiments with several
state-of-the-art algorithms for each of the submodules (pre-processing, feature
extraction, feature selection and classification) of the Signal Processing module
of a BCI system in order to find out what combination of algorithms best
predicts what type of video a person is watching. We found, the best results
(80.0% with 32.32 ms average total execution time per subject) are obtained
when data of channel AF8 are used (i.e. data recorded from the electrode located
at the right frontal lobe of the brain). The combination of algorithms that
achieved this highest average accuracy of 80.0% are FIR Least Squares, Welch
Spectrum, Principal Component Analysis and Adaboost for the submodules
pre-processing, feature extraction, feature selection and classification
respectively.

Keywords: EEG · BCI · HCI · Human factor · Video Category Classification

1 Introduction

Electroencephalography (EEG) signals, due to their non-invasiveness, high time resolution, ease of acquisition, and cost effectiveness compared to other brain signal recording methods, are one of the most widely used brain signal acquisition techniques in existing Brain-Computer Interface (BCI) systems [1].

Advent of off-the-shelf wireless dry electrode EEG recording devices such as NeuroSky's MindWave [2], InteraXon's Muse [3], Emotiv Epoc [4], etc. have created a whole new domain for BCI researchers in recent years especially because of its low-cost, ease of use and portability. Emotion classification [5], attentiveness [6], task discrimination [7, 8], etc. are few examples of areas in which such devices are used in the recent past.

Video tagging from EEG signals is one other area explored by researchers [9, 10]. In [9], 17 participants were shown 49 videos (less than 10 s each) – 7 videos in each of

© Springer International Publishing AG 2017
Y. Zeng et al. (Eds.): BI 2017, LNAI 10654, pp. 39–48, 2017.
https://doi.org/10.1007/978-3-319-70772-3_4

the preselected 7 categories. At the end of each stimulus, a tag was presented for 1 s. Each of the 49 videos were shown twice (98 trials), the first followed by a matching tag and the second followed by an incorrect tag, or vice versa. The authors concluded that the N400 event-related potential (ERP) occurring after an incorrect tag is displayed, is of higher negativity compared to the N400 ERP after a correct tag.

Usage of videos as stimuli are not uncommon in EEG related BCI studies [1, 11, 12]. However, to the best of our knowledge, there are no studies that tries to passively find out whether a person is enjoying a particular genre of videos, specifically from brain signals. To do this, the very first thing that one should be able to classify is, which type of video a person is watching. With the vision of building a BCI based video recommender system, in this paper, we present a novel idea where we experimented to identify which genre of videos does one video belong to by analysing EEG signals of the person watching it.

The rest of the paper is organized as follows. Section 2 discusses the experimental setup and data acquisition techniques, Sect. 3 comprises of the experimental studies and findings, Sect. 4 discusses our findings, and finally Sect. 5 gives a brief concluding remarks.

2 Experimental Setup and Data Acquisition Techniques

2.1 Demographics of Subjects

EEG data were collected from 23 subjects, 15 males and 8 females, all of whom were second or third year undergraduate students at Independent University, Bangladesh (IUB). None of the subjects had any personal history of neurological or psychiatric illness and all of them had normal or corrected-normal vision. We also recorded the whole experiment of each of the subjects using a webcam. After analysing these videos, we excluded data of 2 females and 8 males due to reasons like, excessive blinking, hand or body movements (even after being instructed to move as less as possible), etc. to keep the signal noise at its minimum.

The maximum, minimum, average, and standard deviation age of the 13 subjects (7 males and 6 females) were 23, 20, 21.4 and 0.961 respectively and all the participants were right-handed. All the participants signed informed consent forms specific to this study only after they read and agreed to it.

2.2 EEG Recordings

Electroencephalogram (EEG) data were recorded using the Muse headband by InteraXon [3] which has been used in multiple studies as well [13–15]. This off-the-shelf dry electrode EEG recording device has 5 channels (TP9, AF7, AF8 and TP10 with reference channel at FPz) arranged according to the 10–20 international standard. The data were recorded at a sampling rate of 220 Hz and were transmitted wirelessly to a computer using Bluetooth communication module.

2.3   Experimental Setup

The experiment was designed to show the participants three different types of videos: 1. Calming and informative, 2. Fiction and 3. Emotional. The videos were chosen based on four criteria: 1. the length of the videos must be relatively short, 2. the video is to be understood without any prior specialized knowledge or explanation, 3. the videos must be interesting and/or entertaining so that the participant does not get bored and therefore distracted and 4. to keep the signal noise at its minimum, the videos should not have any content that changes one's facial expressions like smiling, squinting, etc. We carried out a questionnaire based survey on ten subjects (who did not take part in the study) to verify whether our videos meet all the criteria mentioned.

The first and third video clips were further trimmed to keep the experimental period short making sure that there was no break in scene i.e. the trimming did not disrupt the flow of the video or the story.

Each of these three videos were preceded with a five second blank black screen allocated for the resting period of the participants and another five second blank black screen at the very end of the three videos. At the very beginning a message stating "The video will start in 5 s" was shown for two seconds to give a hint of start. The full process of our experiment is of 6 min 43 s and is depicted in Fig. 1. The experiment video can be found in [16]. These videos were displayed on a 21.5-inch LED monitor with a refresh rate of 60 Hz.

Fig. 1. Video presentation sequence

The participants were fully explained to the very details about their task (i.e. just to watch all the videos) in the experiment except for the videos they will be shown. To keep the signal noise at its minimum the participants were instructed to move as less as possible. They were given time to sit comfortably before the experiment started and other than the participant, there was no one else in the room during the experiment.

3   Experimental Study and Findings

3.1   Algorithms and Methods

In our study, we carried out experiments with several state-of-the-art algorithms for each of the submodules (i.e. Pre-processing, Feature Extraction, Feature Selection, and Classification) of the Signal Processing module of a Brain-Computer Interface (BCI) system [1] in order to find out what combination of algorithms for each of the different submodules best predicts what type of video a person is watching.

In the pre-processing submodule, we mainly concentrated on different types of filters as EEG signals are highly prone to different types of noise. Basically, we carried out experiments with two basic family of filters namely, Finite Impulse Response (FIR) and Infinite Impulse Response (IIR) filters.

We designed five filters, two FIR filters and three IIR filters. The EEG signals were bandpass filtered from 5–48 Hz. Details of their configurations are presented in Table 1.

Table 1. Different filters with their design specifications

Filters	Type	Specifications
FIR	1. Equiripple (FE) 2. Least Squares (FLS)	Degree: 256 Sample rate: 220 Hz Stopband Frequency 1: 4 Hz Passband Frequency 1: 5 Hz Stopband Frequency 2: 48 Hz Passband Frequency 2: 50 Hz
IIR	3. Chebyshev I (Passband ripple) (ICS1) 4. Chebyshev II (Stopband ripple) (ICS2) 5. Elliptic (IE)	Degree: Automatic Sample rate: 220 Hz Stopband Frequency 1: 4 Hz Passband Frequency 1: 5 Hz Stopband Frequency 2: 48 Hz Passband Frequency 2: 50 Hz

The goal of the feature extraction module is to convert the raw EEG data to a set of logical useful features using which the Classification submodule can classify with better or more accuracy. The Feature Extraction methods we selected are: Discrete Wavelet Transform (DWT), Fast Fourier Transform (FFT), Welch Spectrum (PWelch), Yule – AR Spectrum (PYAR) and Short Time Fourier Transform (STFT). The parameters for each of these algorithms are presented in Table 2.

Table 2. List of feature extraction methods with their default parameters

Algorithm	DWT	FFT	PWelch	PYAR	STFT
Parameters	Wav.: db1 Dec.: 5	Nfft: 512 Freq. range: 0–110 Hz	Nfft: 512 Freq. range: 0–110 Hz	Nfft: 512 Freq. range: 0–110 Hz AR: 20	Nfft: 512 Freq. range: 0–110 Hz

Feature selection step reduces the dimensions of the features extracted from the previous step and selects only those features which contains most relevant information. In short, it is a dimensionality reduction step. We chose two dimensionality reduction algorithms, Principal Component Analysis (PCA) and Singular Value Decomposition (SVD).

We selected six classifiers to train and test using the EEG features selected in the last step. They are: Adaboost (AB), Support Vector Machines (SVMs), Multi-Class Linear Discriminant Analysis (MLDA), Multiple Linear Regression (MLR), Naïve Bayes (NB) and Decision Trees (MLTREE).

For SVM, we used one-vs-all method for multi-class classification and the parameters chosen were kernel = linear and C = 1. For Adaboost, 100 weak classifiers were used to create all the ensembles. All other classification algorithms and their default parameters relied on the implementations of MATLAB's Statistics and Machine Learning toolbox.

To evaluate the classification accuracy, we chose the 10-fold Cross-Validation approach also known as Leave-One-Out Cross-Validation (LOOCV). In our study, we used the subject-specific approach where the training and test set is generated from the same subject, i.e. we divided a single subject's EEG data into 10 trials, trained the classifier with 9 of them and tested the model generated by the classifier with the remaining one and repeated the whole process for 10 times.

It is important to note that each of the three videos shown to the participants were of different length. Therefore, we selected one minute of raw EEG data from each of these videos to keep the length of the raw EEG data of each of the videos equal to ensure that there was no biasness during classification. This selection of one minute was done based on one single criteria – the part selected should clearly state the main story line of the video. Thus, we selected, the last one minute of the first video, one minute from the exact middle of the second video and finally, the last minute of the third video. Most of the information is stated in the last minute of the first video, the main climax or story exists in the middle one minute of the second video and emotional climax is revealed in the last minute of the third video.

Before the EEG data was used to evaluate classification accuracies, we eliminated all data whose absolute difference was above 2 standard deviation from the mean (which significantly improved our classification accuracies in our pre-experiment results).

All the experiments were run in a computer with 3.4 GHz (Intel Core i7) processor, 16 GB memory and 4 GB (NVDIA GeForce) graphics card.

3.2 Experimental Results

Table 3 presents the top ten combinations of algorithms for each of the submodules of the Signal Processing module of a Brain-Computer Interface (BCI) system for each of the channels.

As per Table 3, the best combination of algorithms when data of channel TP9 are used are ICS2 filter for the pre-processing submodule, PYAR for the feature extraction submodule, PCA for the feature selection submodule and finally, SVM for the classification submodule. Combination of these algorithms achieved an average accuracy of 68.7% and on an average the total execution time taken per subject was 2.05 ms.

A better highest average accuracy of 69.7% is achieved when data of channel AF7 are used. The combination of algorithms is same as that of channel TP9 except for the pre-processing submodule where FLS (1.82 ms) or IE (1.84 ms) achieved the best accuracy.

Similar highest average accuracy of 67.9% is obtained when data of TP10 are used. But the combination of algorithms responsible for this result is quite different than the ones discussed before. They are: FLS, PYAR, SVD and MLR for the submodules

Table 3. Top ten combination of algorithms for each of the submodules for each of the channels

Channel	Filter	Feature extraction	Feature selection	Classifier	Avg. accuracy (%)	Avg. time (msec)
TP9	ICS2	PYAR	PCA	SVM	68.7	2.05
	ICS1	PYAR	PCA	SVM	68.5	2.06
	FLS	PYAR	PCA	MLR	68.5	1.81
	FE	PYAR	SVD	NB	66.9	9.97
	IE	PYAR	PCA	SVM	66.2	1.91
	FE	PYAR	PCA	SVM	65.9	1.80
	IE	DWT	PCA	NB	65.9	6.40
	ICS1	PYAR	PCA	AB	65.9	29.63
	FLS	PYAR	PCA	SVM	65.6	1.84
	ICS1	PYAR	SVD	NB	65.4	10.23
AF7	FLS	PYAR	PCA	SVM	69.7	1.82
	IE	PYAR	PCA	SVM	69.7	1.84
	ICS2	PYAR	PCA	SVM	69.5	2.03
	ICS1	PWELCH	SVD	NB	69.2	11.38
	ICS1	PWELCH	PCA	SVM	69.0	3.21
	FLS	PYAR	SVD	SVM	69.0	1.82
	FLS	PYAR	SVD	NB	69.0	9.86
	FE	PYAR	SVD	SVM	68.7	1.81
	ICS1	PYAR	PCA	SVM	68.7	2.07
	FE	PYAR	SVD	NB	68.7	10.01
AF8	FLS	PWELCH	PCA	AB	80.0	32.32
	IE	PYAR	SVD	NB	79.2	10.53
	ICS2	PYAR	PCA	AB	79.2	31.06
	FLS	PYAR	SVD	NB	79.0	10.57
	ICS1	PYAR	SVD	NB	78.5	10.75
	ICS2	PYAR	PCA	MLDA	78.2	2.98
	FLS	PWELCH	PCA	SVM	78.2	3.25
	ICS1	PYAR	PCA	AB	77.9	31.04
	ICS2	PYAR	PCA	SVM	77.9	2.22
	IE	PYAR	PCA	AB	77.7	31.05
TP10	FLS	PYAR	SVD	MLR	67.9	1.82
	FLS	PYAR	PCA	MLR	67.2	1.85
	FE	PYAR	PCA	MLR	65.4	1.86
	FE	PYAR	SVD	MLR	65.1	1.86
	FLS	PYAR	PCA	SVM	64.1	1.82
	ICS2	PYAR	SVD	MLR	63.8	2.10
	IE	PYAR	SVD	MLR	63.6	1.87
	FLS	PYAR	SVD	AB	63.3	29.53
	ICS1	PYAR	SVD	MLR	63.3	2.10
	IE	PYAR	PCA	AB	63.1	29.38

Table 4. Subject-wise accuracies for each of the channels using top ranked combination of algorithms

Subject	Avg. accuracy (%) of channel TP9	Avg. accuracy (%) of channel AF7	Avg. accuracy (%) of channel AF8	Avg. accuracy (%) of channel TP10
1	76.7	86.7	100.0	53.3
2	56.7	90.0	90.0	56.7
3	76.7	86.7	80.0	73.3
4	46.7	53.3	63.3	70.0
5	70.0	40.0	56.7	56.7
6	76.7	66.7	76.7	90.0
7	63.3	53.3	76.7	53.3
8	80.0	83.3	100.0	63.3
9	50.0	83.3	56.7	66.7
10	66.7	53.3	60.0	50.0
11	76.7	73.3	96.7	83.3
12	66.7	83.3	100.0	93.3
13	86.7	53.3	83.3	73.3
Avg. accuracy (%)	**68.7**	**69.7**	**80.0**	**67.9**

pre-processing, feature extraction, feature selection and classification respectively and the mean total execution time per subject was 1.82 ms.

The best results are obtained when data of channel AF8 are used (i.e. data recorded from the electrode located at the right frontal lobe or more specifically right dorsolateral prefrontal cortex of the brain). Even the tenth highest average classification accuracy (77.7%) is substantially higher than the best accuracies obtained by the data of each of the other three channels. This means, which type of video a person is watching can be best classified from the EEG data of channel AF8. Combination of algorithms which evaluated the highest result of 80.0% are FLS, PWelch, PCA and AB. The average total execution time per subject (32.32 ms) increased significantly mainly due to the usage of AB. However, this total execution time of 32.32 ms can still be considered as an acceptable delay for real-time applications.

Subject wise accuracies for each of the top ranked combination of algorithms for each of the channels is presented in Table 4.

4 Discussion

Many aspects of the brain are triggered when a person is watching a video, i.e. videos have the quality to evoke several types of stimuli – attention, emotion, memory, etc. all of which involves the frontal lobe of the brain. The results (80.0%) we obtained can therefore be induced from several criteria.

The reason for the result we obtained (80.0%) from the data of the channel AF8 (located at the right dorsolateral prefrontal cortex), whose results are far better than any of the other channels, can be because of the influence of working memory in subjects when watching videos. This result, supports the findings of [17] where the authors confirms that right dorsolateral prefrontal cortex plays a crucial role in spatial working memory related tasks.

Many papers [18–20] concluded that signals in the gamma band of the frontal and prefrontal cortices of the brain are heavily involved during working memory load related activity. In our paper, we filtered out the raw signal from 5 Hz to 48 Hz keeping a substantial amount of gamma band in our signal. Taking a larger range of gamma band into account might improve our classification accuracy as videos are involved in evoking working memory in viewers. This hypothesis can be even more supported as gamma band frequencies are also involved in activities requiring cross-modal sensory processing – perception combined from two separate senses, for example from sound and sight [21, 22].

The Video Category Classification (VCC) problem can also be represented as an emotion classification problem. The classification accuracy which we received from our experiment are not even close to that achieved by Wang et al. [12] (best average accuracy of 91.77%). This may be because, if we consider our VCC problem as an emotion classification problem, we considered labels of three unique classes (i.e. we tried to solve a multi-class classification problem) whereas Wang et al.'s study considered a binary classification problem which tried to determine positive and negative emotion only. The other reason for the difference in accuracies between our findings and the findings of [12] might be the use Linear Discriminant Analysis (LDA) as feature dimensionality reduction or feature selection tool which the authors claimed significantly improves emotion classification accuracy.

Other than usage of higher level gamma band frequencies and LDA, the one area which we believe will significantly improve our results is by using some artefact removal tools in the pre-processing step. Almost all EEG related studies considers this as a vital step as EEG signals are highly susceptible to noise and artefacts. Finally, tuning the parameters of each of the algorithms and using multiple combination of signals from several channels to solve the classification problem was out of the scope of this study which will definitely have a huge impact on the average classification accuracies.

5   Conclusion

Among the data of the four channels (TP9, AF7, AF8 and TP10) the highest average classification accuracy of 80.0% (substantially higher than the results of each of the other channels) is obtained when the data of channel AF8 are used (i.e. data recorded from the electrode located at the right frontal lobe of the brain) proving that data of channel AF8 is most fitted for our VCC problem. The mean total execution time taken per subject to achieve this accuracy was 32.32 ms which can be considered as an acceptable delay for real-time applications. The combination of algorithms (for each of the submodules of the Signal Processing module of a BCI system) that achieved this accuracy is FLS filter in the pre-processing step, the PWelch algorithm in the feature extraction step, PCA and AB algorithm in the feature selection and classification step respectively.

References

1. Oikonomou, V.P., Liaros, G., Georgiadis, K., et al.: Comparative evaluation of state-of-the-art algorithms for SSVEP-based BCIs. [1602.00904] (2016). https://arxiv.org/abs/1602.00904. Accessed 13 Aug 2017
2. MindWave. http://store.neurosky.com/pages/mindwave. Accessed 13 Aug 2017
3. MUSE™ | Meditation Made Easy. Muse: the brain sensing headband. http://www.choosemuse.com/. Accessed 13 Aug 2017
4. EMOTIV Epoc - 14 Channel Wireless EEG Headset. In: Emotiv. https://www.emotiv.com/epoc/. Accessed 13 Aug 2017
5. Jalilifard, A., Pizzolato, E.B., Islam, M.K.: Emotion classification using single-channel scalp-EEG recording. In: 38th Annual International Conference of the Engineering in Medicine and Biology Society (EMBC 2016), pp. 845–849. IEEE Press, Orlando (2016). https://doi.org/10.1109/EMBC.2016.7590833
6. Liu, N.-H., Chiang, C.-Y., Chu, H.-C.: Recognizing the degree of human attention using EEG signals from mobile sensors. Sensors **13**(8), 10273–10286 (2013). doi:10.3390/s130810273
7. Nine, M.S.Z., Khan, M., Poon, B., Amin, M.A., Yan, H.: Human computer interaction through wireless brain computer interfacing device. In: 9th International Conference on Information Technology and Applications (ICITA 2014) (2014)
8. Paul, S.K., Zulkar Nine, M.S.Q., Hasan, M., Amin, M.A.: Cognitive task classification from wireless EEG. In: Guo, Y., Friston, K., Aldo, F., Hill, S., Peng, H. (eds.) BIH 2015. LNCS, vol. 9250, pp. 13–22. Springer, Cham (2015). doi:10.1007/978-3-319-23344-4_2
9. Koelstra, S., Mühl, C., Patras, I.: EEG analysis for implicit tagging of video data. In: 2009 3rd International Conference on Affective Computing and Intelligent Interaction and Workshops, pp. 1–6. IEEE Press, Amsterdam (2009). https://doi.org/10.1109/ACII.2009.5349482
10. Soleymani, M., Pantic, M.: Multimedia implicit tagging using EEG signals. In: 2013 IEEE International Conference on Multimedia and Expo (ICME), pp. 1–6. IEEE Press, San Jose (2013). https://doi.org/10.1109/ICME.2013.6607623
11. Hubert, W., Jong-Meyer, R.D.: Autonomic, neuroendocrine, and subjective responses to emotion-inducing film stimuli. Int. J. Psychophysiol. **11**(2), 131–140 (1991). doi:10.1016/0167-8760(91)90005-I
12. Wang, X.W., Nie, D., Lu, B.L.: Emotional state classification from EEG data using machine learning approach. Neurocomputing **129**, 94–106 (2014). doi:10.1016/j.neucom.2013.06.046
13. Abujelala, M., Abellanoza, C., Sharma, A., Makedon, F.: Brain-EE: brain enjoyment evaluation using commercial EEG headband. In: Proceedings of the 9th ACM International Conference on Pervasive Technologies Related to Assistive Environments, p. 33. ACM, Island of Corfu, Greece (2016). https://doi.org/10.1145/2910674.2910691

14. Galway, L., McCullagh, P., Lightbody, G., Brennan, C., Trainor, D.: The potential of the brain-computer interface for learning: a technology review. In: 2015 IEEE International Conference on Computer and Information Technology; Ubiquitous Computing and Communications; Dependable, Autonomic and Secure Computing; Pervasive Intelligence and Computing, pp. 1554–1559. IEEE Press, Liverpool (2015). https://doi.org/10.1109/CIT/IUCC/DASC/PICOM.2015.234

15. Karydis, T., Aguiar, F., Foster, S.L., Mershin, A.: Performance characterization of self-calibrating protocols for wearable EEG applications. In: Proceedings of the 8th ACM International Conference on PErvasive Technologies Related to Assistive Environments, p. 38. ACM, New York (2015). https://doi.org/10.1145/2769493.2769533

16. Experiment 1 Version 2. YouTube (2016). https://www.youtube.com/watch?v=eITcEnCOMc0&feature=youtu.be. Accessed 13 Aug 2017

17. Giglia, G., Brighina, F., Rizzo, S., Puma, A., Indovino, S., Maccora, S., Baschi, R., Cosentino, G., Fierro, B.: Anodal transcranial direct current stimulation of the right dorsolateral prefrontal cortex enhances memory-guided responses in a visuospatial working memory task. Func. Neurol. **29**(3), 189–193 (2014). doi:10.11138/FNeur/2014.29.3.189

18. Howard, M.W., Rizzuto, D.S., Caplan, J.B., Madsen, J.R., Lisman, J., Aschenbrenner-Scheibe, R., Schulze-Bonhage, A., Kahana, M.J.: Gamma oscillations correlate with working memory load in humans. Cereb. Cortex **13**(12), 1369–1374 (2003). doi:10.1093/cercor/bhg084

19. Linden, D.E.J., Oosterhof, N.N., Klein, C., Downing, P.E.: Mapping brain activation and information during category-specific visual working memory. J. Neurophysiol. **107**(2), 628–639 (2011). doi:10.1152/jn.00105.2011

20. Roux, F., Wibral, M., Mohr, H.M., et al.: Gamma-Band Activity in Human Prefrontal Cortex Codes for the Number of Relevant Items Maintained in Working Memory. J. Neurosci. **32**, 12411–12420 (2012). doi:10.1523/jneurosci.0421-12.2012

21. Kanayama, N., Sato, A., Ohira, H.: Crossmodal effect with rubber hand illusion and gamma-band activity. Psychophysiology **44**(3), 392–402 (2007). doi:10.1111/j.1469-8986.2007.00511.x

22. Kisley, M.A., Cornwell, Z.M.: Gamma and beta neural activity evoked during a sensory gating paradigm: effects of auditory somatosensory and cross-modal stimulation. Clin. Neurophysiol. **117**(11), 2549–2563 (2006). doi:10.1016/j.clinph.2006.08.003

Learning Music Emotions via Quantum Convolutional Neural Network

Gong Chen[1], Yan Liu[1(✉)], Jiannong Cao[1], Shenghua Zhong[2], Yang Liu[3], Yuexian Hou[4], and Peng Zhang[4]

[1] Department of Computing, The Hong Kong Polytechnic University,
Hong Kong, China
{csgchen,csyliu}@comp.polyu.edu.hk
[2] College of Computer Science and Software Engineering,
Shenzhen University, Shenzhen, China
[3] Department of Computer Science, Hong Kong Baptist University,
Hong Kong, China
[4] Tianjin Key Laboratory of Cognitive Computing and Application,
School of Computer Science and Technology, Tianjin University, Tianjin, China

Abstract. Music can convey and evoke powerful emotions. But it is very challenging to recognize the music emotions accurately by computational models. The difficulty of the problem can exponentially increase when the music segments delivery multiple and complex emotions. This paper proposes a novel quantum convolutional neural network (QCNN) to learn music emotions. Inheriting the distinguished abstraction ability from deep learning, QCNN automatically extracts the music features that benefit emotion classification. The main contribution of this paper is that we utilize measurement postulate to simulate the human emotion awareness in music appreciation. Statistical experiments on the standard dataset shows that QCNN outperforms the classical algorithms as well as the state-of-the-art in the task of music emotion classification. Moreover, we provide demonstration experiment to explain the good performance of the proposed technique from the perspective of physics and psychology.

Keywords: Music emotion · Convolutional neural network · Quantum mechanics · Superposition collapse

1 Introduction

Music is an artistic form of auditory communication incorporating instrumental or vocal tones in a structured and continuous manner [1]. No matter what style of music may interest you, the ability of arousing powerful emotions might be an important reason behind most people's engagement with music [2,3]. Such an amazing ability of music has fascinated not only the general public but also the researchers from different fields throughout the ages [4].

As defined in Drever's psychology dictionary [5], emotion is a mental state of excitement or perturbation, marked by a strong feeling, and usually an impulse

© Springer International Publishing AG 2017
Y. Zeng et al. (Eds.): BI 2017, LNAI 10654, pp. 49–58, 2017.
https://doi.org/10.1007/978-3-319-70772-3_5

towards a definite form of behavior. Ekman presented a classical categorical model of emotions based on the human facial expressions, which divides emotions into six basic classes: anger, happiness, surprise, disgust, sadness, and fear [6].

For music emotions, psychologists proposed several models specifically. A hierarchical model called Geneva emotional music scale (GEMS-45) is designed by professionals, which includes 40 labels such as moved, and heroic [3]. Turnbull et al. collected a number of user-generated annotations that describe 500 Western popular music tracks, and generated 18 easily understood emotion labels such as calming, and arousing [7]. Unlike the categorical models that represent the musical emotions using a number of classes, another kind of emotion models, called dimensional models, describe the emotions in a Cartesian space with valence and arousal (V-A space) as two dimensions. Here valence means how positive or negative the affect appraisal is and arousal means how high or low the physiological reaction is [8,9]. For instance, happiness is an emotion of positive valence and high arousal, while calm is an emotion of positive valence and low arousal.

It has taken on increasing interests to automate music emotion analysis using computational approaches [10,11]. Considering that one music segment often delivers multiple emotions, Li and Ogihara formulated the music emotion detection as a multi-label classification problem and decomposed it into a set of binary classification problems [12]. Wu et al. proposed a multi-label multi-layer multi-instance multiview learning scheme for music emotion recognition [13]. To discover the nature between music features and human's emotion, multi-label dimensionality reduction techniques are introduced and improved the classification accuracy further [14,15]. Chen et al. proposed an adaption method for personalized emotion recognition in V-A space [16]. Weninger et al. utilized in deep recurrent neural networks for mood regression in V-A space [17].

In summary, most computational models on music emotion analysis utilize artificial intelligence techniques and have achieved good progress in some real world applications, such as music recommendation [18] and music video generation [19]. However, it is very challenging to recognize the music emotions accurately when the music segment delivers multiple and complex emotions. One reason is that these classical artificial intelligence techniques are mainly designed to simulate human's rational thinking and decision making while emotion response is sentimental and spontaneous. Hence, this paper raises a question in the suitability of applying artificial intelligence models to affective computing when most existing models are initially designed for logical reasoning. To address this problem, we want to try a new kind of technology based on quantum theory, which can provide significant cognitive implication in emotion modeling. A brief review of quantum information is provided in Sect. 2. Then we propose a novel quantum convolutional neural network for music emotion analysis in Sect. 3. In Sect. 4, we validate the proposed algorithm on a standard dataset CAL500exp [20]. The paper is closed with conclusion and future work.

2 Related Work on Quantum Information

Quantum information is the study of the information processing tasks that can be accomplished using quantum mechanics, which is a mathematical framework or set of rules for the construction of physical theories. The rules of quantum mechanics are simple, but some of them are counter-intuitive. For example, measurement changes the state of a qubit, is one of the fundamental postulates of quantum mechanics. Mathematically, a qubit can be completely described by a state vector called ket $|\alpha\rangle$. In quantum theory, a qubit can be in a superposition state. For example, $a|0\rangle + b|1\rangle$ indicates the qubit is in a superposition both $|0\rangle$ and $|1\rangle$ under $|a|^2 + |b|^2 = 1$. When a qubit is being measured, the superposition state will collapse into one of the base states either $|0\rangle$ or $|1\rangle$ with certain probability.

Quantum theory has been widely used in information theory since 2004. A pioneering work proposed the quantum language to describe objects and processes [21]. Following this pioneering work, the Quantum Probability Ranking Principle (QPRP) was proposed to rank interdependent documents based on the quantum interference [22]. Inspired by the photon polarization, a Quantum Measurement inspired Ranking model (QMR) was introduced [23]. Zhu et al. proposed a non-parametric quantum theory to rank images [24]. Later, a Quantum Language Model (QLM) was proposed to model term dependencies based on the quantum theory and outperformed the conventional IR model [25]. We have successfully modeled quantum entanglements in quantum language models and improved the performance of re-ranking tasks [26]. Latest work in quantum cognition indicated a new theoretical approach to psychology [27].

3 Quantum Convolutional Neural Network for Music Emotion Analysis

3.1 Rationale

The proposed technique utilizes the measurement postulate to simulate the emotion awareness in music appreciation. The ability of a qubit to be in a superposition state runs counter to our common sense understanding of the physical world. However, quantum theory is much natural to describe the emotion response, and sometime even better than using classical physical language. A kind of emotion, for example, happiness can be defined in a continuum of states between $|happy\rangle$ and $|non - happy\rangle$. When that emotion is observed consciously, just like filling the questionnaire, the superposition collapses into either happy or non-happy probabilistically. It is supported by the psychology that the observation of emotion state has influence on the emotion. We simulate the measurement postulate in the music emotion analysis as supervised learning part.

Note that the superposition of emotion states is different from mixture of emotion states. For example, one music segment may evoke both happy and sad emotions, although they are always considered as the opposite emotions.

More general case is that one music segment conveys cheerful, pleasant, light, and touching simultaneously. Considering the complexity of the emotions, we formulate the music emotion analysis as a multi-label classification problem while each label is represented in a continuum of states and collapsed into a determinate state when measured.

We develop our technique under convolutional neural networks framework because of three considerations. First, the multiple-layer filters can simulate the audio data perception, which has the good cognition implication. Second, convolutional neural networks have demonstrated the good ability of feature extraction and abstraction, which is helpful for finding the features that essentially contribute emotion evoking. Third, convolutional neural networks have shown good performance in multi-label learning problems.

In summary, we propose a novel quantum convolutional neural network for multi-label classification problem.

3.2   Quantum Convolutional Neural Network

The framework of the proposed algorithm is shown in Fig. 1. Given a music segment $M_i \in \{M_1, M_2, ..., M_N\}$, a second order time-frequency feature $X_i \in \mathbb{R}^{T \times F}$ is generated using the short-time Fourier transform (STFT), where N are the number of training data, T and F are the numbers of sampling points in the time order and frequency order, respectively. The corresponding emotion label is denoted as y_i, where $y_i^c = 1$ defines that M_i evokes the cth emotion.

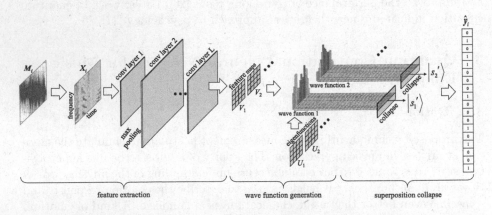

Fig. 1. Illustration of quantum convolutional neural network

The training set $\{(X_1, y_1), ..., (X_N, y_N)\}$ is input to the CNN with three convolutional layers and two pooling operations. The output feature maps can be denoted by $V_k \in \{V_1, V_2, ..., V_K\}$, where K is the number of feature maps. We assume that V_k is a kind of emotion representation when affective state $|\alpha_k\rangle$ is expanded in a certain kind of emotion base $\{|a_k\rangle\}$ as follows:

$$V_k = \langle a_k | \alpha_k \rangle. \tag{1}$$

With the Dirac's notation, a vector $\boldsymbol{\alpha}_k$ corresponding to a quantum state can be expressed as the ket $|\alpha_k\rangle$ and its transpose $\boldsymbol{\alpha}_k{}^{\mathrm{T}}$ is the bra $\langle\alpha_k|$. $\{|a_k\rangle\}$ denotes a set of base kets with J elements in the affective state space.

Following Sakurai's formalism [28], we give the expansion representation for $|\alpha_k\rangle$

$$|\alpha_k\rangle = \sum_{j=1}^{J} |a_k\rangle\langle a_k|\alpha_k\rangle. \tag{2}$$

We define another set of base kets $\{|s_k\rangle\}$ with J elements, and multiply Eq. (2) by the eigenbra $\langle s_k|$ as follows:

$$\langle s_k|\alpha_k\rangle = \sum_{j=1}^{J} \langle s_k|a_k\rangle\langle a_k|\alpha_k\rangle. \tag{3}$$

In wave mechanics [29], $\langle s_k|\alpha_k\rangle$ is usually refered to as the wave function for state $|\alpha_k\rangle$. The wave function represented by the ψ-function [29] with corresponding expansion is as follows:

$$\psi_{\alpha_k}(s_k) = \sum_{j=1}^{J} \langle s_k|a_k\rangle\langle a_k|\alpha_k\rangle. \tag{4}$$

We introduce an eigenfunction with eigenvalue a_k

$$U_k(a_k, s_k) = \langle s_k|a_k\rangle. \tag{5}$$

Intergrating Eqs. (1) and (5), we rewrite Eq. (4) as follows:

$$\psi_{\alpha_k}(s_k) = \sum_{j=1}^{J} V_k U_k(a_k, s_k). \tag{6}$$

For a superposition state $|\alpha_k\rangle$, once a measurement occurs, the superposition state would collapse into a certain state, which is one of all eigenstates [28]. We define the probability $P_{\alpha_k}(s_k)$ with which the affective state $|\alpha_k\rangle$ collapses into the eigenstate $|s_k\rangle$ as follows:

$$P_{\alpha_k}(s_k) = |\psi_{\alpha_k}(s_k)|^2. \tag{7}$$

We calculate the output label $\boldsymbol{y}_i$ according to the probability in Eq. (7). To minimize the difference between output label and the ground truth, we define the objective function as follows:

$$\arg\min_{\boldsymbol{W},\boldsymbol{U}} \sum_{i=1}^{n} |\hat{\boldsymbol{y}}_i(\boldsymbol{X}_i, \boldsymbol{W}, \boldsymbol{U}), \boldsymbol{y}_i|^2, \tag{8}$$

where $\boldsymbol{W}$ is the parameter space of CNN. We learn $\boldsymbol{W}$ and $\boldsymbol{U}$ using stochastic gradient descent method. The computational complexity of QCNN is similar to the one of CNN.

4 Experiments

In this section, we evaluate the performance of the proposed technique on a standard music emotion dataset CAL500exp [20]. This dataset has 500 songs with totally eighteen emotions, including: angry/aggressive, arousing/awakening, bizarre/weird, calming/smoothing, carefree/lighthearted, cheerful/festive, emotional/passionate, exciting/thrilling, happy, laid-back/mellow, light/playful, loving/romantic, pleasant/comfortable, positive/optimistic, powerful/strong, sad, tender/soft, touching/loving. We extract 5580 3-second music segments for training and test. For each music segment, we generate a 64×64 time-frequency matrix as input data using STFT. The emotion ground truth of each music segment is represented by a 18-dimensional vector, in which each component indicates the existence/non-existence of a specific emotion.

The time-frequency matrix is input to the CNN with three convolutional layers. The first convolutional layer filters the input image with 20 kernels of size 5×5. The second convolutional layer takes as input the pooled output of the first convolutional layer and filters it with 50 kernels of size 4×4. And the third convolutional layer has 100 kernels of size 6×6. The CNN is initialized using zero-mean Gaussian distribution with standard derivation 0.01. For simplicity, we set the number of eigenfunctions to 18.

Two groups of criteria are used to evaluate the performance. In the first group, we use the standard label-based metrics, i.e., the precision and F1 score, as the evaluation criteria [15]. Since precision and F1 score are originally designed for binary classification, we use macro average and micro average to evaluate the overall performance across multiple labels. For these four criteria, the larger the metric value the better the performance. In the second group, we use four standard example-based metrics, i.e., average precision, Hamming loss, one-error, and ranking loss, as the evaluation criteria [15]. For average precision, the larger the metric value the better the performance. For Hamming loss, one-error, and ranking loss, the smaller the metric value the better the performance. For both groups, we perform 10-fold cross validation.

We demonstrate the effectiveness of the proposed methods by comparing it with four representative solutions for multi-label problem, including rank-support vector machine (Rank-SVM) [30], backpropagation for multi-label learning

Table 1. Performance evaluation on CAL500exp dataset using label-based metrics

Methods	Macro average		Micro average	
	Precision	F1 score	Precision	F1 score
Rank-SVM	0.572	0.534	0.569	0.544
BP-MLL	0.551	0.504	0.544	0.548
BML-CNN	0.650	0.628	0.639	0.610
ME-SPE	0.513	0.555	0.580	0.534
QCNN	**0.672**	**0.640**	**0.664**	**0.631**

Table 2. Performance evaluation on CAL500exp dataset using example-based metrics

Methods	Average precision	Hamming loss	One-error	Ranking loss
Rank-SVM	0.573	0.408	0.343	0.300
BP-MLL	0.560	0.425	0.440	0.355
ML-CNN	0.650	0.277	0.290	0.248
ME-SPE	0.638	**0.225**	0.294	0.251
QCNN	**0.687**	0.257	**0.281**	**0.246**

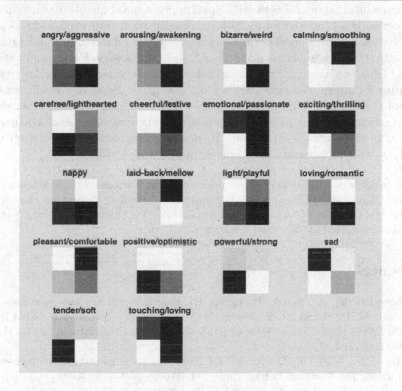

Fig. 2. Visualization of eigenfunctions

(BP-MLL) [31], multi-label convolutional neural networks (ML-CNN), and multi-emotion similarity preserving embedding (ME-SPE) with kNN classifier [15]. As shown in Tables 1 and 2, ML-CNN and QCNN show much better performance under most cases. QCNN always performs better and stably compared with ML-CNN.

To provide the more intuitive impression of quantum information processing, we visualize the eigenfunctions corresponding to the determination of the emotion state in Fig. 2. More bright components indicate more features are used. For some simple emotions, such as emotional/passionate, one component is enough to determine its existence. But for calming/soothing, three components have involved in the emotion determination. We can compare the eigenfunctions of loving/romantic and loving/touching. Although both of them deliver loving emotion, the label romantic is more complex than the label touching, hence, more components involve in the emotion determination for loving/romantic.

5   Conclusions

This is the first work that utilizes quantum theory to analyze music emotions under convolutional neural networks. Specifically, a novel quantum convolutional neural network is proposed for multi-label classification. The proposed technique not only demonstrates good performance on standard dataset but also provides some interesting observations from physics and psychology. Future work will be explored from two aspects. Frist, we will develop novel quantum based regression techniques for music emotion analysis in V-A space. Second, we will consider the label correlation. This paper regards every component in the label vector as independent one. But as indicated in [32], label correction like co-occurrence, is a kind of important information. To describe complex music emotion more accurately, we will integrate label correlation in data modeling.

Acknowledgments. This work was supported by the National Natural Science Foundation of China under Grants 61373122 and Open Funding Project of Tianjin Key Laboratory of Cognitive Computing and Application. We thank Dr. Haidong Yuan for the helpful discussion on quantum mechanics.

References

1. Wieczorkowska, A., Synak, P., Lewis, R., Raś, Z.W.: Extracting emotions from music data. In: Hacid, M.-S., Murray, N.V., Raś, Z.W., Tsumoto, S. (eds.) ISMIS 2005. LNCS (LNAI), vol. 3488, pp. 456–465. Springer, Heidelberg (2005). doi:10.1007/11425274_47
2. Juslin, P.N., Västfjäll, D.: Emotional responses to music: the need to consider underlying mechanisms. Behav. Brain Sci. **31**(05), 559–575 (2008)
3. Zentner, M., Grandjean, D., Scherer, K.R.: Emotions evoked by the sound of music: characterization, classification, and measurement. Emotion **8**(4), 494 (2008)
4. Eerola, T., Vuoskoski, J.K.: A comparison of the discrete and dimensional models of emotion in music. Psychol. Mus. **39**, 18–49 (2010)
5. Drever, J.: A Dictionary of Psychology. Oxford University Press, Oxford (1952)
6. Ekman, P.: Expression and the nature of emotion. In: Approaches to Emotion, vol. 3, pp. 19–344 (1984)
7. Turnbull, D., Barrington, L., Torres, D., Lanckriet, G.: Semantic annotation and retrieval of music and sound effects. IEEE Trans. Audio Speech Lang. Process. **16**(2), 467–476 (2008)
8. Russell, J.A.: A circumplex model of affect. J. Pers. Soc. Psychol. **39**(6), 1161–1178 (1980)
9. Thayer, R.E.: The Biopsychology of Mood and Arousal. Oxford University Press, Oxford (1990)

10. Lu, L., Liu, D., Zhang, H.J.: Automatic mood detection and tracking of music audio signals. IEEE Trans. Audio Speech Lang. Process. **14**(1), 5–18 (2006)
11. Yang, Y.H., Lin, Y.C., Su, Y.F., Chen, H.H.: A regression approach to music emotion recognition. IEEE Trans. Audio Speech Lang. Process. **16**(2), 448–457 (2008)
12. Li, T., Ogihara, M.: Content-based music similarity search and emotion detection. In: IEEE International Conference on Acoustics, Speech, and Signal Processing (ICASSP), Proceedings, vol. 5, p. V-705. IEEE (2004)
13. Wu, B., Zhong, E., Horner, A., Yang, Q.: Music emotion recognition by multi-label multi-layer multi-instance multi-view learning. In: Proceedings of the 22nd ACM International Conference on Multimedia, pp. 117–126. ACM (2014)
14. Panagakis, Y., Kotropoulos, C.: Automatic music tagging via parafac2. In: 2011 IEEE International Conference on Acoustics, Speech and Signal Processing (ICASSP), pp. 481–484. IEEE (2011)
15. Liu, Y., Liu, Y., Zhao, Y., Hua, K.A.: What strikes the strings of your heart? Feature mining for music emotion analysis. IEEE Trans. Affect. Comput. **6**(3), 247–260 (2015)
16. Chen, Y.A., Wang, J.C., Yang, Y.H., Chen, H.: Linear regression-based adaptation of music emotion recognition models for personalization. In: 2014 IEEE International Conference on Acoustics, Speech and Signal Processing (ICASSP), pp. 2149–2153. IEEE (2014)
17. Weninger, F., Eyben, F., Schuller, B.: On-line continuous-time music mood regression with deep recurrent neural networks. In: 2014 IEEE International Conference on Acoustics, Speech and Signal Processing (ICASSP), pp. 5412–5416. IEEE (2014)
18. Wang, X., Rosenblum, D., Wang, Y.: Context-aware mobile music recommendation for daily activities. In: Proceedings of the 20th ACM International Conference on Multimedia, pp. 99–108. ACM (2012)
19. Lin, J.C., Wei, W.L., Wang, H.M.: Demv-matchmaker: emotional temporal course representation and deep similarity matching for automatic music video generation. In: 2016 IEEE International Conference on Acoustics, Speech and Signal Processing (ICASSP), pp. 2772–2776. IEEE (2016)
20. Wang, S.Y., Wang, J.C., Yang, Y.H., Wang, H.M.: Towards time-varying music auto-tagging based on cal500 expansion. In: 2014 IEEE International Conference on Multimedia and Expo (ICME), pp. 1–6. IEEE (2014)
21. Van Rijsbergen, C.J.: The Geometry of Information Retrieval. Cambridge University Press, Cambridge (2004)
22. Zuccon, G., Azzopardi, L.: Using the quantum probability ranking principle to rank interdependent documents. In: Gurrin, C., et al. (eds.) ECIR 2010. LNCS, vol. 5993, pp. 357–369. Springer, Heidelberg (2010). doi:10.1007/978-3-642-12275-0_32
23. Zhao, X., Zhang, P., Song, D., Hou, Y.: A novel re-ranking approach inspired by quantum measurement. In: Clough, P., Foley, C., Gurrin, C., Jones, G.J.F., Kraaij, W., Lee, H., Mudoch, V. (eds.) ECIR 2011. LNCS, vol. 6611, pp. 721–724. Springer, Heidelberg (2011). doi:10.1007/978-3-642-20161-5_79
24. Zhu, S., Wang, B., Liu, Y.: Using non-parametric quantum theory to rank images. In: 2012 IEEE International Conference on Acoustics, Speech and Signal Processing (ICASSP), pp. 1049–1052. IEEE (2012)
25. Sordoni, A., Nie, J.Y., Bengio, Y.: Modeling term dependencies with quantum language models for IR. In: Proceedings of the 36th International ACM SIGIR Conference on Research and Development in Information Retrieval, pp. 653–662. ACM (2013)

26. Xie, M., Hou, Y., Zhang, P., Li, J., Li, W., Song, D.: Modeling quantum entanglements in quantum language models. In: Proceedings of the 24th International Joint Conference on Artificial Intelligence, pp. 1362–1368. AAAI Press (2015)

27. Bruza, P.D., Wang, Z., Busemeyer, J.R.: Quantum cognition: a new theoretical approach to psychology. Trends Cognit. Sci. **19**(7), 383–393 (2015)

28. Sakurai, J.J., Napolitano, J.: Modern Quantum Mechanics. Addison-Wesley, Reading (2011)

29. Bohr, N., et al.: The Quantum Postulate and the Recent Development of Atomic Theory, vol. 3. Printed in Great Britain by R. & R. Clarke, Limited (1928)

30. Elisseeff, A., Weston, J.: A kernel method for multi-labelled classification. In: Advances in neural Information Processing Systems, pp. 681–687 (2001)

31. Zhang, M.L., Zhou, Z.H.: Multilabel neural networks with applications to functional genomics and text categorization. IEEE Trans. Knowl. Data Eng. **18**(10), 1338–1351 (2006)

32. Lo, H.Y., Wang, J.C., Wang, H.M., Lin, S.D.: Cost-sensitive stacking for audio tag annotation and retrieval. In: 2011 IEEE International Conference on Acoustics, Speech and Signal Processing (ICASSP), pp. 2308–2311. IEEE (2011)

Supervised EEG Source Imaging with Graph Regularization in Transformed Domain

Feng Liu[1], Jing Qin[2], Shouyi Wang[1](✉), Jay Rosenberger[1], and Jianzhong Su[3]

[1] Department of Industrial, Manufacturing and Systems Engineering,
University of Texas at Arlington, Arlington, TX, USA
feng.liu@mavs.uta.edu, {shouyiw,jrosenbe}@uta.edu
[2] Department of Mathematical Sciences, Montana State University,
Bozeman, MT, USA
jing.qin@montana.edu
[3] Department of Mathematics, University of Texas at Arlington,
Arlington, TX, USA
su@uta.edu

Abstract. It is of great significance to infer activation extents under different cognitive tasks in neuroscience research as well as clinical applications. However, the EEG electrodes measure electrical potentials on the scalp instead of directly measuring activities of brain sources. To infer the activated cortex sources given the EEG data, many approaches were proposed with different neurophysiological assumptions. Traditionally, the EEG inverse problem was solved in an unsupervised way without any utilization of the brain status label information. We propose that by leveraging label information, the task related discriminative extended source patches can be much better retrieved from strong spontaneous background signals. In particular, to find task related source extents, a novel supervised EEG source imaging model called Graph regularized Variation-Based Sparse Cortical Current Density (GVB-SCCD) was proposed to explicitly extract the discriminative source extents by embedding the label information into the graph regularization term. The graph regularization was derived from the constraint that requires consistency for all the solutions on different time points within the same class. An optimization algorithm based on the alternating direction method of multipliers (ADMM) is derived to solve the GVB-SCCD model. Numerical results show the effectiveness of our proposed framework.

Keywords: EEG source imaging · Discriminative source · Graph regularization · Total variation (TV) · Alternating direction method of multiplier (ADMM)

1 Introduction

Electroencephalography (EEG) is a non-invasive brain imaging technique that records the electric field on the scalp generated by the synchronous activation of

© Springer International Publishing AG 2017
Y. Zeng et al. (Eds.): BI 2017, LNAI 10654, pp. 59–71, 2017.
https://doi.org/10.1007/978-3-319-70772-3_6

neuronal populations. It has been previously estimated that if as few as one in a thousand synapses become activated simultaneously in a region of about $40 \, \text{mm}^2$ of cortex, the generated signal can be detected and recorded by EEG electrodes [1,2]. Compared to other functional neuroimaging techniques such as functional magnetic resonance imaging (fMRI) and positron emission tomography (PET), EEG is a direct measurement of real-time electrical neural activities, so EEG can provide the answer on exactly *when* different brain regions are involved during different processing [3]. PET and fMRI, by contrast, measure brain activity indirectly through associated metabolic or cerebrovascular changes which are slow and time-delayed [2–4].

Although EEG has the advantages of easy portability, low cost and high temporal resolution, it can only measure the electrical potential from on the scalp instead of measuring directly the neural activation on the cortex. Reconstructing the activated brain sources from the recorded EEG data is called inverse problem (also known as source localization). Due to its significance in clinical applications and scientific investigations on understanding how our brain is functioning under different cognitive tasks, numerous methods have been proposed to solve the inverse problem [5–8]. Furthermore, source localization can serve as a preliminary step for brain connectivity network analysis [9,10], as the calculation of connectivity between two brain regions is based on the measurement of how "closely" related of the time series from those two regions, using Pearson correlation, transfer entropy etc. as summarized in [11]. The next step is to analyze the brain network using complex networks properties such as small-worldness or clustering coefficient [12–17], as we saw a trend that more attention has been focused on in neuroscience community from traditional "segregation" perspective to "integration" perspective where the functional and effective connectivity are intensively studied [18] in the past decades.

It is very challenging to solve the EEG inverse problem since it is highly ill-posed due to the fact that the number of dipoles is much larger than that of electrodes. To seek a unique solution, regularization technique could be applied to incorporate prior knowledge of sources. The most traditionally used priors are based on minimum energy, resulting in the ℓ_2 norm based minimum norm estimate (MNE) inverse solver [19]. By replacing ℓ_2 by ℓ_1, minimum current estimate (MCE) [20] can overcome the disadvantage of over-estimation of active area sizes incurred by the ℓ_2 norm. Bio-inspired algorithms such as genetic algorithm [21], Particle Swarm Optimization (PSO) is also used in EEG source estimation [22,23]. Pascual-Marqui et al. presented standardized low-resolution brain electromagnetic tomography (sLORETA) [7] that enforces spatial smoothness of the neighboring sources and normalizes the solution with respect to the estimated noise level; Gramfort et al. proposed the Mixed Norm Estimates (MxNE) which imposes sparsity over space and smoothness over time using the $\ell_{1,2}$-norm regularization [24]. In [25], the same authors proposed time-frequency mixed-norm estimate (TF-MxNE) which makes use of structured sparse priors in time-frequency domain for better estimation of the non-stationary and transient source signal. Li et al. presented the graph Fractional-Order Total Variation (gFOTV) [26], which enhances the spatial smoothness of the source signal by utilizing the sparsity of the fractional order TV defined on a triangular mesh. Liu et al. first proposed to combine the EEG inverse problem and the classification problem into a joint framework which is solved using a sparse dictionary learning technique [27]. In recent years, it has been found that by imposing sparseness on the original domain is inappropriate for the extended source estimation, source extents can be obtained by enforcing sparseness in transformed domains with total variation regularization [6,8,26,28], which makes more sense considering the true sources are not distributed in an isolated and independent way.

It is worth noting that the aforementioned algorithms for solving the EEG inverse problem used different prior assumptions of sources configurations. To the best of our knowledge, the traditional algorithms solved the inverse problem in an unsupervised way and did not use the brain status label information such as happiness, sadness or calm. Due to the fact that brain spontaneous sources contribute to the majority of EEG data, label information plays a key role to find the discriminative task-related sources [27,29,30]. The remaining question is why not utilizing label information to better reconstruct discriminative source and how to use those information and fuse the label information into traditional inverse models.

In this research, we proposed a graph regularized EEG inverse model that uses label information to estimate source extents with sparsity constraint in the transformed domain. The graph regularization was derived from the assumption that requires intra-class consistency for the solutions at different time points. The proposed graph regularized model was solved by the alternating direction method of multipliers (ADMM). The proposed model is tested to find discriminative source extents and its effectiveness is illustrated by numerical experiments.

2   Inverse Problem

EEG data are mostly generated by pyramidal cells in the gray matter with an orientation perpendicular to the cortex. It is well established to assume the orientation of cortex sources is perpendicular to the surface [31]. With a so-called lead field matrix L that describes the superposition of linear mappings from the cortex sources to the EEG recording sensors, the forward model can be described as

$$X = LS + \varepsilon, \tag{1}$$

where $X \in \mathbb{R}^{N \times T}$ is the EEG data measured at N electrodes on the scalp for T time samples and ε represents the noise , each column of lead field matrix L ($L \in \mathbb{R}^{N \times D}$) represents the electrical potential mapping pattern of a cortex dipole to the EEG electrodes. D is the number of dipoles. Figure 1 illustrates the process of building a realistic brain model, the cortex sources or dipoles are represented with triangle meshes, each triangle represents a dipole. Here $S \in \mathbb{R}^{D \times T}$ represents the electrical potentials in D source locations for all the T time points that are transmitted to surface of the scalp. As L is a wide matrix with much more columns than rows, the inverse problem of inferring S from

X becomes ill-posed. Mathematically speaking, to seek a unique solution, a regularization technique could be applied. An estimate of S can be calculated by minimizing the following objective function:

$$\underset{S}{\operatorname{argmin}} \; \|X - LS\|_F^2 + \lambda R(S), \qquad (2)$$

the first term is called fidelity term measuring the fitting error and the second term is called regularization term. Here $\|\cdot\|_F$ is the Frobenius norm which sums up the squared values of all entries in a matrix. The regularization term $R(S)$ is to encourage spatially or temporally smooth source configurations and to enforce neurophysiologically plausible solutions. For example, to restrict the total number of total activated sources, ℓ_0 norm can be ideally used as it measures the cardinality of sources. However, the ℓ_0-norm regularized problem is NP-hard from the perspective of computational complexity. Instead of ℓ_0-norm, many convex or quasi-convex norms have been used as $R(\cdot)$, such as ℓ_1. More precisely, source localization can be obtained by solving the following ℓ_1-regularized model.

$$\underset{s_i}{\operatorname{argmin}} \; \|x_i - Ls_i\|_2^2 + \gamma \|s_i\|_1, \qquad (3)$$

where s_i is the sparse coding for all the dipoles, i.e., the i-th column of S, and the nonzero entry of s_i represents the activated source. We want to find the best s_i to fit the observed data x_i while maintaining sparse configurations.

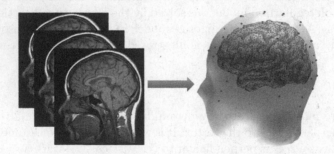

Fig. 1. From MRI images to realistic brain model. After gathering the MRI scans of the head, tissue segmentation is conducted followed by mesh generation. After assigning conductivity values to different tissues and electrodes co-registered with the meshing model, boundary element method (BEM) was used to solve the forward model. Each triangle represents a brain source, the direction of the source is assumed to be perpendicular to the triangular surface.

3   Graph Regularized EEG Source Imaging in Transformed Domain

In this section, we first introduce the sparsity related regularization in the transformed domain which promotes the discovery of extended source patches rather than isolated sources on the cortex. Our proposed graph regularized model in the transformed domain is presented in detail.

3.1   EEG Source Imaging in Transformed Domain

As the cortex sources are discretized with meshing triangles, simply distributed source imaging using the ℓ_1 norm on S will cause the calculated sources distributed in an isolated way instead of grouped as source extents. In order to encourage source extents estimation, Ding [6] proposed to used sparse constraint in the transformed domain and the model was termed as Variation-Based Sparse Cortical Current Density (VB-SCCD). In [8], Zhu *et al.* proposed to use multiple priors including variation-based and wavelet-based constraints. The VB-SCCD model can be extended by adding sparse regularization in the original source domain named Sparse VB-SCCD (SVB-SCCD) [32]. The common yet most important term is the TV term, which considers the total sparsity of the gradients of the source distribution so that the spatial smoothness can be guaranteed. The total variation was defined to be the ℓ_1 norm of the transformed domain using a linear transform characterized with the matrix $V \in \mathbb{R}^{P \times N}$ whose entries are

$$\begin{cases} v_{ij} = 1; v_{ik} = -1; \text{ if voxels } j,k \text{ share edge } i; \\ \qquad v_{ij} = 0; \text{ otherwise.} \end{cases} \tag{4}$$

where and N is the number of voxels, P is the number of edges from the triangular grid. The motivation of the total variation regularization is illustrated in Fig. 2, when the red triangle is estimated to be activated, the neighboring blue voxels should also be activated.

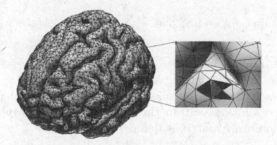

Fig. 2. Illustration of V matrix design purpose. When one voxel (in red) is activated, the neighbor voxels (in blue) are encouraged to be activated to achieve smaller goal value in Eq. 5 (Color figure online)

We can see that the matrix VS contains all differences of amplitudes of any two adjacent voxels. Then the model for the VB-SCCD has the following form:

$$\langle S \rangle = \underset{S}{\operatorname{argmin}} \|X - LS\|_F^2 + \lambda \|VS\|_{1,1}. \tag{5}$$

3.2 Discriminative Source Reconstruction with Graph Regularization

Previous studies [18,33] indicated that the brain spontaneous sources contribute most part of the EEG signal. The neurons in our brain are still active even the subjects are in closed-eye resting state. As a result, the source solution given by traditional EEG inverse algorithm is likely to be corrupted by background noise. A simple example is that suppose $x_1 = L(s_0 + s_1)$ and $x_2 = L(s_0 + s_2)$, and s_0 is the spontaneous common source across different classes: brain status 1 and status 2. Here s_1 is the discriminative source for class 1, and s_2 is the discriminative source for class 2. Without the label information, traditional methods such as MNE, MCE, MxNE are trying to estimate the overall source activation pattern instead of the task related ones. Even worse, when the magnitude of s_0 is much greater than s_1 and s_2, the traditional method will be more likely to fail.

Inspired by graph regularization in computer vision community for discovering discriminators of images [34–37], the proposed model employs a Laplacian graph regularization term in the original VB-SCCD and is termed as GVB-SCCD in this paper. Although we can consider the graph regularized SVB-SCCD model which involves sparsity regularization imposed on both the original source signal domain and TV transformed domain [8], the parameter tuning process will be very complicated since it involves three parameters that balance the trade off between data fidelity, sparsity on original domain, sparsity on the transformed domain, and graph regularization, and its effectiveness will be evaluated in our future work. The graph regularization tries to eradicate the spurious sources that are not consistent intra-class. The common source inter-class are decomposed as the first step using Voting Orthogonal Matching Pursuit (VOMP) algorithm proposed in [29]. The GVB-SCCD model is given below:

$$\langle S \rangle = \underset{S}{\text{argmin}} \, \frac{1}{2} \|X - LS\|_F^2 + \frac{\beta}{2} \sum_{i,j=1}^{N} \|s_i - s_j\|_2^2 M_{ij} + \lambda \|VS\|_{1,1}. \qquad (6)$$

Here $\|\cdot\|_{1,1}$ is the ℓ_1 norm notation for a matrix, equal to the sum of absolute values of all entries from a matrix. $X \in \mathbb{R}^{N \times T}$ the EEG data, where T is the number of samples from different classes. The second term that penalizes the inconsistent source solutions within the same class is the graph regularization term. The definition of M matrix is defined as:

$$M_{ij} = \begin{cases} +1, & \text{if } (s_i, s_j) \text{ belong to the same class;} \\ 0, & \text{if } (s_i, s_j) \text{ otherwise.} \end{cases} \qquad (7)$$

The goal of this formulation is to find discriminative sources by decomposing the common source while maintaining the consistency of intra-class reconstructed sources. By defining D as a diagonal matrix whose entries are row sums of the symmetric matrix M, $D_{ii} = \sum_j M_{ij}$ and the graph Laplacian $G = D - M$ [34], the second term of Eq. (6) is further expanded as:

$$\sum_{i,j=1}^{N} \|s_i - s_j\|_2^2 M_{ij} = \sum_{i,j=1}^{N} (s_i^T s_i + s_j^T s_j - 2s_i^T s_j) M_{ij} = 2\text{Tr}(SGS^T). \qquad (8)$$

As a result, Eq. (6) is written as

$$\langle S \rangle = \underset{S}{\text{argmin}} \frac{1}{2} \|X - LS\|_F^2 + \beta(\text{Tr}(SGS^T)) + \lambda \|VS\|_{1,1}, \tag{9}$$

where the second term is called graph regularization. Using variable splitting, Eq. (9) is equivalent to

$$\min_S \frac{1}{2} \|X - LS\|_F^2 + \lambda \|Y\|_{1,1} + \beta(\text{Tr}(SGS^T))$$
$$s.t. \quad Y = VS. \tag{10}$$

The new formulation makes the objective function separable with respect to the variables S and Y. For Problem (10), S can also be written in a separable form as

$$\min_{s_i} \frac{1}{2} \|x_i - Ls_i\|_2^2 + \lambda \|y_i\|_1 + \beta G_{ii} s_i^T s_i + s_i^T h_i$$
$$s.t. \quad y_i = V s_i, \tag{11}$$

where $h_i = 2\beta(\sum_{j \neq i} G_{ij} s_j)$, and x_i, s_i, y_i and z_i are the i-th column of the matrices X, S, Y and Z, respectively, G_{ij} is the (i,j)-th entry of the matrix G.

4 Optimization with ADMM Algorithm

After reformulating Problem (11) to unconstrained augmented Lagrangian function, it can be solved using ADMM [38]:

$$L_p(s_i, y_i, u_i) = \frac{1}{2} \|x_i - Ls_i\|_2^2 + \lambda \|y_i\|_1 + \beta G_{ii} s_i^T s_i + s_i^T h_i \tag{12}$$
$$+ u_i^T (V s_i - y_i) + \frac{\rho}{2} \|V s_i - y_i\|_2^2.$$

The variable s_i, y_i, u_i are updated sequentially, with the hope that each subproblem has a closed form solution or can be calculated efficiently. In short, ADMM results in two sub-problems Eqs. (13) and (14) plus a variable update Eq. (15),

$$s_i^{(k+1)} := \underset{s}{\text{argmin}} \, L_\rho(s, y_i^{(k)}, u_i^{(k)}) = \underset{s}{\text{argmin}} \frac{1}{2} \|x_i - Ls\|_2^2 + \beta G_{ii} s^T s + s^T h_i$$
$$+ \frac{\rho}{2} \left\| Vs - y_i^{(k)} + \frac{u_i^{(k)}}{\rho} \right\|_2^2 \tag{13}$$

$$y_i^{(k+1)} := \underset{y}{\text{argmin}} \, L_\rho(s_i^{(k+1)}, y, u_i^{(k)}) = \underset{y}{\text{argmin}} \lambda \|y\|_1 + \frac{\rho}{2} \left\| V s_i^{(k+1)} - y + \frac{u_i^{(k)}}{\rho} \right\|_2^2 \tag{14}$$

$$u_i^{(k+1)} := u_i^{(k)} + \rho(V s_i^{(k+1)} - y_i^{(k+1)}) \tag{15}$$

The update of $s_i^{(k+1)}$ has a closed form solution, which is

$$s_i^{(k+1)} = P^{-1}[L^T x_i - h_i + \rho V^T(y_i^{(k)} - \frac{u_i^{(k)}}{\rho})], \tag{16}$$

where $P = L^T L + 2\beta G_{ii} I + \rho V^T V$. The update of $y_i^{(k+1)}$ involves the proximal operator of the ℓ_1-norm, and its update can be expressed as:

$$y_i^{(k+1)} = \text{shrink}(V s_i^{(k+1)} + \frac{u_i^{(k+1)}}{\rho}, \frac{\lambda}{\rho}), \tag{17}$$

where the shrinkage function is defined as

$$\text{shrink}(v, \mu) = (|v| - \mu)_+ \, \text{sgn}(v). \tag{18}$$

Here $(x)_+$ is x when $x > 0$, otherwise 0, and $\text{sgn}(v)$ is the sign of v. The shrinkage function provides an efficient way to solve the ℓ_1-regularized minimization problem due to its calculation is element-wise.

The procedure for solving problem (10) is summarized in Algorithm 1. Although it is time-consuming to update all s_i's, the separability of s_i's suggests further improvement with the help of parallel computing techniques.

Algorithm 1. ADMM framework for solving goal function (10)

INPUT: Lead field matrix L, EEG signal matrix X, Laplacian Graph G, total variation matrix V, parameter α and β, λ
OUTPUT: Source matrix S
while the stopping criteria is not met
    for $i = 1, \ldots, N$
    Alternating update until converge:
    $s_i^{(k+1)} = P^{-1}[L^T x_i - h_i + \rho V^T(y_i^{(k)} - \frac{u_i^{(k)}}{\rho})]$,
    $y_i^{(k+1)} = \text{shrink}(V s_i^{(k+1)} + \frac{u_i^{(k+1)}}{\rho}, \frac{\lambda}{\rho})$,
    $u_i^{(k+1)} = u_i^{(k)} + \rho(V s_i^{(k+1)} - y_i^{(k+1)})$
    end for
end while

---

5   Numerical Experiment

A realistic head model called "New York Head" [39] and synthetic data with known ground truth is used for validation of our method. The dimension of lead field matrix is 108 by 2004 for the head model. We sampled 1 s of the data with 200 Hz frequency for each class of brain status. The number of classes is fixed to be 3. An extended common source patch for all 3 classes with 4,

6, 8 neighboring sources are assumed to be activated respectively. The task related discriminative source extents for each class also have the same number of neighboring sources activated. The magnitude of common source that represents spontaneous activation pattern is set as 0.8 with a standard derivation of 0.1 and the discriminative source is assigned to be 0.2 with a standard derivation to be 0.05. To mimic the noise from other source locations, we set the magnitude of 10 spurious sources as 0.35 or 0.5 in the simulation experiments. The spurious sources are randomly chosen in each time point but not consistent intra-class. Under the current experiment settings, the actual activated number of source can be 18, 22 and 26. According to the result in [40], the recovery rate drops quickly when the number of dipoles increases. When the number of activated sources is 20, the recovery rate is about 40% since the lead field matrix has a very high coherence across columns. The parameters λ and β in Eq. (9) is set be 0.1 and 1 respectively. Each configuration was repeated 5 times and the ground truth source localization is randomly simulated in 8 different Regions of Interest (ROI) defined in [41]. The initialization of S in ADMM algorithm was done by solving the ℓ_1 constrained problem using Homotopy Algorithm.

The accuracy is based on the shortest path distance (in mm) from the ground truth location to the reconstructed source location along the cortex surface. The final results under different simulation settings are summarized in Table 1. In Table 1, PAcc represents the primary common source location accuracy (in mm), and DAcc represents the discriminative source location accuracy averaged from 3 classes. The simulation result illustrated our proposed method can perform better than all benchmark methods for discriminative source under different configurations, and also performs well for the primary common source. Moreover, from the comparison of the benchmark methods, the ℓ_1 based methods within MCE framework is better than the ℓ_2 based methods such as sLORETA and MNE. The ℓ_2 based methods give an overestimation of sources and ignore the task related discriminative sources. The ℓ_1 constrained MCE can recover both the primary source well, however, the discriminative source extent is corrupted by spurious sources. Our method can provide a sparse and precise reconstruction

Table 1. Accuracy summary

SSM	Patch size = 4				Patch size = 6				Patch size = 8			
	0.35		0.5		0.35		0.5		0.35		0.5	
Method	PAcc	DAcc	PAcc	DAcc	PAcc	DAcc	PAcc	DAcc	PAcc	DAcc	PAcc	DAcc
Homotopy	4.76	33.32	6.97	46.56	4.25	22.43	14.69	47.45	5.08	22.72	12.55	30.48
DALM	4.75	34.04	7.94	46.98	4.08	22.67	15.03	46.85	**4.65**	22.58	12.93	30.97
FISTA	5.63	39.04	5.78	58.25	6.20	34.08	13.66	49.56	8.84	43.16	**8.09**	33.84
sLoreta	9.75	160.33	14.44	154.7	5.07	179.31	13.98	133.15	5.99	166.53	11.26	152.43
MNE	9.51	140.73	22.16	136.33	6.06	161.86	26.28	90.32	7.69	151.67	22.31	126.56
GVB-SCCD	**2.26**	**7.85**	**3.88**	**3.46**	**3.69**	**8.99**	**7.92**	**13.67**	4.81	**9.05**	10.24	**12.18**

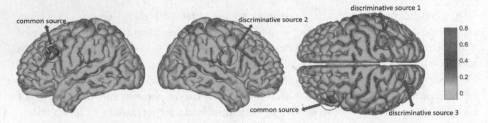

Fig. 3. Ground truth for all 3 classes aggregated in one figure with a common source and 3 discriminative sources

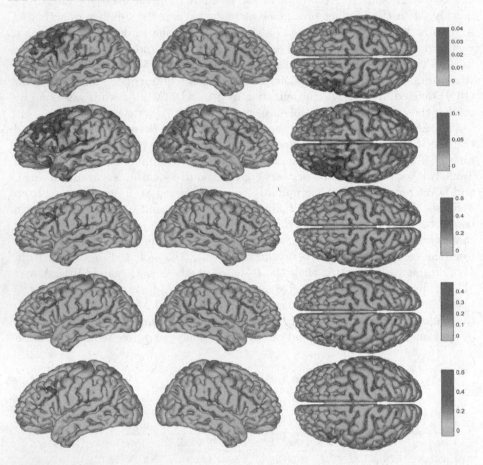

Fig. 4. Illustration of primary source reconstruction and discriminative source reconstruction by difference methods. The first row is source solution provided by MNE, the second row is from the solution of sLORETA, the third and fourth row are DALM and FISTA method within the MCE framework, the last row is our proposed method GVB-SCCD.

of both primary source and discriminative source. For illustration, the ground truth source and reconstructed source from different algorithms are given in Figs. 3 and 4 respectively. Those illustration results are from the case when spurious source magnitude (SSM) is equal to 0.5 and the patch size is equal to 6. The brain imaging figures in Fig. 4 also show that the ℓ_2 based methods gave a diffuse solution and failed to recover the discriminative sources. Our proposed method can recover the primary source as well as discriminative source by encouraging intra-class consistency imposed by graph regularization.

6 Conclusion

The EEG inverse problem is usually solved independently under different brain states. In this research, we used label information to find discriminative source patches by fusing it into a graph regularization term. The proposed model called GVB-SCCD has the advantage to better find the task related activation pattern than traditional methods. The additional graph regularization can promote intra-class consistency thus eliminate the spurious sources. The proposed ADMM algorithm is given to solve the GVB-SCCD model with better performance than the benchmark methods validated in the numerical experiments. One of the drawbacks for the proposed framework is the TV term only allow smoothness for the first spatial derivative, future work will consider a smoother higher-order TV to enhance smoothness of reconstructed sources.

References

1. Hämäläinen, M., Hari, R., Ilmoniemi, R.J., Knuutila, J., Lounasmaa, O.V.: Magnetoencephalographytheory, instrumentation, and applications to noninvasive studies of the working human brain. Rev. Modern Phys. **65**(2), 413 (1993)
2. Lamus, C., Hämäläinen, M.S., Temereanca, S., Brown, E.N., Purdon, P.L.: A spatiotemporal dynamic solution to the meg inverse problem: An empirical bayes approach. arXiv preprint (2015). arXiv:1511.05056
3. Michel, C.M., Murray, M.M., Lantz, G., Gonzalez, S., Spinelli, L., de Peralta, R.G.: EEG source imaging. Clin. Neurophysiol. **115**(10), 2195–2222 (2004)
4. Haufe, S., Nikulin, V.V., Ziehe, A., Müller, K.R., Nolte, G.: Combining sparsity and rotational invariance in EEG/MEG source reconstruction. Neuroimage **42**(2), 726–738 (2008)
5. Costa, F., Batatia, H., Chaari, L., Tourneret, J.Y.: Sparse EEG source localization using bernoulli laplacian priors. IEEE Trans. Biomed. Eng. **62**(12), 2888–2898 (2015)
6. Ding, L.: Reconstructing cortical current density by exploring sparseness in the transform domain. Phys. Med. Biol. **54**(9), 2683 (2009)
7. Pascual-Marqui, R.D., et al.: Standardized low-resolution brain electromagnetic tomography (sloreta): technical details. Methods Find. Exp. Clin. Pharmacol. **24**(Suppl D), 5–12 (2002)
8. Zhu, M., Zhang, W., Dickens, D.L., Ding, L.: Reconstructing spatially extended brain sources via enforcing multiple transform sparseness. Neuroimage **86**, 280–293 (2014)

9. Mahjoory, K., Nikulin, V.V., Botrel, L., Linkenkaer-Hansen, K., Fato, M.M., Haufe, S.: Consistency of EEG source localization and connectivity estimates. Neuroimage **152**, 590–601 (2017)
10. Yang, Y., Aminoff, E., Tarr, M., Robert, K.E.: A state-space model of cross-region dynamic connectivity in MEG/EEG. In: Advances in Neural Information Processing Systems, pp. 1234–1242 (2016)
11. Liu, F., Xiang, W., Wang, S., Lega, B.: Prediction of seizure spread network via sparse representations of overcomplete dictionaries. In: Ascoli, G.A., Hawrylycz, M., Ali, H., Khazanchi, D., Shi, Y. (eds.) BIH 2016. LNCS (LNAI), vol. 9919, pp. 262–273. Springer, Cham (2016). doi:10.1007/978-3-319-47103-7_26
12. Guan, Z.H., Liu, F., Li, J., Wang, Y.W.: Chaotification of complex networks with impulsive control. Chaos: an interdisciplinary. J. Nonlinear Sci. **22**(2), 023137 (2012)
13. Newman, M.E.: The structure and function of complex networks. SIAM Rev. **45**(2), 167–256 (2003)
14. Watts, D.J., Strogatz, S.H.: Collective dynamics of small-world networks. Nature **393**(6684), 440–442 (1998)
15. Zhang, H., Guan, Z.H., Li, T., Zhang, X.H., Zhang, D.X.: A stochastic sir epidemic on scale-free network with community structure. Physica A Statist. Mech. Appl. **392**(4), 974–981 (2013)
16. Zhang, H., Shen, Y., Thai, M.T.: Robustness of power-law networks: its assessment and optimization. J. Comb. Optim. **32**(3), 696–720 (2016)
17. Chen, G., Cairelli, M.J., Kilicoglu, H., Shin, D., Rindflesch, T.C.: Augmenting microarray data with literature-based knowledge to enhance gene regulatory network inference. PLoS Comput. Biol. **10**(6), e1003666 (2014)
18. Hipp, J.F., Hawellek, D.J., Corbetta, M., Siegel, M., Engel, A.K.: Large-scale cortical correlation structure of spontaneous oscillatory activity. Nature Neurosci. **15**(6), 884–890 (2012)
19. Hämäläinen, M.S., Ilmoniemi, R.J.: Interpreting magnetic fields of the brain: minimum norm estimates. Med. Biol. Eng. Comput. **32**(1), 35–42 (1994)
20. Uutela, K., Hämäläinen, M., Somersalo, E.: Visualization of magneto encephalographic data using minimum current estimates. Neuroimage **10**(2), 173–180 (1999)
21. Han, Z., Wang, D., Liu, F., Zhao, Z.: Multi-AGV path planning with double-path constraints by using an improved genetic algorithm. PLoS One **12**(7), e0181747 (2017)
22. McNay, D., Michielssen, E., Rogers, R., Taylor, S., Akhtari, M., Sutherling, W.: Multiple source localization using genetic algorithms. J. Neurosci. Methods **64**(2), 163–172 (1996)
23. Shirvany, Y., Mahmood, Q., Edelvik, F., Jakobsson, S., Hedstrom, A., Persson, M.: Particle swarm optimization applied to EEG source localization of somatosensory evoked potentials. IEEE Trans. Neural Syst. Rehabil. Eng. **22**(1), 11–20 (2014)
24. Gramfort, A., Kowalski, M., Hämäläinen, M.: Mixed-norm estimates for the M/EEG inverse problem using accelerated gradient methods. Phys. Med. Biol. **57**(7), 1937 (2012)
25. Gramfort, A., Strohmeier, D., Haueisen, J., Hämäläinen, M.S., Kowalski, M.: Time-frequency mixed-norm estimates: sparse M/EEG imaging with non-stationary source activations. Neuroimage **70**, 410–422 (2013)
26. Li, Y., Qin, J., Hsin, Y.L., Osher, S., Liu, W.: s-SMOOTH: sparsity and smoothness enhanced EEG brain tomography. Front. Neurosci. **10**, 543 (2016)

27. Liu, F., Wang, S., Rosenberger, J., Su, J., Liu, H.: A sparse dictionary learning framework to discover discriminative source activations in EEG brain mapping. In: AAA, vol. 1, pp. 1431–1437 (2017)
28. Sohrabpour, A., Lu, Y., Worrell, G., He, B.: Imaging brain source extent from EEG/MEG by means of an iteratively reweighted edge sparsity minimization (ires) strategy. Neuroimage **142**, 27–42 (2016)
29. Liu, F., Hosseini, R., Rosenberger, J., Wang, S., Su, J.: Supervised discriminative EEG brain source imaging with graph regularization. In: Descoteaux, M., Maier-Hein, L., Franz, A., Jannin, P., Collins, D.L., Duchesne, S. (eds.) MICCAI 2017. LNCS, vol. 10433, pp. 495–504. Springer, Cham (2017). doi:10.1007/978-3-319-66182-7_57
30. Liu, F., Rosenberger, J., Lou, Y., Hosseini, R., Su, J., Wang, S.: Graph regularized EEG source imaging with in-class consistency and out-class discrimination. IEEE Trans. Big Data (2017)
31. Becker, H., Albera, L., Comon, P., Gribonval, R., Wendling, F., Merlet, I.: Localization of distributed EEG sources in the context of epilepsy: a simulation study. IRBM **37**(5), 242–253 (2016)
32. Becker, H., Albera, L., Comon, P., Gribonval, R., Merlet, I.: Fast, variation-based methods for the analysis of extended brain sources. In: 2014 Proceedings of the 22nd European Signal Processing Conference (EUSIPCO), pp. 41–45. IEEE (2014)
33. Raichle, M.E.: The brain's dark energy. Science **314**(5803), 1249–1250 (2006)
34. Cai, D., He, X., Han, J., Huang, T.S.: Graph regularized nonnegative matrix factorization for data representation. IEEE Trans. Pattern Anal. Mach. Intell. **33**(8), 1548–1560 (2011)
35. Guo, H., Jiang, Z., Davis, L.S.: Discriminative dictionary learning with pairwise constraints. In: Lee, K.M., Matsushita, Y., Rehg, J.M., Hu, Z. (eds.) ACCV 2012. LNCS, vol. 7724, pp. 328–342. Springer, Heidelberg (2013). doi:10.1007/978-3-642-37331-2_25
36. Lu, X., Wang, Y., Yuan, Y.: Graph-regularized low-rank representation for destriping of hyperspectral images. IEEE Trans. Geosci. Remote Sensing **51**(7), 4009–4018 (2013)
37. Ramirez, I., Sprechmann, P., Sapiro, G.: Classification and clustering via dictionary learning with structured incoherence and shared features. In: CVPR, pp. 3501–3508. IEEE (2010)
38. Boyd, S., Parikh, N., Chu, E., Peleato, B., Eckstein, J.: Distributed optimization and statistical learning via the alternating direction method of multipliers. Found. Trends Mach. Learn. **3**(1), 1–122 (2011)
39. Huang, Y., Parra, L.C., Haufe, S.: The New York Head - a precise standardized volume conductor model for EEG source localization and tES targeting. Neuroimage **140**, 150–162 (2016). Transcranial electric stimulation (tES) and Neuroimaging
40. Costa, F., Batatia, H., Oberlin, T., D'giano, C., Tourneret, J.Y.: Bayesian EEG source localization using a structured sparsity prior. Neuroimage **144**, 142–152 (2017)
41. Haufe, S., Ewald, A.: A simulation framework for benchmarking EEG-based brain connectivity estimation methodologies. Brain Topogr., 1–18 (2016)

Insula Functional Parcellation from FMRI Data via Improved Artificial Bee-Colony Clustering

Xuewu Zhao[1,2], Junzhong Ji[1(✉)], and Yao Yao[1]

[1] Beijing Municipal Key Laboratory of Multimedia and Intelligent Software,
Faculty of Information Technology, Beijing University of Technology, Beijing, China
zhaoxuewuyonghu@163.com, jjz01@bjut.edu.cn, yaoyao1314@emails.bjut.edu.cn
[2] College of Software, Nanyang Normal University, Nanyang, China

Abstract. The paper presents a novel artificial bee colony clustering (ABCC) algorithm with a self-adaptive multidimensional search mechanism based on difference bias for insula functional parcellation, called as DABCC. In the new algorithm, the preprocessed functional magnetic resonance imaging (fMRI) data was mapped into a low-dimension space by spectral mapping to reduce its dimension in the initialization. Then, clustering centers in the space were searched by the search procedure composed of employed bee search, onlooker bee search and scout bee search, where a self-adaptive multidimensional search mechanism based on difference bias for employed bee search was developed to improve search capability of ABCC. Finally, the experiments on fMRI data demonstrate that DABCC not only has stronger search ability, but can produce better parcellation structures in terms of functional consistency and regional continuity.

Keywords: Insula functional parcellation · Spectral mapping · Artificial bee colony clustering · Self-adaptive multidimensional search mechanism

1 Introduction

Insula is a triangular island under the lateral fissure and in the depths of the sulcus [1], which involves in many kinds of brain activity, such as smell, taste process, motor sense, interoception and motivation [2]. Moreover, related researches have also shown that insula is associated with Alzheimer's disease, Parkson's disease and anxiety disorder. However, we have not yet reached a unified cognition about insula functional organization up to now. Thus, exploring it further both contributes to understand Human brain cognition mechanism deeply and is beneficial to grasp the pathogenesis of cerebral diseases more precisely.

Human brain functional parcellation is a crucial method for functional organization of Human brain by dividing the whole brain or local regions into some parcels/subregions [3,4]. In recent years, some researchers attempted to investigate functional parcellation of insula based on fMRI data. However, this method

© Springer International Publishing AG 2017
Y. Zeng et al. (Eds.): BI 2017, LNAI 10654, pp. 72–82, 2017.
https://doi.org/10.1007/978-3-319-70772-3_7

has a strong dependence on experimental data collected in BrainMap. Literature [2] clustered functional connectivity between voxels in insula with K-means and three functional subregions were obtained. But K-means is sensitive to initial values and easily traps into local optimum. In 2012, Cauda et al. performed hierarchical clustering after building meta-analysis connectivity on data about insula from BrainMap, revealing insula functional organization at different granularity [5]. Although hierarchical clustering can show the whole process of clustering, it needs bigger storage space and it is difficult to determine the final parcellation result. In 2015, Vercelli et al. applied fuzzy clustering to insula functional parcellation and twelve functional subregions were obtained in left and right insula respectively [6]. However, fuzzy clustering easily falls into local optimum. In 2016, literature [7] employed independent component analysis (ICA) to divide insular into four functional subregions. Its advantage is to easily find the potential data structure and independent components obtained by ICA are unstable. These studies are a helpful exploration of insula functional organization. However, most of them applied existed methods to insula functional parcellation, which could not address challenges of the high dimensionality and low signal-to-noise ratio of fMRI data well and results in poor functional parcellation of insula in terms of functional consistency and regional continuity.

Researches suggest that swarm intelligence algorithms have strong search ability and some robustness and clustering methods based on them have proved to be superior to traditional clustering methods [8]. Especially, artificial bee colony (ABC) algorithm has the virtues of simplicity, fewer parameters and ease of implementation and artificial bee colony clustering (ABCC) algorithms that are based on ABC have strong competitiveness in terms of search and clustering [9].

Driven by the above motivation, we propose an improved ABCC for insula functional parcellation. To reduce the dimension of fMRI data, DABCC projected it into a low-dimension space by spectral mapping in initialization. Then, the search procedure of improved ABCC, composed of employed bee search, onlooker bee search and scout bee search, was employed to search cluster centers in the space. For employed bee search, a self-adaptive multidimensional search mechanism based on difference bias was developed to improve the search capability of ABCC. The experiments on fMRI data show that DABCC has stronger search capability and can produce parcellation results with better functional consistency and regional continuity in comparison with some clustering algorithms.

2   Related Content

2.1   Insula Functional Parcellation Based on FMRI Data

During fMRI data acquisition, a fMRI scanner divides a brain into voxels by magnetic field gradient in it for imaging and records changes of magnetic resonance signal intensity of each voxel at different time points. As a result, time series of every voxel are obtained. Taking time series of voxels in insula as data, insula functional parcellation based on fMRI data segments insula into some mutually disjoint functional subregions (parcels) with strong functional consistency in space. More intuitively, it assigns voxels in insula to different functional subregions/parcels according to the similarity of the corresponding time series. Generally, a parcellation result with stronger functional consistency and spatial continuity is preferred.

2.2   Artificial Bee Colony (ABC) Algorithm

ABC, proposed by Karaboga in 2005, is a novel meta-heuristic search algorithm by simulating the intelligence foraging behavior of honey bees [10]. In the initialization operation, some parameters are initialized, such as the population size N, the max iterative count N_C and the maximum mining times thr. Then, the population (food sources), solutions of some problem to solve, is initialized in the light of (1):

$$x_{ij} = lb_j + rand(0, 1) * (ub_j - lb_j) \tag{1}$$

where $i = 1, 2, \cdots, N$, $j = 1, 2, \cdots, d$ (d is the dimension of a food source). x_{ij} denotes the j-th dimension of the i-th food source $\mathbf{x}_i = \{x_{i1}, x_{i2}, \cdots, x_{id}\}$. lb_j and ub_j represent the lower and upper bounds of the j-th dimension, respectively. Meanwhile, each food source has a variable, named as $limit$, to evaluate its mining potential. After initialization, search operations are performed.

(1) Employed bee search. The number of employed bees is equal to that of the population and each employed bee associates with a food source in ABC. For each food source, the corresponding employed bee performs neighborhood search in accordance with (2):

$$v_{ij} = x_{ij} + \phi_{ij} * (x_{rj} - x_{ij}) \tag{2}$$

where x_{rj} and v_{rj} denotes the j-th dimension of a neighbor food source $\mathbf{x}_r$ and a candidate food source $\mathbf{v}_i$. $\phi_{ij} \in [-1, 1]$. If the fitness of $\mathbf{v}_i$ is better than that of $\mathbf{x}_i$, $\mathbf{x}_i$ is replaced with $\mathbf{v}_i$ and the corresponding $limit$ is set to 0; otherwise, it is increased by one. After employed bee search, the probability of each food source is computed according to (3):

$$p_i = fit_i / \sum_{j=1}^{N} fit_j \tag{3}$$

where fit_j is the fitness of the j-th food source. The probabilities and fitness information are used in the following onlooker bee search.

(2) Onlooker bee search. After getting the above information, an onlooker bee selects a food source according to some selection strategy. Then, it searches a new food source according to (2). Likewise, the greedy selection between them is done and the $limit$ is also updated similarly.

(3) Scout bee search. After onlooker bee search, if *limit* of a food source is greater than *thr*, the employed bee associated with it changes into a scout bee to randomly search a new food source according to (1) and the scout bee becomes into an employed bee again. Theoretically, a scout bee can get any point in search space.

From the above, an employed bee and an onlooker bee perform exploitation search, while a scout bee does exploration search. However, an employed bee and an onlooker bee adopt the similar search pattens in the basic ABC and only one dimension is searched with (2) in a iteration, limiting search performance of artificial bees.

3   DABCC Algorithm

3.1   Food Source Representation

DABCC first performed spectral mapping and then searched cluster centers in a low-dimension space. Suppose that the dimension of a low-dimension space is K after spectral mapping. A cluster center is a K-dimensional vector, so a cluster solution consisted of K clusters can be represented as a K^2-dimensional vector. Thus, a food source can be repressed as a K^2-dimensional vector composed of K cluster centers.

3.2   Initialization

For the low signal-noise ratio and the high dimension of the preprocessed fMRI data, we did not directly cluster it, but first performed a mapping for it in the initialization. Here, spectral mapping based on spectral graph theory was adopted for the purpose. Its pseudo-code is shown in Algorithm 1. After spectral mapping, some parameters were initialized. Subsequently, the population was assigned randomly and their fitness was computed.

Algorithm 1. spectral mapping

Input: D: Dataset, K: cluster number.

Output: A: connectivity matrix, **Y**: a matrix in a K-dimensional space.

1: Compute the functional connectivity matrix **A** of **D** by Pearson's correlation coefficient;

2: Calculate the diagonal matrix $\mathbf{B}(B_{ii} = \sum_j \mathbf{A}_{ij})$ and construct $\mathbf{L} = \mathbf{B}^{-1/2}\mathbf{A}\mathbf{B}^{-1/2}$;

3: Select the K largest eigenvectors of **L** and form the matrix **X** by arranging them in columns;

4: Normalize each of **X**'s rows, denoted as **Y**;

5: **Return A, Y**;

3.3   Self-adaptive Multidimensional Search Mechanism Based on Difference Bias for Employed Bee Search

In the basic ABC, an employed bee searches a new food source by only modifying some dimension of its associated with food source to guide it "fly" forward. Figure 1a illustrates this search when the dimension is 3: an employed bee associated with $\mathbf{x}$ can only move along the coordinate axis x_1, x_2 or x_3 when it searches a new food source. However, with the increasing of the dimension of a food source, the search pattern makes the search ability of ABC poorer. The reasons are as follows: (i) One dimension is only modified in one iteration, bringing a weak effect on the improvement of a food source. (ii) The unidimensional search pattern has some blindness. (iii) The search pattern does not take full advantage of difference information between two food sources. To increase the dimension of search may alleviate this issue to some extent.

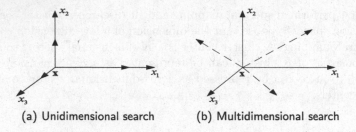

(a) Unidimensional search   (b) Multidimensional search

Fig. 1. Search in different dimensions

Along the line of the thought, a self-adaptive multidimensional search mechanism based on difference bias is proposed. Its core idea is as follows: in the light of the maximum difference principle, m dimensions in a searched food source are selected to generate a candidate food source. When the dimension is 3 with $m = 2$, the search is shown in Fig. 1b: an employed bee associated with $\mathbf{x}$ could move along any direction in a plane composed of any two of the three coordinate axes x_1, x_2 and x_3, increasing the diversity of search directions. The formulas for the search mechanism are defined as follows:

$$m = max\{\lfloor K^2 * (1 - \frac{fitness_{best}}{fitness_i})\rfloor, 1\} \qquad (4)$$

$$\mathbf{v}_i(j_1, j_2, \cdots, j_m) = \mathbf{x}_i(j_1, j_2, \cdots, j_m) + \phi_i(\mathbf{x}_b(j_1, j_2, \cdots, j_m) - \mathbf{x}_i(j_1, j_2, \cdots, j_m)) \quad (5)$$

where $fitness_{best}$ and $fitness_i$ denote the fitness of the current best $\mathbf{x}_b$ and the food source $\mathbf{x}_i$. $j_1, j_2, \cdots, j_m$ are m subscripts corresponding to the dimensions on which the m differences between them are the biggest with $\phi_i \in (0, 1)$. From (4) and (5), we can see that: the searched dimension m for each food is adaptively determined by its fitness and the current best, avoiding the adverse effect brought by too big or small m. In addition, no parameter was introduced in the mechanism, reducing artificial blind intervention during experiments. In the new mechanism, the current best food source is employed instead of a randomly selected food source to guide search with the purpose of encouraging food sources to move towards it.

3.4 Algorithm Description

DABCC mainly consists of an initialization phase and a search phase. In the former phase, the preprocessed fMRI data was projected into a low-dimension space by spectral mapping. In the latter phase, employed bee search, onlooker bee search and scout bee search were orderly performed in one iteration. The procedure of DABCC is shown in Algorithm 2.

4 Experimental Results and Analysis

In order to verify the performance of the proposed algorithm DABCC, we conduct experiments on real fMRI data with comparison to other typical algorithms, such as spectral clustering (SC) and sparse-representation-based spectral clustering (SSC) [11] and K-means. All the following algorithms were implemented with matlab. The parameters are set as follows: $N_C = 10000$, $N = 50$, $thr = K^2 \times N$.

4.1 Data Description and Preprocessing

We got the real fMRI data from the public website[1], where fifty-seven subjects were scanned in resting state. The scanning parameters of functional images and structural images are listed in Table 1. Fun. Vol and stru. Vol represent functional image and structural image. TR denotes scanning time for a whole bran. No.S and No.V represent the number of slice in a volume and that of the sampled whole brain, respectively.

Table 1. The scanning parameters for fMRI data

Image	Sequence	TR	No.S	FOV	No.V
Fun. Vol	EPI	2000 ms	33	200×200	200
Stru. Vol	MPRAGE	2530 ms	144	256×256	1

The fMRI data was preprocessed with DPARSF[2] and the specific process is as follows: the structural images were divided into white matter, gray matter and cerebral spinal fluid and normalized to the MNI normalized brain atlas. The

---

[1] http://www.yonghelab.org/downloads/data.
[2] http://rfmri.org/DPARSF.

functional images belonging to first 10 volumes (frames) of every subject were discarded to remove the effects of a fMRI scanner and the adaptive process of the subjects and they were slice timing corrected, motion corrected, written to MNI space at $3 \times 3 \times 3\,\mathrm{mm}^3$ resolution. The final fMRI data was restricted to gray matter by regressing out 24 nuisance signals. At last, the time series were filtered with passing band 0.01 hz–0.10 hz. A Gaussian kernel of FWHM 4 mm was employed for spatial smoothing and the linear trend was removed from all signals. In order to extract signals from the left insula in a subject, the mask of the left insula was made in the following: an elementary mask was obtained by extracting the 29-th area in the automated anatomical labeling atlas and then the final mask was gotten by intersecting it and the grey mask of the

---

Algorithm 2. DABCC

Input: D: Dataset, K: cluster number.
Output: groups, fit^*, $dunn$.
1: **A) Initialization:**
2: obtain the functional connectivity matrix **A** and the mapped matrix **Y** according to algorithm 1.
3: Initialize N: the population size, N_C: the max iterative count, thr: the max mining times, $populations$: the population. $pop\_fits = f(\mathbf{Y}, populations)$; $fit^* = min\{pop\_fits\}$; $fd^* = populations[argmin\{pop\_fits\}]$;
4: **B) Search phase:**
5: **repeat**
6:    a) **employed bee search:**
7:    **for** $i = 1; i <= N; i++$ **do**
8:       perform search for the i-th individual in accordance with (4)~(5), denoted as $temp\_i$; $fit\_i = f(\mathbf{Y}, temp\_i)$;
9:       Select between $temp_i$ and $population[i]$ greedily and update $limits[i]$;
10:   **end for**
11:   b) **onlooker bee search:**
12:   **for** $i = 1; i <= N; i++$ **do**
13:      randomly select a food source and the corresponding index is denoted as pos; perform search for it according to (2), denoted as $temp\_i$; $fit\_i = f(\mathbf{Y}, temp\_i)$;
14:      Select between $temp_i$ and $population[i]$ greedily and update $limits[i]$;
15:   **end for**
16:   **if** $fit^* > min(pop\_fits)$ **then**
17:      $fit^* = min(pop\_fits)$; $fd^* = populations[argmin(pop\_fits)]$;
18:   **end if**
19:   c) **scout bee search:**
20:   **for** $i = 1; i <= N; i++$ **do**
21:      **if** $limits[i] > thr$ **then**
22:         randomly generate a new food source $newf\_i$; $populations[i] = newf\_i$; $pop\_fits[i] = f(\mathbf{Y}, newf\_i)$; $limits[i] = 0$;
23:      **end if**
24:   **end for**
25: **until** the iterative count reaches N_C
26: calculate clustering labels **groups** of fd^* with SSE; $dunn = dunns(\mathbf{A}, \mathbf{groups})$; $SI = SIs(\mathbf{A}, \mathbf{groups})$;
27: **return groups**, fit^*, $dunn$;

---

corresponding subject. We randomly selected a subject and took its left insula for the following experiments.

4.2   Evaluation Metrics

In order to compare with K-means conveniently, the sum of within-cluster squared error (SSE) was used. As a common index in clustering, it is defined in (6):

$$f(\mathbf{D}, \mathbf{Z}) = \sum_{i=1}^{N} \min_{j \in \{1,2,\cdots,K\}} \{dist(\mathbf{x}_i, \mathbf{z}_j)\} \tag{6}$$

where K is the number of clusters and $dist(\mathbf{x}_j, \mathbf{z}_i)$ represents squared Euclidean distance between data point $\mathbf{x}_i \in \mathbf{D}$ and the j-th center $\mathbf{z}_j \subset \mathbf{Z}$. Obviously, the smaller its SSE, the better a clustering result.

From the perspective of Human brain functional parcellation, functional consistency is a very important index measured the performance of a parcellation method. Here, Silhoutte index is often used and its definition is as follows:

$$SI(\mathbf{C}) = \frac{1}{K} \sum_{k=1}^{K} \frac{a_k - b_k}{\max\{a_k, b_k\}} \tag{7}$$

$$a_k = \frac{1}{n_k(n_k - 1)} \sum_{i,j \in c_k, i \neq j} s(v_i, v_j) \tag{8}$$

$$b_k = \frac{1}{n_k(N_v - n_k)} \sum_{i \in c_k} \sum_{j \notin c_k} s(v_i, v_j) \tag{9}$$

where $\mathbf{C}$ denotes a parcellation result composed of K parcels/clusters and $\mathbf{c}_k$ represents the k-th parcel. $s(v_i, v_j)$ is the similarity between voxel v_i and voxel v_j, which is computed by Pearson correlation. N_v expresses the total number of voxels. The bigger SI of a parcellation, the higher its functional consistency.

4.3   Search Capability

In order to examine search capability of DABCC, we conduct experiments on the selected subject. With convenience of comparison, SC, ABCC and DABCC were run 20 times on it. SSE values in low dimension space were record. The results are shown in Table 2, where the form $\mu \pm \sigma$ is adopted (The form is used in all the following tables). K-means searches clusters in unreduced high dimension space spanned by functional matrix and SSC performs clustering in the space spanned by sparse coefficient matrix, so they are not included in Table 2. It can be seen that: (i) SSE values from ABCC is lower than them obtained by SC. The reason is mainly that K-means in SC is a single-path greedy search and easily falls into local optimum, while ABCC is a multi-path search technique based on population and has stronger spatial search capability. (ii) The DABCC is the best in terms of SSE. The main explanation for it is that the self-adaptive multidimensional search mechanism based on difference bias updates many dimensions of a food source according to its fitness and provides more suitable search direction.

Table 2. SSE of different cluster sizes for three algorithms

No.	SC	ABCC	DABCC
2	39.76 ± 7.29E-15	39.76 ± 7.29E-15	39.76 ± 2.42E-14
3	64.90 ± 0	64.90 ± 0	64.88 ± 2.42E-14
4	83.62 ± 0	83.61 ± 4.99E-14	83.37 ± 4.86E-14
5	97.25 ± 2.22E-01	97.09 ± 2.55E-06	96.91 ± 5.76E-06
6	100.05 ± 3.8E-03	100.04 ± 3E-03	100.01 ± 2.98E-03
7	101.92 ± 4.9E-03	101.90 ± 7.5E-03	101.72 ± 8.78E-03
8	101.72 ± 2.52E-2	101.71 ± 1.1E-03	101.47 ± 8.48E-04
9	91.12 ± 3.23E-14	91.11 ± 1.7E0-03	91.02 ± 1.20E-03
10	99.04 ± 1.88E-01	99.00 ± 1.72E-02	98.79 ± 2.19E-02
11	100.48 ± 1.13	99.86 ± 2.07E-02	99.65 ± 2.03E-02
12	104.56 ± 8.8E-01	104.45 ± 2.53E-01	104.23 ± 7.49E-02

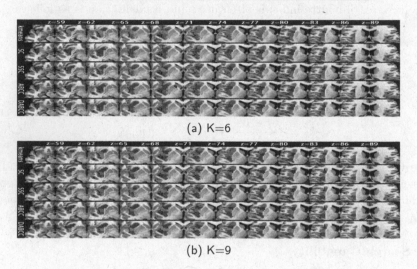

(a) K=6

(b) K=9

Fig. 2. Parcellation results on different parcellation numbers (A color represents a functional subregion) (Color figure online)

4.4   Parcellation Results

To evaluate spatial structure of parcellation results, we also made a comparison in terms of functional parcellation maps of parcellation that are the closest to the corresponding averaged SSE. From Fig. 2, we can see that spatial continuity of parcellation results from K-means and SSC is poorer. The main reasons are that both K-means has poor search ability and sparse representation in SSC is error. The results from DABCC have the best spatial continuity and the most smooth edges. It suggests that the self-adaptive multidimensional search mechanism could improve the search capability of DABCC.

4.5   Functional Consistency

The aim of Human brain functional parcellation is to get some subregions with stronger functional consistency, so we computed functional consistency of a parcellation result with SI from 2 to 12 clusters. The algorithms in Table 3 were respectively run 20 times. We can see that SI values of K-means are the lowest and the corresponding standard deviations also are the biggest. The functional consistency from SC is bigger than them of K-means, because SC searches cluster centers in a low-dimension space. As the pacellation number increases, the results obtained by SSC gradually become poor. It may be the reasons that sparse representation is error and the similarity between coefficients of nearby time series is very high. Since ABCC has stronger search capability than K-means and some robustness, ABCC has the second highest SI values. The results from DABCC are the best in terms of SI. The explanation is that our self-adaptive multidimensional mechanism captures difference information between food sources and provides more suitable search directions.

Table 3. SI of different cluster sizes for five algorithms

No.	K-means	SSC	SC	ABCC	DABCC
2	$0.5652 \pm 1.44E-01$	$0.5758 \pm 2.28E-16$	$0.8745 \pm 2.28E-16$	$0.8746 \pm 2.28E-16$	$0.8777 \pm 2.28E-16$
3	$0.6797 \pm 5.43E-02$	$0.7206 \pm 3.42E-16$	0.8799 ± 0	0.8799 ± 0	$0.8801 \pm 2.28E-16$
4	$0.7558 \pm 4.29E-02$	$0.7670 \pm 3.42E-16$	$0.8881 \pm 1.14E-16$	$0.8890 \pm 1.14E-16$	$0.8897 \pm 5.70E-16$
5	$0.8077 \pm 3.59E-02$	$0.7857 \pm 4.53E-16$	$0.8959 \pm 4.2E-03$	0.8988 ± 0	$0.8997 \pm 3.42E-16$
6	$0.8230 \pm 1.93E-02$	$0.8282 \pm 4.45E-16$	$0.9057 \pm 2E-04$	$0.9058 \pm 1.35E-04$	$0.9009 \pm 2.28E-16$
7	$0.8414 \pm 1.94E-02$	$0.8367 \pm 1.14E-16$	$0.8962 \pm 5E-04$	$0.8965 \pm 5.94E-04$	$0.8972 \pm 6.67E-03$
8	$0.8525 \pm 1.42E-02$	$0.8380 \pm 9E-04$	$0.9067 \pm 5.32E-16$	$0.9076 \pm 1.67E-04$	$0.9084 \pm 3.82E-03$
9	$0.8695 \pm 1.26E-02$	$0.8434 \pm 2.5E-03$	$0.9136 \pm 5.06E-16$	$0.9137 \pm 2.71E-05$	$0.9146 \pm 2.84E-03$
10	$0.8738 \pm 1.21E-02$	$0.8121 \pm 3.1E-03$	$0.9118 \pm 7E-04$	$0.9122 \pm 4.74E-04$	$0.9150 \pm 3.94E-03$
11	$0.8815 \pm 8.9E-03$	$0.8227 \pm 2E-03$	$0.9171 \pm 3.5E-03$	$0.9187 \pm 3.92E-04$	$0.9199 \pm 4.08E-03$
12	$0.8865 \pm 8.6E-03$	$0.8314 \pm 2.7E-03$	$0.9179 \pm 2.8E-03$	$0.9180 \pm 8.64E-04$	$0.9201 \pm 3.04E-03$

5   Conclusion

The paper developed an new algorithm DABCC for insula functional parcellation. DABCC performed spectral mapping to reduce the dimension of fMRI data in the initialization. Then, the search procedure based on ABCC was employed to search clustering centers in a low-dimension space, where a self-adaptive multidimensional search mechanism based on difference bias was adopted for employed bee search. Finally, the results on real fMRI data demonstrate that DABCC not only has stronger search capability, but can generate better insula functional parcellation in terms of functional consistency and regional continuity. Moreover, DABCC is also applied to functional parcellation of other regions. So this work both deepens our understanding of insula functional organization further and promotes the development of Human brain functional parcellation method.

Acknowledgments. The work is partly supported by the NSFC Research Program (61375059, 61672065), the National "973" Key Basic Research Program of China (2014CB744601), Nanyang Normal University - level Young Teacher Project (QN2017040), the scientific and technological project in Henan Province of China (142102210588, 172102310702), and the Science and Technology Foundation of Henan Educational Committee of China (17A520049, 17A630046).

References

1. Zaccarella, E., Friederici, A.D.: Reflections of word processing in the insular cortex: a sub-regional parcellation based functional assessment. Brain Lang. **142**(1), 1–7 (2015)
2. Deen, B., Pitskel, N.B., Kevin, A.P.: Three systems of insular functional connectivity identified with cluster analysis. Cereb. Cortex **21**(7), 1498–506 (2011)
3. Zhao, X.W., Ji, J.Z., Liang, P.P.: The human brain functional parcellation based on fMRI data. Chin. Sci. Bull. **61**(18), 2035–2052 (2016)
4. Zhao, X.W., Ji, J.Z., Yao, Y.: Insula functional parcellation by searching Gaussian mixture model using immune clonal selection algorithm. J. Zhejiang Univ. (Eng. Sci.) **51**(12), 1–12 (2017)
5. Franco, C., Tommaso, C., Torta, D.M., Katiuscia, S., Federico, D., Sergio, D., Giuliano, G., Fox, P.T., Alessandro, V.: Meta-analytic clustering of the insular cortex: characterizing the meta-analytic connectivity of the insula when involved in active tasks. Neuroimage **62**(1), 343–355 (2012)
6. Vercelli, U., Diano, M., Costa, T., Nani, A., Duca, S., Geminiani, G., Vercelli, A., Cauda, F.: Node detection using high-dimensional fuzzy parcellation applied to the insular cortex. Neural Plast. **5–6**, 1–8 (2016)
7. Nomi, J.S., Farrant, K., Damaraju, E., Rachakonda, S., Calhoun, V.D., Uddin, L.Q.: Dynamic functional network connectivity reveals unique and overlapping profiles of insula subdivisions. Hum. Brain Mapp. **37**(5), 1770–1787 (2016)
8. İnkaya, T., Kayalıgil, S., Özdemirel, N.E.: Swarm intelligence-based clustering algorithms: a survey. In: Celebi, M.E., Aydin, K. (eds.) Unsupervised Learning Algorithms, pp. 303–341. Springer, Cham (2016). https://doi.org/10.1007/978-3-319-24211-8_12
9. Karaboga, D., Ozturk, C.: A novel clustering approach: Artificial Bee Colony (ABC) algorithm. Appl. Soft Comput. **11**(1), 652–657 (2011)
10. Karaboga, D.: An idea based on honey bee swarm for numerical optimization. Technical Report TR06, Erciyes University, Engineering Faculty, Computer Engineering Department, Kayseri, Turkiye (2005)
11. Zhang, Y., Caspers, S., Fan, L.Z., Fan, Y., Song, M., Liu, C.R., Mo, Y., Roski, C., Eickhoff, S., Amunts, K.: Robust brain parcellation using sparse representation on resting-state fMRI. Brain Struct. Funct. **220**(6), 3565–3579 (2015)

EEG-Based Emotion Recognition via Fast and Robust Feature Smoothing

Cheng Tang[1], Di Wang[2(✉)], Ah-Hwee Tan[1,2], and Chunyan Miao[1,2]

[1] School of Computer Science and Engineering, Nanyang Technological University,
Singapore 639798, Singapore
[2] Joint NTU-UBC Research Center of Excellence in Active Living for the Elderly,
Nanyang Technological University, Singapore 639798, Singapore
`tang0314@e.ntu.edu.sg`, {`wangdi,asahtan,ascymiao`}`@ntu.edu.sg`

Abstract. Electroencephalograph (EEG) signals reveal much of our brain states and have been widely used in emotion recognition. However, the recognition accuracy is hardly ideal mainly due to the following reasons: (i) the features extracted from EEG signals may not solely reflect one's emotional patterns and their quality is easily affected by noise; and (ii) increasing feature dimension may enhance the recognition accuracy, but it often requires extra computation time. In this paper, we propose a feature smoothing method to alleviate the aforementioned problems. Specifically, we extract six statistical features from raw EEG signals and apply a simple yet cost-effective feature smoothing method to improve the recognition accuracy. The experimental results on the well-known DEAP dataset demonstrate the effectiveness of our approach. Comparing to other studies on the same dataset, ours achieves the shortest feature processing time and the highest classification accuracy on emotion recognition in the valence-arousal quadrant space.

Keywords: Emotion recognition · EEG · DEAP · Feature smoothing

1 Introduction

Emotion is the subjective experience that reflects our mental states and can significantly affect our cognitive function and action tendencies [10]. With the advances in artificial intelligence (AI) and brain-computer interface (BCI) technologies, the ability for computer applications to recognize human emotions can provide us more intelligent services, such as style-adjusting e-learning system [1], driver's fatigue detection [6], e-healthcare assistance [4], etc.

In efforts to recognize human emotions using machines, researchers mainly rely on the following three types of data: (i) behavioral patterns such as facial expressions, (ii) physiological signals from peripheral nervous system such as electrooculography (EOG), and (iii) physiological signals from central nervous system such as electroencephalograph (EEG). Compared to the other two types of signals, EEG is more informative for high-level brain activities [7]. Moreover,

© Springer International Publishing AG 2017
Y. Zeng et al. (Eds.): BI 2017, LNAI 10654, pp. 83–92, 2017.
https://doi.org/10.1007/978-3-319-70772-3_8

studies showed that EEG exhibits promising characteristics in revealing the subject's emotional states [17]. Thanks to the emerging non-invasive brain-computer interfacing devices, EEG has become one of the most prevalent signals being used to recognize human emotions.

With the aid of machine learning algorithms, many prior studies used different features extracted from raw EEG signals to decode the underlying emotion. However, the recognition accuracy is hardly ideal because of the following reasons: (i) EEG is a mixture of fluctuations induced by many neuronal activities in the brain and is susceptible to interference [13]. The features extracted from EEG may vary drastically within short periods but human emotions are relatively stable, which means the features may not directly reflect the emotional patterns; and (ii) increasing the feature dimension may improve the recognition accuracy, but this approach usually introduces more computational cost in feature extraction, classifier training, and the classification task.

To address the aforementioned feature instability problem without increasing the feature dimension, in this paper, we propose a fast and robust feature smoothing method, which can be applied on the extracted EEG features to improve the emotion recognition accuracy.

Without feature smoothing, emotion-irrelevant patterns make the extracted features less distinctive, thus reduce the classification accuracy. This problem can be alleviated by applying moving average smoothing on the extracted EEG features. This simple feature smoothing method will not increase the feature dimension nor add in much time to the total feature processing process. We evaluate our proposed approach on the widely studied DEAP dataset [8]. Specifically, we extract six statistical features from the EEG signals and apply moving average smoothing on all the extracted features. Using support vector machine (SVM) as the classifier, we obtain 82.3% accuracy in recognizing four classes of emotions, which is higher than four prior studies using the same dataset. Moreover, the processing time of our feature set is the shortest.

The rest of this paper is organized as follows. In Sect. 2, we review related work on EEG-based emotion recognition. In Sect. 3, we present the motivation of feature smoothing and the details of our methodology. In Sect. 4, we show our experimental results on the DEAP dataset with comparisons and discussions. Finally, we conclude and propose future work in Sect. 5.

2   Related Work

Various prior studies have been conducted to explore how to extract better feature sets for EEG-based emotion recognition. Some studies investigated the characteristics of EEG signals in the frequency domain. Heraz and Frasson [5] used the amplitude of four frequency bands to obtain an averaged accuracy of 74%, 74% and 75% on 17 subjects in the valence, arousal and dominance dimensions, respectively. Bos [2] explored arithmetic combinations of the power on frequency bands to obtain the highest accuracy of 92% in both arousal and valence dimensions on five subjects. Some studies investigated other feature sets such as discrete wavelet coefficients (84.67% for happy and sad on five subjects) [23], fractal dimension (around 90% for arousal on twelve subjects) [16], and higher order crossing (50.13% for four emotions on DEAP dataset) [9]. However, majority of the afore-reviewed results are based on binary emotion classifications, which may not be enough to capture the emotional variations in our daily life. Moreover, as a general observation from the literature, the recognition accuracy decreases, sometimes significantly, if the models need to recognize more classes of emotion. Vyzas et al. [12] managed to recognize six emotions with a remarkable accuracy of 81% on a single subject, but they incorporated other physiological signals such as blood pressure and heart rate besides EEG.

The difficulty in accurately recognizing different emotions merely from EEG signals lies in the non-stability of EEG features. It has been found that feature smoothing can reduce such non-stability and improve recognition accuracy. Shi and Lu first proposed a linear dynamical system (LDS) approach to estimate the latent states of vigilance [15] and later used this model for feature smoothing in emotion recognition (91.77% for positive/negative) [21]. Although LDS is effective in enhancing recognition accuracy, the expectation-maximization (EM) algorithm incorporated in the smoothing process makes the overall approach computationally expensive. Pham et al. [11] used the Savitzky-Golay method, which is based on local least-squares polynomial approximation, to smooth EEG features. Although their proposed method improved the recognition accuracy in the valence dimension, the improved accuracy of 77.38% is not high among similar binary classification problems. In this paper, we apply moving average feature smoothing on six statistical features extracted from the raw EEG signals. As such, the smoothing method does not introduce much computational cost.

3   Moving Average Smoothing on Statistical Feature Set

We adopt the common process to recognize human emotions using EEG signals, i.e., extract features from the raw data, then train the classifier to perform emotion recognition. In addition, we apply feature smoothing on the extracted features before training the classifier. We introduce each of these key steps of our proposed approach with details in the following subsections.

3.1   Feature Extraction

In our proposed approach, we only extract six statistical features, which have been widely adopted in prior studies. Vyzas et al. [18] showed that these six features are strongly correlated to emotions. Moreover, it is computationally inexpensive to extract these simple statistical features.

Let X_n denote an EEG signal value at the nth time stamp, where $n = 1, 2, \ldots, N$ and N denotes the total number of data samples. Moreover, let $\bar{X}_n$ denote the corresponding normalized signal with zero mean and unit variance. Then we extract the following six statistical features:

1. μ, mean of the raw signal over time N:

$$\mu = \frac{1}{N} \sum_{n=1}^{N} X_n. \tag{1}$$

2. σ, standard deviation of the raw signal:

$$\sigma = \sqrt{\frac{1}{N-1} \sum_{n=1}^{N} (X_n - \mu)^2}. \tag{2}$$

3. δ, mean of the absolute values of the first differences of the raw signal:

$$\delta = \frac{1}{N-1} \sum_{n=1}^{N-1} |X_{n+1} - X_n|. \tag{3}$$

4. $\bar{\delta}$, mean of the absolute values of the first differences of the normalized signal:

$$\bar{\delta} = \frac{1}{N-1} \sum_{n=1}^{N-1} |\bar{X}_{n+1} - \bar{X}_n| = \frac{\delta}{\sigma}. \tag{4}$$

5. γ, mean of the absolute values of the second differences of the raw signal:

$$\gamma = \frac{1}{N-2} \sum_{n=1}^{N-2} |X_{n+2} - X_n|. \tag{5}$$

6. $\bar{\gamma}$, mean of the absolute values of the second differences of the normalized signal:

$$\bar{\gamma} = \frac{1}{N-2} \sum_{n=1}^{N-2} |\bar{X}_{n+2} - \bar{X}_n| = \frac{\gamma}{\sigma}. \tag{6}$$

3.2   Moving Average Smoothing on Extracted Features

Within short time periods, the emotional states of human are relatively stable, but the features obtained from EEG signals may have strong variation in time due to the impact of emotion-irrelevant activities and random fluctuations [13]. To make the features more robust for emotion recognition, we propose to use the moving average method to smooth the features in time sequence. Specifically, we first divide EEG data into non-overlapping windows and extract features from each window. Let f_i denote a single feature f extracted from the ith time window, where $i = 1, 2, \ldots, I$ and I denotes the total number of the non-overlapping windows. The smoothed feature $\bar{f}_i$ is then computed as follows:

$$\bar{f}_i = \frac{1}{T} \sum_{\lfloor i-2/T \rfloor}^{\lfloor i+2/T \rfloor} f_i, \tag{7}$$

where T is the size of the moving average smoother.

3.3 Classification Algorithm

In this paper, we use SVM as the classifier due to its well-known generalization property. In particular, we use the one-vs-all scheme for multiclass classification of the LIBSVM package [3].

For binary classification, given training samples $\{x_i, y_i\}$, where $i = 1, 2, \ldots, l$, $x_i \in \mathbb{R}^d$ and $y_i \in \{-1, 1\}$, SVM solves the following optimization problem:

$$\underset{w, b, \xi}{\text{minimize}} \quad \frac{1}{2}||w||^2 + C \sum_{i=1}^{l} \xi_i,$$
$$\text{subject to} \quad y_i(w^T \phi(x_i) + b) \geq 1 - \xi_i, \quad \xi_i \geq 0, \tag{8}$$

where C denotes the cost parameter indicating the penalty of error and ξ_i denotes the tolerance of error. Kernel function $\phi(x_i)$ maps the feature vector x_i into another feature space. In this paper, we use radial basis function (RBF) kernel, which is represented as:

$$K(x_i, x_j) \equiv \phi(x_i)^T \phi(x_j) = \exp(-\gamma||x_i - x_j||^2), \gamma > 0, \tag{9}$$

where γ defines the steepness of the decision boundary.

4 Emotion Recognition on DEAP Dataset

To assess the performance of our feature extraction and feature smoothing strategy, we use the well-known DEAP dataset for evaluations. DEAP dataset was collected by Koelstra et al. [8] for human emotion analysis. EEG signals of 32 subjects were elicited using multimodal stimuli and recorded on 32 channels using the Biosemi ActiveTwo system[1]. In the preprocessed dataset provided[2], each subject has 40 min recordings of EEG signals. Moreover, ratings of valence, arousal and dominance were labeled by the subjects after each trial. The EEG data were down-sampled to 128 Hz, filtered by a bandpass filter of 4–45 Hz, and normalized with respect to the common reference in each channel. In this paper, we take the subjects' labels in the valence and arousal dimensions as the ground truth of the EEG data. Actually, we are following the circumplex model of affect proposed by Russel [14]. In his widely adopted model (e.g., applied in [8,19,20]), emotions are represented in a two-dimensional space, where the two axes represent valence and arousal, respectively.

4.1 Experimental Setup

In our experiments, we segment all EEG data given in the DEAP dataset into non-overlapping windows of one second, where each window consists of 128 data samples. Therefore, the total number of observations/windows of a subject is

---

[1] http://www.biosemi.com.
[2] http://www.eecs.qmul.ac.uk/mmv/datasets/deap/.

2400 (40 videos times 60 s). Moreover, for every observation, we extract the six statistical features from each EEG channel. Because data in DEAP were collected using a 32-channel device, the size of our feature set is 192.

In DEAP, the ratings of valence and arousal were given as decimals in the [1, 9] interval. Therefore, we choose 5 as the threshold for class labeling in the valence-arousal space. In other words, we use the ratings provided by the subjects in DEAP as the ground truth to define four classes of emotion for assessments. These four classes are $V_L A_L, V_L A_H, V_H A_L$ and $V_H A_H$, where V denotes valence, A denotes arousal, L denotes low value (<5), and H denotes high value (≥ 5).

To split the training and testing samples, we did not choose the k-fold cross-validation scheme because feature smoothing should only be applied on continuous time sequence that segmenting a continuous feature sequence into k parts will break down the continuity. Instead, we choose 80%/20% splitting strategy to preserve the continuity in features. Specifically, for samples in each minute, we use the first 80% for training and the rest 20% for testing. Feature smoothing is then applied separately on the two sets. This splitting strategy is depicted in Fig. 1(a). Furthermore, we normalize training samples (referring to the extracted features rather than the raw signals) to zero-mean and unit variance to train the classifier, and then use the normalization parameters obtained from training samples to normalize testing samples before performing classification.

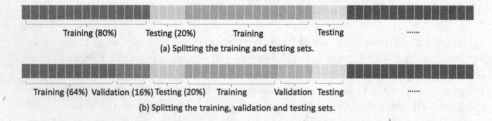

(a) Splitting the training and testing sets.

(b) Splitting the training, validation and testing sets.

Fig. 1. 80%/20% splitting strategy for obtaining training, validation and testing sets.

In feature smoothing, the window size T (see (7)) greatly affects the performance. To obtain the best value of T, we further split a validation set from the training set using the same splitting ratio as illustrated in Fig. 1(b). Table 1 shows the accuracy obtained on the validation set based on classifiers trained using the training set (64%) with respect to different T values. As shown in Table 1, we obtain the best accuracy when $T = 11$. As such, we use $T = 11$ during subsequent feature smoothing on both the training (80%) and testing datasets.

On the other hand, to find the appropriate parameter settings of SVM, we perform grid search on both the cost parameter C (see (8)) and the penalty parameter γ (see (9)) for $\{n \times 10^m\}$, where $n = 1, 2, \ldots, 9$ and $m = -3, -2, \ldots, 3$. When $T = 11$, we find the best performing values as $C = 10$ and $\gamma = 0.005$. This combination of parameter values are used in all the experiments presented in this paper.

Table 1. Classification accuracy on the validation set with varying T values

T	1	2	3	4	5	6	7	8	9	10	11	12	13	14	15	16
Acc (%)	59.83	68.23	73.55	76.15	77.86	78.96	80.36	80.58	81.13	81.06	**81.31**	80.81	80.77	80.67	80.90	79.49

4.2 Results and Discussions

Figure 2 shows the four-class emotion recognition results of each individual subject before and after applying our moving average feature smoothing method. The average accuracy (across all subjects) improves significantly from $61.73\% \pm 7.07\%$ to $82.30\% \pm 8.44\%$ after applying feature smoothing. This finding strongly support the high effectiveness of our feature smoothing method.

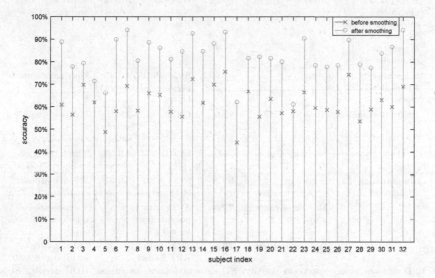

Fig. 2. Emotion recognition accuracy before and after feature smoothing.

Besides 80%/20%, we also conduct experiments on other splitting ratios and then present the results in Table 2. Similar to Fig. 1(b), for each splitting ratio R, we further split a validation set from the training set based on R to obtain the corresponding optimal T value. As clearly shown in Table 2, our feature smoothing method always improves the recognition accuracy. It is more encouraging to see that even when $R = 20\%/80\%$, our approach can obtain an accuracy of 62.43% on four-class emotion recognition. This satisfactory performance obtained on low splitting ratio well demonstrates the generalization capability or robustness of our approach. To be more elaborate, for a longitudinal study with stable emotion transitions, say each session lasts one minute (same as DEAP), our approach only requires the EEG signals collected in the first twelve seconds to be labeled for training, hereafter, it can already achieve 62.43% accuracy on four-class emotion recognition.

Table 2. Performance of feature smoothing using different train/test splitting ratios

Train/Test (%)	Acc before smoothing (%)	T value	Acc after smoothing (%)	Acc improvement (%)
20/80	51.37	5	62.43	11.06
40/60	55.67	7	72.41	16.74
60/40	58.26	11	78.36	20.10
80/20	61.73	11	82.30	20.57

Table 3. Comparison with prior studies on DEAP

Study	Feature Set	Feature Smoothing	Time (s)	Classifier	Subjects	Performance
[8]	PSD (32 channels)+ difference between 14 pairs of symmetric channels	None	166.2	Gaussian Naive Bayes	All 32 subjects	Valence: 57.6%, arousal: 62%
[9]	Six statistics + FD + HOC (32 channels)	None	>1000	SVM with polynomial kernel	All 32 subjects	Four emotions: 80%
[22]	PSD (16 channels)	None	82.6	CNN	22 selected subjects	Valence: 76.63%
[11]	PSD (32 channels)	Savitzky-Golay	207.4	SVDD	All 32 subjects	Three levels of valence: 71.75%
Our work	Six statistics (32 channels)	None	**67.8**	SVM with RBF kernel	All 32 subjects	Valence-arousal quadrant: 61.73%
		Moving average	**77.4**	SVM with polynomial kernel		Valence-arousal quadrant: 67.90%
		Moving average	**77.4**	SVM with RBF kernel		Valence-arousal quadrant: **82.30%**

Note: PSD denotes power spectrum density, FD denotes fractal dimension, HOC denotes higher order crossing, CNN denotes convolutional neural network, and SVDD denotes support vector data description.

To further assess the performance of our proposed method, we compare our results with some prior studies using the same DEAP dataset. In Table 3, [8] used leave-one-video-out scheme for classification, [22] conducted 11-fold cross-validation on 22 selected subjects, [9,11] used 5-fold cross-validation. Although our data splitting strategy is different from all the benchmarking studies (similar but conducted four times less than 5-fold CV) and some models adopted different number of classes for emotion recognition, we still compare all the results in the same table. Nonetheless, we use the same computer (2.20 GHz CPU with 8 GB RAM) and the same programming language (MATLAB) to obtain all the feature sets shown in Table 3 and report their processing time (feature extraction and smoothing if applicable). It is encouraging to see that our approach achieves the highest accuracy of 82.3%, which is even higher than those recognizing lesser number of emotional classes. Moreover, our moving average approach does not add in much computational time to the overall feature processing procedure $((77.4 - 67.8)/77.4 = 12.4\%)$. Compared to other benchmarking models, either with or without feature smoothing, our approach has the shortest feature processing time.

5   Conclusion

In this paper, we propose a fast and robust EEG-based emotion recognition approach that applies the simple moving average feature smoothing method on the six extracted statistical features. To assess the effectiveness of our approach, we apply it on the well-known DEAP dataset to perform emotion recognition in the valence-arousal quadrant space. The results are more than encouraging. First of all, the average accuracy (when using 80%/20% splitting ratio) significantly improves from 61.73% to 82.3% after feature smoothing is applied. Secondly, we show the robustness of our approach that it always significantly improves the recognition accuracy for various data splitting ratios. Last but most importantly, our approach achieves the best performance in terms of both feature processing time and recognition accuracy among all the benchmarking models.

In the future, we will further test the robustness of our approach by conducting experiments on own collected and other datasets. We will also look into the theoretical insights of why the simple moving average method may significantly improve the emotion recognition accuracy.

Acknowledgments. This research is supported in part by the National Research Foundation, Prime Minister's Office, Singapore under its IDM Futures Funding Initiative. This research is also partially supported by the NTU-PKU Joint Research Institute, a collaboration between Nanyang Technological University and Peking University that is sponsored by a donation from the Ng Teng Fong Charitable Foundation.

References

1. Asteriadis, S., Tzouveli, P., Karpouzis, K., Kollias, S.: Estimation of behavioral user state based on eye gaze and head pose-application in an e-learning environment. Multimedia Tools Appl. **41**(3), 469–493 (2009)
2. Bos, D.O.: EEG-based emotion recognition-the influence of visual and auditory stimuli. Capita Selecta (MSc course) (2006)
3. Chang, C.C., Lin, C.J.: LIBSVM: a library for support vector machines. ACM Trans. Intell. Syst. Technol. **2**, 27:1–27:27 (2011). http://www.csie.ntu.edu.tw/~cjlin/libsvm
4. Goldberg, D.: The detection and treatment of depression in the physically ill. World Psychiatry **9**(1), 16–20 (2010)
5. Heraz, A., Frasson, C.: Predicting the three major dimensions of the learners emotions from brainwaves. Int. J. Comput. Sci. **2**(3), 183–193 (2007)
6. Jap, B.T., Lal, S., Fischer, P., Bekiaris, E.: Using EEG spectral components to assess algorithms for detecting fatigue. Expert Syst. Appl. **36**(2), 2352–2359 (2009)

7. Kandel, E.R., Schwartz, J.H., Jessell, T.M., Siegelbaum, S.A., Hudspeth, A.J.: Principles of Neural Science. Mc Graw Hill, New York (2012)
8. Koelstra, S., Muhl, C., Soleymani, M., Lee, J.S., Yazdani, A., Ebrahimi, T., Pun, T., Nijholt, A., Patras, I.: DEAP: a database for emotion analysis using physiological signals. IEEE Trans. Affect. Comput. **3**(1), 18–31 (2012)
9. Liu, Y., Sourina, O.: EEG databases for emotion recognition. In: International Conference on Cyberworlds, pp. 302–309. IEEE (2013)
10. Mauss, I.B., Robinson, M.D.: Measures of emotion: a review. Cogn. Emot. **23**(2), 209–237 (2009)
11. Pham, T.D., Tran, D., Ma, W., Tran, N.T.: Enhancing performance of EEG-based emotion recognition systems using feature smoothing. In: Arik, S., Huang, T., Lai, W.K., Liu, Q. (eds.) ICONIP 2015. LNCS, vol. 9492, pp. 95–102. Springer, Cham (2015). doi:10.1007/978-3-319-26561-2_12
12. Picard, R.W., Vyzas, E., Healey, J.: Toward machine emotional intelligence: Analysis of affective physiological state. IEEE Trans. Pattern Anal. Mach. Intell. **23**(10), 1175–1191 (2001)
13. Pijn, J.P., Van Neerven, J., Noest, A., da Silva, F.H.L.: Chaos or noise in EEG signals' dependence on state and brain site. Electroencephalogr. Clin. Neurophysiol. **79**(5), 371–381 (1991)
14. Russell, J.A.: A circumplex model of affect. J. Pers. Soc. Psychol. **39**(6), 1161–1178 (1980)
15. Shi, L.C., Lu, B.L.: Off-line and on-line vigilance estimation based on linear dynamical system and manifold learning. In: International Conference on Engineering in Medicine and Biology, pp. 6587–6590. IEEE (2010)
16. Sourina, O., Liu, Y.: A fractal-based algorithm of emotion recognition from EEG using arousal-valence model. In: BIOSIGNALS, pp. 209–214 (2011)
17. Takahashi, K.: Remarks on emotion recognition from multi-modal bio-potential signals. In: International Conference on Industrial Technology, vol. 3, pp. 1138–1143. IEEE (2004)
18. Vyzas, E., Picard, R.W.: Affective pattern classification. In: Emotional and Intelligent: The Tangled Knot of Cognition, pp. 176–182 (1998)
19. Wang, D., Tan, A.H., Miao, C.: Modelling autobiographical memory in humanlike autonomous agents. In: International Conference on Autonomous Agents and Multiagent Systems, pp. 845–853. ACM (2016)
20. Wang, D., Tan, A.H.: Mobile humanoid agent with mood awareness for elderly care. In: International Joint Conference on Neural Networks, pp. 1549–1556. IEEE (2014)
21. Wang, X.W., Nie, D., Lu, B.L.: Emotional state classification from EEG data using machine learning approach. Neurocomputing **129**, 94–106 (2014)
22. Yanagimoto, M., Sugimoto, C.: Recognition of persisting emotional valence from EEG using convolutional neural networks. In: International Workshop on Computational Intelligence and Applications, pp. 27–32. IEEE (2016)
23. Yohanes, R.E., Ser, W., Huang, G.B.: Discrete wavelet transform coefficients for emotion recognition from EEG signals. In: International Conference on Engineering in Medicine and Biology, pp. 2251–2254. IEEE (2012)

Human Information Processing Systems

Stronger Activation in Widely Distributed Regions May not Compensate for an Ineffectively Connected Neural Network When Reading a Second Language

Hao Yan[1,2(✉)], Chuanzhu Sun[3], Shan Wang[3], and Lijun Bai[3]

[1] Department of Linguistics, Xidian University, Xi'an 710071, China
yanhao@xidian.edu.cn
[2] Center for Language and Brain, Shenzhen Institute of Neuroscience,
Shenzhen 518057, China
[3] Department of Biomedical Engineering, School of Life Science
and Technology, Xi'an Jiaotong University, Xi'an 710049, China

Abstract. Even though how bilinguals process the second language (L2) still remain disputable, it is agreed that L2 processing involve more brain areas and activate common regions more strongly. It interested us to probe why heavier manipulation of cortical regions did not guarantee a high language proficiency. Since the responses of individual brain regions were inadequate to explain how the brain enabled behavior, we sought to explore this question at the neural network prospect via the Psychophysiological interaction (PPI) analysis. We found that Chinese English bilinguals adopted the assimilation/accommodation strategy to read L2, and English activated common brain areas more strongly. However, the whole brain voxel-wise analysis of effective connectivity showed that these brain areas formed a less synchronized network, which may indicate an ineffective neural network of L2. Our findings provided a possible explanation why the proficiency level of L2 was always lower than L1, and suggested that future fMRI studies may better explore language issues by depicting functional connectivity efficacy.

Keywords: Functional connectivity · Neural modulation effect · Chinese English bilinguals · Psychophysiological interaction (PPI) analysis · FMRI

1 Introduction

It is agreed that the acquisition age of a second language (L2) affects neural activities during language processing, which is always associated with low language proficiency of L2 [1]. The "single system hypothesis" (also known as the Convergence Hypothesis) advocates that late bilingual learners have already formed a consolidated and entrenched linguistic system in place, and late bilinguals would manipulate neural literacy of L1 to perform L2, but with additional reliance on cognitive control due to unmatched language proficiency [2]. A situation the "single system hypothesis" ignored is that it failed to explain how bilinguals processed L2-specific features. A more comprehensive

© Springer International Publishing AG 2017
Y. Zeng et al. (Eds.): BI 2017, LNAI 10654, pp. 95–106, 2017.
https://doi.org/10.1007/978-3-319-70772-3_9

theoretical framework proposed that L2 processing relied on combined neural activity of assimilation (refers to using the existing reading network in the acquisition of a new writing system) and accommodation (refers to using new procedures for reading the new writing system) [3]. They considered that reading the second language was associated with similar brain areas to that of L1, but with differences in certain regions that were likely to reflect specific properties of a language. Even though theoretical disputes still remain, particularly at the sentence level, both the "single system hypothesis" and the "assimilation/accommodation theory" proposed that L2 processing was characterized by more areas and a stronger activation intensity. Hereafter, it interested us to probe why involvement of more brain regions and stronger activation did not mean high language proficiency.

It may be inadequate, at least insufficient, to illustrate complex links between structure and function only on condition of neural responses of individual brain regions to tasks. Due to various anatomical inputs to a single region, any specific region can be involved in several cognitive processes and cooperate with many other brain areas in a wide spatial distribution. Different functions of an area made it challenging to understand how the brain enables behavior based on functional segregation viewpoint [4, 5]. Only when a brain region cooperated with other regions to form an effective neural network for a cognitive process, may we claim that it was necessarily involved in that process. Since determining the function of a region depends on its interactions with other brain regions, solving this challenge rests on how we understood functional connectivity of cooperated areas [6]. Brain connectivity is now being explored by depicting neuronal coupling between brain regions through various techniques [7]. We highlighted how responses in individual brain regions can be effectively integrated through functional connectivity (FC).

Effective connectivity quantifies directed relationships between brain regions and controls for confounds that limit functional connectivity—features that facilitate insight into functional integration [5]. It overcomes important pitfalls of functional connectivity that limit our understanding of neuronal coupling. For example, functional connectivity elicited by an interested task could involve functional synchronization with other cognitive processes, observational noise, or neuronal fluctuations [8]. One popular approach to make effective connectivity analysis is psychophysiological interactions (PPI), which allows simple synchronization analysis (between two regions) [9]. This approach measures in a psychological context how one brain region (a seed region) contributes to another brain region (a target region). It explains neural responses in one brain area in terms of the interaction between influences of another brain region and a cognitive/sensory process, indicating a condition specific change of coupling between brain areas [9].

Chinese and English are two different languages but share common features at the sentence level. For example, Chinese is an analytic language, and depends on word order to express meaning [10], in which changing only one word position may keep the sentence grammatically correct but generate different meanings. Its verbs remain invariably uninflected, regardless of difference in tense, gender or number, and nuances are expressed by other words. In contrast, English is a synthetic language, which marks tense through morphological inflections. However, Chinese and English both exhibit an SVO (subject verb object) word order. The two languages are suitable to test the assimilation/accommodation theory at the sentence level, and we hoped to explore the working mechanism of reading L2 sentences from the perspective of effectively connected neural networks.

2 Methods

2.1 Participants

To reduce intersubject variabilities, the current study selected 21 Chinese-English right-handed bilinguals (12 females and 9 males; mean age of 22.3 years, ranging from 21 to 27) as experiment subjects. They are graduates or undergraduates, and all have received a minimum of 10 years of formal pedagogical English training until the experiment (according to self-report). There were four criteria of inclusion in this study: (a) high English proficiency in reference to scores of College English Test at the advanced level (CET, a national standardized English as a Foreign Language test in the People's Republic of China); (b) right-handedness based on self-report; (c) normal or corrected-to-normal vision; (d) no known history of neurological impairment. Informed consents were obtained prior to scanning, and participants would be paid after the experiment.

2.2 Materials

Since we intended to explore syntactic contrasts between Chinese (L1) and English (L2) in the bilingual brain, we adopted a grammar acceptability judgment task. To make the word order effect more prominent, we added an indefinite frequency adverb in the SVO structure and used active simple present sentences of the subject-adverbial-verb-object (SAVO) structure. Indefinite frequency adverbs (often, always, and usually, etc.) usually appear before the main verb (except the main verb "to be"), and occupy the middle position in English sentences, which reminds the subject-verb agreement check of English sentences in the simple present tense (involving syntax-semantic integration). In contrast, Chinese frequency adverbs are more (over 50) and have a very flexible position in sentences. They can appear at the head, the end of a sentence (usually reduplicated frequency adverbs), or before the main verb, but can never position after the main verb. When Chinese natives judge grammatical canonicity of a sentence, they have to tentatively check all possible positions of the frequency adverb (involving working memory). Here, we used a frequency adverb of "chang (常)" before the main verb in grammatically correct Chinese sentences, and inserted it after the main verb to form incorrect Chinese sentences. In contrast, incorrect English sentences only violated subject-verb agreement rules (subjects were all third person pronouns). Therefore, Chinese and English sentences in the current study would differ only in interested aspects. To make experiment materials comprehensive, the current study also adopted negative sentences, in which the frequency adverbs were replaced by a negator.

In all, there were four test conditions of Chinese affirmative sentences, Chinese negative sentences, English affirmative sentences, and English negative sentences.

Each condition contained 30 sentence items. Each sentence consisted of a pronoun as the subject, an adverb (a frequency adverb or a negation marker with matched auxiliary in the contracted form), a main verb, and an object. In all, there were 120 verb phrases. Predicate phrases were randomly selected to be used in Chinese or English sentences. Since there lacks a word frequency corpus based on Chinese-English bilinguals' language practice, we chose predicates from textbooks appointed by College Entrance Examination or College English Test of basic level in order to assure that experiment materials were very familiar to subjects. The baseline condition was composed of word strings. Constituent words were selected from formal experiment materials. Care was taken to avoid a meaningful verb phrase or noun phrase in baseline word strings. Participants were asked to judge if word strings were composed of the same words, as in the inconsistent condition the last word was a different one from the preceding words. The number of words in the baseline word string was matched with that in English or Chinese sentences. All sentences were rechecked by well-experienced English teachers for the sake of naturalness and acceptability.

2.3   Experimental Procedure

The whole experiment was composed of 4 sessions (the baseline condition consisting of 60 trials was tested in the last session). Participants may have a short rest lasting 1 min between sessions, motionless and eye closed. The task was implemented using E-Prime 2.0 (Psychology Software Tools, Inc). Experiment materials were projected on a screen and displayed in random. Each trial began with a 500 ms "+" fixation, followed by the sentence stimuli for 2500 ms. To make sure participants were timely sufficient to respond, there was a response blank after the probe for the maximum of 1500 ms. Participants were guaranteed to see the stimuli presented in the middle of the vision field. They were instructed to respond as quickly and accurately as possible at the sight of the sentence stimuli. Two accessory key buttons (one for each hand) were used to record response time and error rates. Response hand assignments were counterbalanced across subjects. Before scanning, subjects were asked to complete a short practice session of 24 training trials. Only after lowering the error rate to 5%, were they eligible to start the formal experiment. Practice trials were not reused during the formal experiment.

2.4   Data Acquisition

MRI images were acquired on a Siemens Trio 3T MRI Scanner. A custom-built head holder was used to prevent head movements. Thirty-two axial slices (FOV = 240 mm × 240 mm, matrix = 64 × 64, thickness = 5 mm), parallel to the AC-PC plane and covering the whole brain were obtained using a T2*-weighted single-shot, gradient-recalled echo planar imaging (EPI) sequence (TR = 2000 ms, TE = 30 ms, flip angle = 75°). Prior to the functional run, high-resolution structural images were also acquired using 3D MRI sequences with a voxel size of 1 millimeter cube for anatomical localization (TR = 2700 ms, TE = 2.19 ms, matrix = 256 × 256, FOV = 256 mm 256 mm, flip angle = 7°, slice thickness = 1 mm).

2.5 Data Processing

Functional imaging analyses were carried out using statistical parametric mapping (SPM8, http://www.fil.ion.ucl.ac.uk/spm/). The images were first slice-timed and then realigned to correct for head motions (none of the subjects had head movements exceeding 2 mm on any axis and head rotation greater than 2°). The image data was further processed with spatial normalization based on the MNI space and re-sampled at 2 mm × 2 mm × 2 mm. Finally, the functional images were spatially smoothed with a 6 mm full-width-at-half maximum (FWHM) Gaussian kernel.

At the first level, data from each subject were entered into a general linear model using an event-related analysis procedure. At the group level, each condition was first contrasted with its baseline, and then each regressor entered a flexible factorial design matrix configured under two conditions of Chinese and English. Regions of interest (ROI) were defined by the conjunction analysis of Chinese and English conditions. All the whole brain activation results and conjunction analysis result were reported at $p < 0.05$ (FDR corrected) with cluster >20 voxels. The statistics were color-coded in MNI space, while activation results were estimated from Talairach and Tournoux (to be reported in accordance with the Brodmann atlas) after adjustments for differences between MNI and Talariach coordinates with a nonlinear transform.

To undertake the PPI analysis, a design matrix containing three columns of variables are established, such as (1) a psychological variable that reflects the experimental paradigm, (2) a time-series variable representing the time course of the seed region, and (3) an interaction variable that represents the interaction between (1) and (2) [11]. Separate PPI analyses were conducted for each ROI. The psychological variable used was a vector coding for the specific task (English and Chinese) convolved with the HRF. The individual time series for the ROIs was extracted from all raw voxel time series in a sphere (8 mm radius) centered on the coordinates of the subject-specific activations. The physiological factor (the deconvolved extracted time series of ROIs) was then multiplied with the psychological factor to constitute the interaction term (referred to as "PPI regressor"). These approaches were run using the generalized psychophysiological interaction (gPPI) toolbox [12] implemented in SPM8. The interaction term, the psychological factor, and the physiological factor as regressors were entered in a first-level model of effective connectivity. For the second-level connectivity analysis, we built similar models for each ROI via a flexible factorial model, with regression coefficient for the interaction term giving a measure of PPI. These modeled the same conditions as for the activation analysis, but focused on the psychophysiological interaction. For the connectivity analysis, a statistical threshold of uncorrected $p < 0.001$, cluster size (k) > 20 was used.

3 Results

By contrasting sentences and the corresponding word string, we identified regions involved in reading Chinese and English sentences respectively (see Tables 1 and 2; Fig. 1). We found that reading Chinese (L1) activated areas of the left middle frontal gyrus (MFG, BA9), the bilateral superior temporal gyrus (STG, BA22/41), and the Caudate.

Table 1. Regions involved in reading Chinese SAVO sentences

		Talairach			F	V
		x	y	z	value	voxels
MFG BA9	L	−48	15	25	5.36	44
MeFG BA8/6	L	−10	33	43	4.64	22
PCG BA6/4	R	50	−9	23	4.49	26
STG	L	−44	−29	5	4.95	45
BA41/22/38	R	50	−13	4	5.22	90
PocG BA2/3	L	−55	−13	15	7.16	357
Caudate	L	−14	17	19	3.68	49

Table 2. Regions involved in reading English SAVO sentences

		Talairach			F	V
		x	y	z	value	voxels
IFG BA9	R	50	5	24	4.82	47
MFG BA9	L	−48	15	25	5.98	120
MeFG BA8/6	L	−2	1	63	4.18	23
PCG BA6/4	L	−50	−8	43	5.66	26
	R	53	−5	22	6.18	85
STG	L	−44	−29	5	4.84	29
BA41/22/38	R	50	−13	4	6.08	111
IPL BA 40	L	−46	−39	39	4.90	26
PoCG BA2/3	L	−55	−13	15	7.16	313
Caudate	L	−10	22	15	5.66	43
Putamen	R	20	13	−4	4.19	23

Reading English also activated these three areas (the same peak coordinates), but the activation was much stronger and extended to a larger extent. Besides, reading English additionally elicited neural responses in more widely distributed spatial regions, such as the right inferior frontal gyrus (IFG, BA9), the left inferior parietal lobule (IPL, BA40), and the putamen.

The contrast of English > Chinese showed that English elicited significantly more activation in common areas of the left MFG and bilateral STG (Table 3), but also recruited additional areas of the right IFG, the right inferior temporal gyrus (ITG, BA37), the bilateral middle temporal gyrus (MTG, BA20/21), the left angular gyrus (AG, BA39), the right supramarginal gyrus (SMG, BA40), the anterior cingulate cortex (ACC), and the posterior cingulate cortex (PCC). The contrast of Chinese > English displayed no significant activation cluster.

Conjunction analysis revealed that activation areas in processing English and Chinese were significantly overlapped in the left MFG and bilateral STG, which were determined as seed ROIs for PPI analysis. Results of functional connectivity for the English and Chinese condition were presented in Fig. 2. For the Chinese language, the

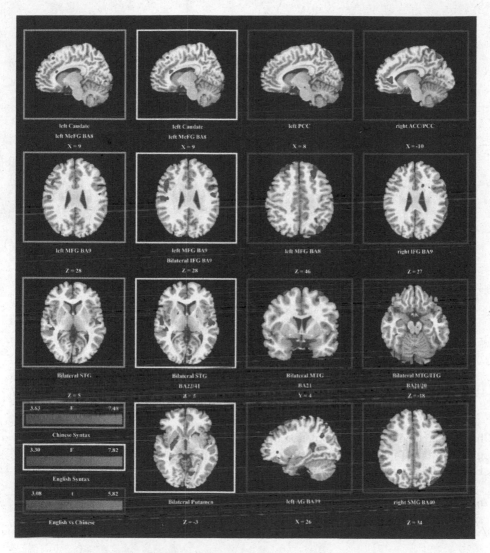

Fig. 1. Activated Regions in reading Chinese, English, and their contrast. Results of reading Chinese, English, and their contrast were presented in the pink box, white box, and green box respectively. (Color figure online)

voxel-wise PPI results showed that the left MFG was effectively connected with the left MTG, the bilateral putamen and the right cuneus; the right STG was effectivelycon-nected with the right paracentral gyrus, left parahippocampal gyrus, left MTG, left MFG, bilateral STG, left lingual gyrus, left ACC, and the right postcentral gyrus (PoCG). For the English language, however, no area was found significantly connected with seed ROIs in the whole brain.

Between-group PPI analysis revealed that reading Chinese involved increased effective connectivity between the right STG and right SFG than English. On the other

Table 3. Regions by the English vs Chinese contrast

		Talairach			t	V
		x	y	z	value	voxels
IFG BA9	R	59	9	24	5.13	99
MFG BA9	L	−32	39	40	4.70	643
	R	32	5	61	4.55	505
MeFG BA8/6	L	−4	59	10	4.15	215
SFG BA8/10	R	24	29	43	5.46	833
PCG BA6/4	L	−28	−12	67	3.70	38
	R	26	−14	67	4.39	328
ITG BA37	R	51	−60	−2	5.14	918
MTG BA20/21	L	−53	−7	−16	4.11	102
	R	50	−9	−16	5.5	137
STG	L	−46	−44	15	3.88	34
BA41/22/38	R	55	3	−14	4.29	52
PoCG BA2/3	L	−55	−23	45	3.97	54
	R	63	−16	30	4.78	73
SPG BA7	R	32	−65	55	5.67	507
AG BA39	L	−28	−59	34	4.06	72
SMG BA40	R	55	−55	32	4.21	163
ACC	R	12	39	2	5.82	2264
PCC	L	−8	−44	11	3.71	78
	R	10	−48	17	3.52	45

hand, increased effective modulation of the right STG to regions of the right precentral gyrus and right insular, and of the left STG to right PCG were identified by contrasting English EC against that of Chinese.

We made further correlation analysis between activation, connectivity strength, and behavioral performance (both response and accuracy rate). However, we found no significant correlation.

4 Discussion

The current study explored how late bilinguals read L1 and L2 sentences. Activation results were consistent with the theoretical framework of the assimilation/accommodation theory that reading L2 activated similar regions like L1 but at an increased activation level. However, the whole brain voxel-wise analysis of functional connectivity showed that commonly activated brain areas were differently integrated with other components in the language network when reading a particular language. Bilinguals did not form an effectively interconnected neural network of L2, which may provide a possible explanation about lower proficiency of L2 than L1. Weak effective interaction of neural network components, and decreased neural modulation effects of ROIs in reading English than Chinese suggested less effectiveness and the compensation

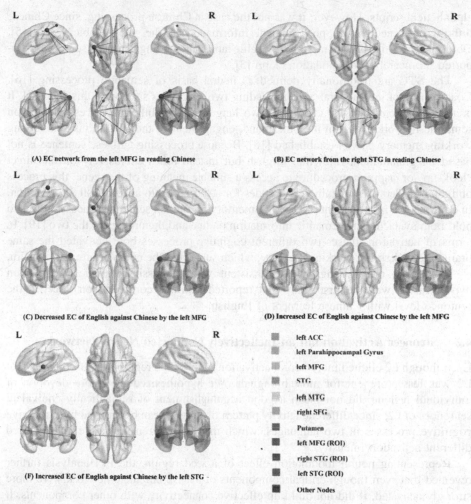

(A) EC network from the left MFG in reading Chinese   (B) EC network from the right STG in reading Chinese

(C) Decreased EC of English against Chinese by the left MFG   (D) Increased EC of English against Chinese by the left MFG

(F) Increased EC of English against Chinese by the left STG

left ACC
left Parahippocampal Gyrus
left MFG
STG
left MTG
right SFG
Putamen
left MFG (ROI)
right STG (ROI)
left STG (ROI)
Other Nodes

Fig. 2. Effectively connected networks in reading Chinese, and the compensated English networks. The effectively connected network of English was not established even in proficient English learners. Both increased and decreased neural modulation effects were identified in reading English, which reflected a compensation mechanism at the neural network level.

mechanism of the L2 network. Meanwhile, increased effective connectivity of ROIs in reading L2 reflected a kind of compensation mechanism at the neural network level.

4.1   Assimilated and Accommodated Neural Network for L2

Somatosensory cortex has been demonstrated critical in phonological motor processing [13]. In tasks involving both overt and covert articulation, the postcentral gyrus appears to represent a significant component of the speech motor network [14]. It is reported that the IPL and PoCG played an important role in phoneme-level manipulation in alphabetical scripts. However, it was not the case in Chinese processing, since Chinese orthography encodes no phonological information at the sub-syllabic level [15]. Obviously, the involvement of L2-specific areas in reading English sentences supported the neural accommodation claim [3].

The STG are traditionally defined as neural basis of semantic processing [16]. Coactivation of the bilateral STG in reading two languages supported this proposal. It is also consistent with the claim that two languages of a bilingual access a common semantic network [17]. On the other hand, engagement of the PFC in tasks involving working memory is well established [18]. Because processing Chinese sentence is not based on grammatical rules like English but instead relies exclusively on the direct "look-up" or mapping procedure to access a suitable meaning of the scene, this process obligated readers to maintain verbal codes for a short term to accomplish the selection. In contrast, the involvement of working memory in English sentence processing was to hold both syntactic and semantic information on line and then integrate the two [19]. In terms of activation, theses two different cognitive processes both activated the same brain area underlying working memory, which supported the neural assimilation claim.

Obviously, our findings were consistent with the assimilation/accommodation hypothesis. It was the first study that reported neural accommodation effect at the sentence level with Chinese learners of English.

4.2   Stronger Activation but an Ineffectively Connected Neural Network

Even though L2 elicited more neural activation of the same region, it was common that L2 was less proficient for most bilinguals. We hypothesized that more devotion of individual regions did not mean a good accomplishment of semantically equivalent sentences of L2, since different activity pattern of a region can be aroused by distinctive cognitive processes in two languages, which may be different in nature and required different activation intensity.

Representing neural modulation effect of a seed region, the PPI analysis further revealed that even though critical components of the L2 language network were more strongly activated, it did not build up effective connectivity with other components. It demonstrated that more devotion of a single area cannot guarantee high proficiency of the second language. We postulated that stronger activation of an individual region may reflect specific requirements of L2 language features per se, and suggested that bilinguals employed spotted brain areas, whose function were versatile and flexible in nature, to accomplish the current task [20]. However, a complex cognitive task depended on cooperation of a constellation of brain areas. Lower network efficacy hurdled normal operation of the L2 neural network, causing low language proficiency.

It was obvious that in the current study the working memory system functioned differently in processing English and Chinese sentences. However, by activation tomography analysis, no clear-cut difference was revealed except intensity change. In contrast, the PPI analysis revealed that the two languages were processed by distinctive working mechanisms in the network aspect. Our findings reinforced that observing only the intensity and activation topography could not fully represent cognitive processes of a task. We proposed that functional connectivity analysis was more feasible to examine sentence reading than the simple spatial activation index.

In sum, even though reading L2 relied on L1 neural literacy and additionally involved more brain areas, these areas were not effectively connected, which reduced the efficacy of L2 neural network. It implied that language proficiency may depend on functional connectivity between critical hub areas and many other brain regions. To implement a complex cognitive task, such as language, we could not solely rely on robuster devotion of single brain areas. It provided a possible explanation of lower proficiency level of L2 than L1. We suggested that future study should discuss complex cognitive processes at the functionally connected network prospect.

Acknowledgments. This paper is supported by the National Natural Science Foundation of China (31400962, 81571752, 81371630), China Postdoctoral Science Foundation Funded Project (2015M582400), the Fundamental Research Funds for the Central Universities (RW150401), Shenzhen Peacock Plan (KQTD2015033016104926), Shanxi Nova Program (2014KJXX-34), and the Shenzhen Science and Technology Program (JCYJ20160608173106220). There are no conflicts of interest.

References

1. Wartenburger, I., Heekeren, H.R., Abutalebi, J., Cappa, S.F., Villringer, A., Perani, D.: Early setting of grammatical processing in the bilingual brain. Neuron **37**, 159–170 (2003). doi:10.1016/S0896-6273(02)01150-9
2. Abutalebi, J.: Neural aspects of second language representation and language control. Acta Psychol. **128**, 466–478 (2008). doi:10.1016/j.actpsy.2008.03.014
3. Perfetti, C.A., Liu, Y., Tan, L.H.: The Lexical Constituency Model: Some implications of research on Chinese for general theories of reading. Psycho. Rev. **12**, 43–59 (2005). doi:10.1037/0033-295X.112.1.43
4. Friston, K.J.: Models of brain function in neuroimaging. Annu. Rev. Psychol. **56**, 57–87 (2005). doi:10.1146/annurev.psych.56.091103.070311
5. Park, H.J., Friston, K.: Structural and functional brain networks: From connections to cognition. Science **342**, 1238411 (2013). doi:10.1126/science.1238411
6. Fox, P.T., Friston, K.J.: Distributed processing; distributed functions? Neuroimage **61**, 407–426 (2012). doi:10.1016/j.neuroimage.2011.12.051
7. Sporns, O.: Contributions and challenges for network models in cognitive neuroscience. Nat. Neurosci. **17**, 652–660 (2014). doi:10.1038/nn.3690
8. Friston, K.J.: Functional and effective connectivity: a review. Brain Connect. **1**, 13–36 (2011). doi:10.1089/brain.2011.0008
9. Friston, K.J., Buechel, C., Fink, G.R., Morris, J., Rolls, E., Dolan, R.J.: Psychophysiological and modulatory interactions in neuroimaging. Neuroimage **6**, 218–229 (1997). doi:10.1006/nimg.1997.0291
10. Luke, K.K., Liu, H.L., Wai, Y.Y., Wan, Y.L., Tan, L.H.: Functional anatomy of syntactic and semantic processing in language comprehension. Hum. Brain Mapp. **16**, 133–145 (2002). doi:10.1002/hbm.10029
11. Hayashi, A., Okamoto, Y., Yoshimura, S., Yoshino, A., Toki, S., Yamashita, H., Matsuda, F., Yamawaki, S.: Visual imagery while reading concrete and abstract Japanese kanji words: an fMRI study. Neurosci. Res. **79**, 61–66 (2014). doi:10.1016/j.neures.2013.10.007

12. McLaren, D.G., Ries, M.L., Xu, G., Johnson, S.C.: A generalizedform of context-dependent psychophysiological interactions(Gppi): a comparison to standard approaches. Neuroimage **61**, 1277–1286 (2012). doi:10.1016/j.neuroimage.2012.03.068

13. Dhanjal, N.S., Handunnetthi, L., Patel, M.C., Wise, R.J.S.: Perceptual systems controlling speech production. J. Neurosci. **28**, 9969–9975 (2008). doi:10.1523/JNEUROSCI.2607-08. 2008

14. Pulvermüller, F., Huss, M., Kherif, F., del Prado Martin, F.M., Hauk, O., Shtyrov, Y.: Motor cortex maps articulatory features of speech sounds. Proc. Natl. Acad. Sci. U.S.A. **103**, 7865– 7870 (2006). doi:10.1073/pnas.0509989103

15. Cao, F., Kim, S.Y., Liu, Y., Liu, L.: Similarities and differences in brain activation and functional connectivity in first and second language reading: evidence from Chinese learners of English. Neuropsychologia **63**, 275–284 (2014). doi:10.1016/j.neuropsychologia.2014. 09.001

16. Vigneau, M., Beaucousin, V., Herve, P.Y., Duffau, H., Crivello, F., Houdé, O., Mazoyer, B., Tzourio-Mazoyer, N.: Meta-analyzing left hemisphere language areas: phonology, semantics, and sentence processing. Neuroimage **30**, 1414–1432 (2006). doi:10.1016/j. neuroimage.2005.11.002

17. Ruschemeyer, S.A., Fiebach, C.J., Kempe, V., Friederici, A.D.: Processing lexical semantic and syntactic information in first and second language: fMRI evidence from German and Russian. Hum. Brain Mapp. **25**, 266–286 (2005). doi:10.1002/hbm.20098

18. Braver, T.S., Barch, D.M., Kelley, W.M., Buckner, R.L., Cohen, N.J., Miezin, F.M., Snyder, A.Z., Ollinger, J.M., Akbudak, E., Conturo, T.E., Petersen, S.E.: Direct comparison of prefrontal cortex regions engaged by working and long-term memory tasks. Neuroimage **14**, 48–59 (2001). doi:10.1006/nimg.2001.0791

19. Hagoort, P., Indefrey, P.: The neurobiology of language beyond single words. Annu. Rev. Neurosci. **37**, 347–362 (2014). doi:10.1146/annurev-neuro-071013-013847

20. Reuter-Lorenz, P.A., Cappell, K.A.: Neurocognitive aging and the compensation hypothesis. Curr. Dir. Psychol. Sci. **17**, 177–182 (2008). doi:10.1111/j.1467-8721.2008.00570.x

Objects Categorization on fMRI Data: Evidences for Feature-Map Representation of Objects in Human Brain

Sutao Song[1], Jiacai Zhang[2], and Yuehua Tong[1(✉)]

[1] School of Education and Psychology, University of Jinan, Jinan 250022, China
sep_tongyh@ujn.edu.cn
[2] College of Information Science and Technology, Beijing Normal University, Beijing 100875, China

Abstract. Brain imaging studies in humans have reported each object category was associated with different neural response pattern reflecting visual, structure or semantic attributes of visual appearance, and the representation of an object is distributed across a broader expanse of cortex rather than a specific region. These findings suggest the feature-map model of object representation. The present object categorization study provided another evidence for feature-map representation of objects. Linear Support Vector Machine (SVM) was used to analyze the functional magnetic resonance imaging (fMRI) data when subjects viewed four representative categories of objects (house, face, car and cat) to investigate the representation of different categories of objects in human brain. We designed 6 linear SVM classifiers to discriminate one category from the other one (1 vs. 1), 12 linear SVM classifiers to discriminate one category from other two categories (1 vs. 2), 3 linear SVM classifiers to discriminate two categories of objects from the other two categories (2 vs. 2). Results showed that objects with visually similar features have lower classification accuracy under all conditions, which may provide new evidences for the feature-map representation of different categories of objects in human brain.

Keywords: Feature-map representation · Support vector machine · fMRI

1 Introduction

The representation of objects in human brain is a matter of intense debate, and in the domain of neuroimaging study, three major models exist [1, 2]: category specific model, process-map model and feature-map model.

The category specific model proposes that ventral temporal cortex contains a limited number of areas that are specialized for representing specific categories of stimuli. Evidences from patients with brain lesion showed that patient with lesions in one paticular brain area lost the ability to recognize facial expressions or other objets [3, 4]. One study found that a farmer with brain lesion no longer recognized his own cows [4]. The study of event-related-potentials (ERP) and magnetic encephalography (MEG) supported the face specificity in visual processing, human faces elicited a negative component peaking at about 170 ms from stimulus onset (N170 or M170) [5–7].

© Springer International Publishing AG 2017
Y. Zeng et al. (Eds.): BI 2017, LNAI 10654, pp. 107–115, 2017.
https://doi.org/10.1007/978-3-319-70772-3_10

Functional magnetic resonance imaging (fMRI) studies described specialized areas for faces and some specific objects: the fusiform face area (FFA) for human faces, the parahippocampal place area (PPA) for scenes and the "extrastriate body area" (EBA) for visual processing of the human body [8–13].

The process-map model [1, 14–16] proposes that different areas in ventral temporal cortex are specialized for different types of perceptual processes. The studies from Gauthier et al. showed that FFA was not just specialized for faces, but for expert visual recognition of individual exemplars from any object category. For example, for bird experts, FFA shows more activity when they view the pictures of bird, and for car experts, FFA shows more activity when they view car than bird. Study also showed that the acquisition of expertise with novel objects (such as greebles, one kind of man-made object) led to increased activation in the right FFA [14].

For feature-map model, it proposes that the representations of faces and different categories of objects are widely distributed and overlapping [2, 17–21]. In the study Haxby et al. [2], fMRI data of ventral temporal cortex was recorded when subjects viewed faces, cats, five categories of man-made objects, and nonsense pictures. A correlation-based distance measure was used to predict the object categories and the prediction result indicates that the representations of faces and objects in ventral temporal cortex are widely distributed and overlapping.

The evidences for the three models came mainly from the neuroimaging study of healthy subjects or patients with brain lesion. Generally, for the analysis of the neuroimaging data, univariate method was used, such as general linear model. However, fMRI was multi-variate in nature. In recent years, multi-variate pattern analysis (MVPA) methods have been widely used in fMRI analysis [22–26]. Compared with the traditional univariate method, MVPA method takes the correlation among neurons or cortical locations into consideration and is more sensitive and informative. In this study, we further investigate the representation of objects in human brain using Support Vector Machine (SVM). As one representative MVPA method, SVM is effective in digging the information behind fMRI data. Four representative objects (house, face, car and cat) were selected as stimulus, which can be grouped in the following ways: face vs. other objects, animate vs. inanimate objects. SVM was applied to predict the label of brain states, i.e. which kind of stimulus the subject was viewing, and 6 classifiers were trained to classify one object category versus the other category (house vs. face, house vs. car, house vs. cat, face vs. car, face vs. cat, car vs. cat). To further investigate the representation of objects in human brain, 15 other classifiers were trained to cover the possible combinations of the 2-class classification problem for the four categories of objects (1 vs. 2, 2 vs. 2).

2   Method

2.1   Subjects and fMRI Data Acquisition

The data came from one of our previous study [26]. Fourteen healthy college students participated in this study (6 males, 8 females). Subjects gave written informed consent. A 3-T Siemens scanner equipped for echo planar imaging (EPI) at the Brain Imaging

Center of Beijing Normal University was used for image acquisition. Functional images were collected with the following parameters: repeat time (TR) = 2000 ms; echo time (TE) = 30 ms; 32 slices; matrix size = 64 × 64; acquisition voxel size = 3.125 × 3.125 × 3.84 mm³; flip angle (FA) = 90°; field of view (FOV) = 190 ~ 200 mm. In addition, a high-resolution, three-dimensional T1-weighted structural image was acquired (TR = 2530 ms; TE = 3.39 ms; 128 slices; FA = 7°; matrix size = 256 × 256; resolution = 1 × 1 × 1.33 mm³).

2.2 Stimuli and Experimental Procedure

The experiment was designed in a blocked fashion. Subject participated in 8 runs and each run consisted of 4 task blocks and 5 control blocks. During each task block lasted for 24 s, 12 gray-scale images belonging to one category (houses, faces, cars or cats) were presented which were chosen randomly from 40 pictures of that particular category, and subject had to press a button with left or right thumb as long as images were repeated consecutively. Two identical images were displayed consecutively 2 times randomly during each task block. Each stimulus was presented for 500 ms followed by a 1500 ms blank screen. Control blocks were 12 s fixation in the beginning of a run and at the end of every task block (Fig. 1). Each kind of objects block was presented once during each run, and the order of them was counterbalanced in the whole session which lasted 20.8 min.

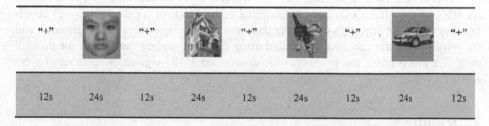

Fig. 1. The experimental procedure for one run.

2.3 Data Preprocessing

The preprocessing steps were the same as our previous study [26]. SPM2 (http://www.fil.ion.ucl.ac.uk/spm/) was used to finish the preprocessing job. It mainly contains 3 steps: realignment, normalization and smoothing. Subjects were preprocessed separately. In the beginning, the first 3 volumes were discarded as the initial images of each session showed some artifacts related to signal stabilization (according to the SPM2 manual). Images were realigned to the first image of the scan run and were normalized to the Montreal Neurological Institute (MNI) template. The voxel size of the normalized images was set to be 3 * 3 * 4 mm. At last, images were smoothed with 8 mm full-width at half maximum (FWHM) Gaussian kernel. The baseline and the low frequency components were removed by applying a regression model for each voxel [23]. The cut-off period chosen was 72 s.

2.4   Voxel Selection

Voxels that activated for any kind of object within the whole brain were selected for further analysis (family-wise error correction, p = 0.05) (Fig. 2).

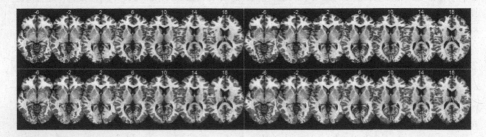

Fig. 2. The voxel selection for one representative subject (slices $-8 \sim 20$), (Red: house; Blue: Face; Green: Car; Yellow: Cat). (Color figure online)

2.5   SVM Method

LibSVM (http://www.csie.ntu.edu.tw/~cjlin/libsvm) was used to predict the brain states. The data of first 4 runs was used to train the model, and the data of last 4 runs was used to test the model. To reduce the number of features, principle component analysis (PCA) procedure was conducted over the features and PCs accumulatively accounting for 95% of the total variance of the original data were kept for the subsequent classification. Then the attributes of training data was scaled to the range $[-1, 1]$ linearly; and the attributes of the test data was scaled using the same scaling function of the training data. To compensate the hemodynamic delays, the fMRI signals of each voxel were shifted by 4 s.

3   Results

For all the 21 combinations of two-class classification problems of the four categories of objects (1 vs. 1, 1 vs. 2, 2 vs. 2), the classification accuracies were all above the chance level (Kappa coefficients: 0.73 ± 0.13, M $\pm$ SD).

3.1   Classification Results for One vs. One Classifiers

Classification performances for discriminating one category from another category were shown in Fig. 3. In this situation, 6 classifiers were trained (house vs. face, house vs. cat, face vs. car, car vs. cat). Significant differences were found among the 6 classifiers ($F(3.2, 42) = 11.88, p < 0.001, \eta^2 = .478$.). And two groups, house vs. car, face vs. cat, have the lowest classification accuracy. The lower performance in distinguishing houses from car (or face from cat) suggests that houses (or face) share more common activity with car (or cat) and therefore less dissociable.

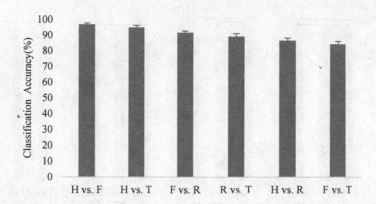

Fig. 3. Classification results for One vs. One Classifiers (M ± SE) (H: house; F: Face; R: Car; T: Cat).

3.2   Classification Results for One vs. Two Classifiers

Classification performances for discriminating one category from the other two categories of objects were shown in Fig. 4. In this situation, 4 groups, and totally 12 classifiers were trained, each group contains three categories of objects (e.g., group one: face vs. house and car; house vs. face and car; car vs. house and face). Significant difference were found among the 3 classifiers for each group ($F(2, 26) = 31.22, p < 0.001, \eta^2 = .706; F(1.39, 18.07) = 22.88, p < 0.001, \eta^2 = .638; \quad F(2, 26) = 11.18, p < 0.001, \eta^2 = .462; F(2, 26) = 28.80, p < 0.001, \eta^2 = .689$). When distinguishing car (or cat) from the other two categories of objects, the classifier performed worst, which suggests that car (or cat) share more common activity with the other categories of objects. Take group one for example, the classifier performed worst to discriminate car from house and face, which implies the similar spatial activity of car with house and (or) face. To look the three 2-class classifiers that involved the three objects (house vs. face, house vs. car, car vs. cat in Fig. 4 further, the two classifiers included car had lower classification accuracy. The results were similar for group 2, 3 and 4.

3.3   Classification Results for Two vs. Two Classifiers

Classification performances for discriminating two categories from the other two categories of objects were shown in Fig. 5. In this situation, 3 classifiers were trained (house and car vs. cat and face; house and cat vs. face and car; house and face vs. car and cat). Significant differences were found among the 3 classifiers ($F(2, 26) = 39.59, p < 0.001, \eta^2 = .753$). When discriminating house and car from face and cat, the classifier performed best, which suggests the dissociative spatial pattern may exist.

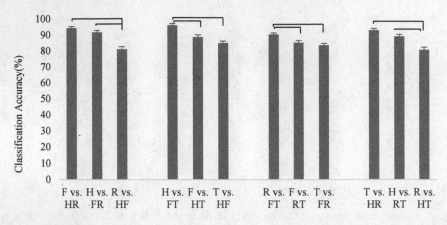

Fig. 4. Classification results for One vs. Two Classifiers (M ± SE) (H: house; F: Face; R:Car; T: Cat).

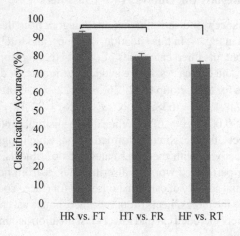

Fig. 5. Classification results for Two vs. Two Classifiers (M ± SE) (H: house; F: Face; R: Car; T: Cat).

3.4   Classification Results for Regions Maximally Responsive to One Category of Objects

The classification accuracies for One vs. One and Two vs. Two classifiers were also provided when the voxels that responded maximally to one category were chosen as features (Fig. 6). Again, significant differences were found (all *ps* < .001). And similar patterns were observed across voxel selection schemes.

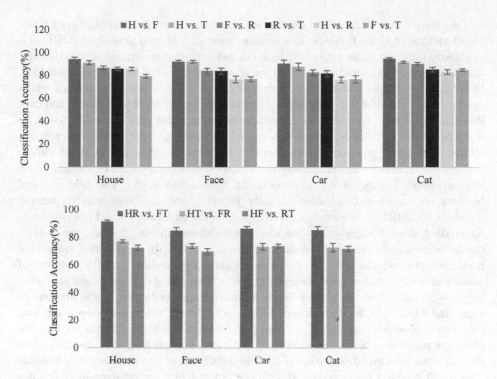

Fig. 6. Classification results for One vs. One (up) and Two vs. Two (below) classifiers when the voxels that responded maximally to one category (i.e. House, face, car or cat) were chosen as features (M ± SE) (H: house; F: Face; R: Car; T: Cat).

4 Discussion and Conclusions

In this study, one MVPA method, SVM was used to analyze the fMRI data when subjects viewed faces and other objects. We investigated the possibility to classify the brain states by various groups. This study selected four representative objects to study the representation of objects in human brain in large scale (i.e. the scale of fMRI technology). Totally 21 classifiers were trained to cover most of the possible combinations of the four objects (the 1 vs. 3 classifiers were not included, as they provide no useful information about the representation way of objects in human brain). Results showed that objects with visually similar features have lower classification accuracy under all conditions, which may provide new evidence for the feature-map representation of different category of objects in human brain.

The current analysis applied linear SVM to predict the categories of objects that the subject viewed. SVM finds a linear combination of features which characterize or separate two or more classes of objects or events. Thus, the higher the classification accuracy is, the less in common the spatial activities are, and vice versa. As one multi-variate analysis method, SVM is powerful in digging information behind fMRI data. However, the use of multi-variate analysis method in fMRI study when subjects viewed the pictures of faces and objects was not new. Haxby et al. applied correlation based method (it is the first time multi-variate method was used to analyze fMRI data) to classify the brain states evoked by face, cat and five other man-made objects (houses, shoes, scissors, bottles and chairs), and the result supported the feature-map model [2]. Different from Haxby's study, we chose four objects (house, face, car and cat), which can be further classified as animate (face and cat) vs. inanimate objects (house and car). Besides, intuitively, face and cat contain information about face processing (such as features related with eyes, mouth and ears), and house and car contain information related with scene processing. The result of Fig. 3 shows that it is most difficult to classify the brain activities elicited by the following two groups, face vs. cat and house vs. car, which is more likely to support the feature-map model. The similar visual features are represented adjacent spatially in brain, and the brain activity patterns recorded by fMRI are adjacent or overlapped, as the patterns of voxel activities corresponding to each category on the whole brain shown in Fig. 2. Thus, the classification accuracy for linear SVM is low. Besides, when the voxels that responded maximally to one category of objects were chosen as features, similar patterns of classification accuracies were observed as that shown in Figs. 2 and 5, and the accuracies were all above the chance level, indicating the overlapped representations of faces and objects. If the definition of feature is not clear, when we grouped any two categories of objects as one class, the classification result (Fig. 5) shows that the classifier performed best when discriminating house and car from face and cat, while the classifier performed worst when discriminating house and face from car and cat. This result indicates house and car share more features, face and cat share more features in common, and thus have similar brain activity pattern. In other situations, results also showed that objects with visually similar features achieved lower classification accuracy (Fig. 4), which further supports the feature-map representation of different category of objects in human brain.

In conclusion, MVPA methods and fMRI technology provide new way to under the representation of different categories of objects in human brain. The current study shows new evidence for feature-map representation of objects.

Acknowledgments. This research was financially supported by Young Scientist Fund of National Natural Science Foundation of China (NSFC) (31300924), NSFC General program (61375116), the Fund of University of JiNan (XKY1508, XKY1408).

References

1. Gauthier, I.: What constrains the organization of the ventral temporal cortex? Trends Cogn. Sci. **4**(1), 1–2 (2000)
2. Haxby, J., et al.: Distributed and overlapping representations of faces and objects in ventral temporal cortex. Science **293**(5539), 2425–2430 (2001)
3. Hecaen, H., Angelergues, R.: Agnosia for faces (prosopagnosia). Arch. Neurol. **7**(2), 92 (1962)
4. Assal, G., Favre, C., Anderes, J.: Nonrecognition of familiar animals by a farmer. Zooagnosia or prosopagnosia for animals. Revue Neurologique **140**(10), 580 (1984)

5. Carmel, D., Bentin, S.: Domain specificity versus expertise: factors influencing distinct processing of faces. Cognition **83**(1), 1–29 (2002)
6. Rossion, B., et al.: The N170 occipito-temporal component is delayed and enhanced to inverted faces but not to inverted objects: An electrophysiological account of face-specific processes in the human brain. NeuroReport **11**(1), 69 (2000)
7. Liu, J., et al.: The selectivity of the occipitotemporal M170 for faces. NeuroReport **11**(02), 337 (2000)
8. Kanwisher, N., McDermott, J., Chun, M.: The fusiform face area: a module in human extrastriate cortex specialized for face perception. J. Neurosci. **17**(11), 4302 (1997)
9. McCarthy, G., et al.: Face-specific processing in the human fusiform gyrus. J. Cogn. Neurosci. **9**(5), 605–610 (1997)
10. Kanwisher, N.: Domain specificity in face perception. Nat. Neurosci. **3**, 759–763 (2000)
11. Fodor, J.A.: The Modularity of Mind. MIT, Cambridge (1981)
12. Downing, P.E., et al.: A cortical area selective for visual processing of the human body. Science **293**(5539), 2470 (2001)
13. Epstein, R., Kanwisher, N.: A cortical representation of the local visual environment. Nature **392**(6676), 598–601 (1998)
14. Tarr, I.G.M.J., et al.: Activation of the middle fusiform 'face area' increases with expertise in recognizing novel objects. Nat. Neurosci. **2**(6), 569 (1999)
15. Gauthier, I., et al.: Expertise for cars and birds recruits brain areas involved in face recognition. Nat. Neurosci. **3**, 191–197 (2000)
16. Tarr, M.J., Gauthier, I.: FFA: a flexible fusiform area for subordinate-level visual processing automatized by expertise. Nat. Neurosci. **3**, 764–770 (2000)
17. Ishai, A., et al.: Distributed representation of objects in the human ventral visual pathway. Nat. Acad. Sci. **96**, 9379–9384 (1999)
18. Ishai, A., et al.: The representation of objects in the human occipital and temporal cortex. J. Cogn. Neurosci. **12**(Supplement 2), 35–51 (2000)
19. Haxby, J.V., Hoffman, E.A., Gobbini, M.I.: The distributed human neural system for face perception. Trends Cogn. Sci. **4**(6), 428–432 (2000)
20. Chao, L.L., Haxby, J.V., Martin, A.: Attribute-based neural substrates in temporal cortex for perceiving and knowing about objects. Nat. Neurosci. **2**, 913–919 (1999)
21. Tanaka, K.: Infcrotemporal cortex and object vision. Ann. Rev. Neurosci. **19**(1), 109–139 (1996)
22. Zhang, H., et al.: Face-selective regions differ in their ability to classify facial expressions. NeuroImage **130**, 77–90 (2016)
23. Wegrzyn, M., et al.: Investigating the brain basis of facial expression perception using multi-voxel pattern analysis. Cortex **69**, 131–140 (2015)
24. Kragel, P.A., Labar, K.S.: Multivariate neural biomarkers of emotional states are categorically distinct. Soc. Cogn. Affect. Neurosci. **10**(11), 1437–1448 (2015)
25. Cowen, A.S., Chun, M.M., Kuhl, B.A.: Neural portraits of perception: Reconstructing face images from evoked brain activity. NeuroImage **94**(1), 12–22 (2014)
26. Song, S., et al.: Comparative study of SVM methods combined with voxel selection for object category classification on fMRI data. PLoS ONE **6**(2), e17191 (2011)

Gender Role Differences of Female College Students in Facial Expression Recognition: Evidence from N170 and VPP

Sutao Song[1(✉)], Jieyin Feng[2], Meiyun Wu[1], Beixi Tang[1], and Gongxiang Chen[1]

[1] School of Education and Psychology, University of Jinan, Jinan 250022, China
sep_songst@ujn.edu.cn
[2] Institute of Cognitive Neuroscience, East China Normal University, Shanghai 200062, China

Abstract. Previous studies have extensively reported an advantage of females over males in facial expression recognition. However, few studies have concerned the gender role differences. In this study, gender role differences on facial recognition were investigated by reaction time and the early event-related potentials (ERPs), N170 and Vertex Positive Potential (VPP). A total of 466 female college students were investigated by gender role inventory, and 34 of them were chosen as subjects, with equal numbers in masculinity and femininity. Subjects were asked to discriminate fearful, happy and neutral expressions explicitly in two emotional states: neutral and fearful. First, N170 and VPP showed greater activity in femininities than in masculinities. Second, subjects showed a predominance of negative face processing, as VPP was more positive in response to fearful expressions than neutral and happy expressions, but no gender role difference was found. Third, in fearful state, the reaction time was shorter, especially for fear expression, and N170 showed enhanced negativity, suggesting that fearful state could promote individuals to recognize expressions, and there was no gender role difference. In conclusion, gender role differences exist in the early stage of facial expressions recognition and femininities are more sensitive than masculinities. Our ERP results provide neuroscience evidence for differences in the early components of facial expression cognition process between the two gender roles of females.

Keywords: Gender role differences · ERPs · Facial expression recognition · N170 · VPP · Fearful state

1 Introduction

Sustained literature and reviews report gender differences in facial processing, indicating that females are more emotionally sensitive, especially to negative emotions [1, 2], and this view is also supported by neuroimaging [3] and event-related potentials (ERPs) studies [4–6].

According to findings from a meta-analysis of neuroimaging study, compared with men, women responded strongly to negative emotions with greater activation in left

© Springer International Publishing AG 2017
Y. Zeng et al. (Eds.): BI 2017, LNAI 10654, pp. 116–125, 2017.
https://doi.org/10.1007/978-3-319-70772-3_11

amygdala [3]. Several ERPs studies found women exhibited larger N170 amplitudes to facial stimuli in comparison with men [4, 5]. Interestingly, a counterpart of N170 is detected at the fronto-central electrode and peaks within 200 ms following stimulus onset named Vertex Positive Potential (VPP), which is a positive-going ERP and is considered to be concomitant of N170 [7]. Both N170 and VPP have been found to be more sensitive to faces than objects, particularly to threatening faces, with a larger enhancement to fearful and angry faces than the neutral or pleasant [7–9].

Although there are many interesting findings in the field of the physiological sex difference in facial expression recognition, few have concerned the psychological gender role difference. According to Bem's sex role theory, gender role is suggested as a personality type, including two dimensions, masculinity and femininity. An individual could be high or low on each dimension, determining four gender role types, androgynous, undifferentiated, traditional men and traditional women [10, 11]. The androgynous are supposed be high on femininity and masculinity, while the differentiated act both low on those two dimensions. Traditional women type could be high on femininity but low on masculinity, showing more feminine, which we name as the short term "femininity". By contrast, traditional men type could be low on femininity but high on masculinity, tending to be more masculine, which we name as "masculinity". Bem argued that beliefs of gender role could guide and frame our behaviors in social interactions. Whereas, stuck with social norms and public pressure or gender identity, we tend to act consistently to our biological gender [12], leaving real thoughts and behaviors depressive or covert, which are framed by gender role. Ridgeway et al. suggested that gender role played as a cognitive background and implicitly influences our minds and behaviors [13]. Therefore, it is necessary for us to study the gender role differences on face cognition. We pay close attention to two gender role types of females in this study, traditional women and traditional men, or femininity and masculinity.

Furthermore, in our daily life, facial expression never shows up in isolation, but almost always appears within a situational context [14–16], such as the surrounding visual scene of the expressions, or the emotional states of the subjects. However, to the best of our knowledge, very few studies have investigated the effect of the emotional states on subjects with different gender roles in expressions cognition. This issue therefore remains unclear.

Taken together, the goal of our study was to clarify gender role differences on face cognition in N170 and VPP and the impact of emotional states. We adopted 2*2*3 mixed design, with the reaction time and ERPs as dependent variables and three independent variables: gender roles, emotional states and facial expressions. The between-subject factor is gender role, including two levels of masculinity and femininity. One within-subjects factor is emotional state, including the fearful and neutral states, and the other is facial expression: fearful, neutral, and happy. We hypothesized that femininity, compared to masculinity, would show greater N170 in expression discrimination task, and VPP would show similar result; an interaction between gender roles and emotional state may exist, femininity would react faster and show larger N170 and VPP than masculinity in fearful state.

2   Materials and Methods

2.1   Participants

Participants with different gender roles were selected from a pool of 466 female undergraduate students based on the response to "New 50-item gender role inventory for Chinese college students" (CSRI-50) [17]. The CSRI-50 yields measures of masculinity and femininity. Participants whose feminine score was greater than 5 and masculine score less than 4.8, or whose masculine score was greater than 4.8 and feminine score less than 5, comprised the feminine and masculine group respectively. And they were invited to participate in the formal experiment. Finally, 17 subjects (femininity scale: M = 5.43, SD = 0.60, masculinity scale: M = 3.89, SD = 0.54) from the feminine group and 17 subjects (femininity scale M = 4.06, SD = 0.51, masculinity scale M = 5.20, SD = 0.42) from the masculine group are willing to take part in our formal experiment. Therefore, 34 healthy participants (mean age = 20.3, SD = 2.1) gave informed consent, and performed task while electroencephalogram (EEG) was recorded.

2.2   Stimuli

Two fearful film clips from the Filmstim collection [18], named "Shining" and "Scream" (with a mean duration of 3.92 min), were used in the formal experiment for emotion induction.

Emotional facial expression stimuli came from the NIMSTIM set [19]. One hundred and ninety-two images, specifically, 32 different actors (16 males) portrayed each of three expressions: neutral, fearful and happy. Each expression included separate open- and closed-mouth versions. Participants were seated in a comfortable chair with their eyes approximately 70 cm away from a 17-in screen, and the viewing angle was 5.25° × 6.06°.

2.3   Experimental Procedure

The participants were instructed to do a facial expression discrimination task by pressing three different but adjacent buttons with their right hand during the experiment before and after watching movies.

The experimental procedure was described briefly in Fig. 1 (a) which consisted of two parts. In the first part, participants did facial expression discrimination task lasting for about 4 min, during which the expressions of 16 actors (8 female, totally 96 pictures) were shown, which resulted in 96 trails. After a rest of about 5 min, participants continued to finish the second part of the discrimination task. In the second part, they initially watched a fearful movie clip, then, the expressions of the other 16 actors were shown to them. To ensure the subjects experienced intense fear emotion, the discrimination tasks were separated into two parts, and each lasted about 2 min, during which the expressions of 8 actors (4 females, totally 48 pictures or 48 trials) were shown, after watching two fearful movie clips respectively. Each face was presented once, and the order of face presentation was randomized for each participant. Totally, the experiment consisted of 192 trials, with 96 trials in neutral and fearful state respectively. After the experiment, the subjects were instructed to relax.

An overview for one experimental trial was shown in Fig. 1(b). Faces were presented for 1000 ms when subjects were asked to do facial expression discrimination task, and a fixation mark (+) was presented during the inter-trial interval, which varied randomly between 1000 and 1500 ms [20]. Each trial started with a fixation mark (+) lasted for 500 ms.

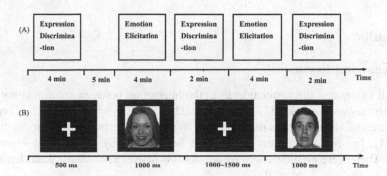

Fig. 1. Overview of the experiment procedure (A) and one example trial (B).

2.4 Behavioral Data Analysis

The reaction times (RTs) were analyzed for subjects whose discrimination accuracy was higher than 87.73% (144/192). Referenced to this criterion, the RTs (only for the trials responded correctly) of 33 out of 34 subjects were assessed. All data were analyzed by three-way ANOVAs with emotional states (neutral state, fearful state), facial expressions (happy, neutral, fearful), as within-subject factors, and gender roles (masculinity, femininity) as between-subject factors.

2.5 EEG Recordings and Analysis

The EEG data were recorded by 64 scalp electrodes mounted on an electrode cap (Compumedics Neuroscan's Quick-cap). Electrooculogram (EOG) generated from blinks and eye movements was recorded. All scalp electrodes were referenced to the vertex Cz. Resistance of all electrodes were kept lower than 10 kΩ. The EEG signals were recorded with a band-pass filter of 0.05-400 Hz and a sampling rate of 1000 Hz.

Offline, using curry7 (Compumedics Neuroscan USA, Ltd), the raw EEG data were re-referenced to average reference. Afterwards, all data were band-pass filtered between 0.1 and 30 Hz. Trials with EOG artifacts (amplitudes of EOG exceeding ± 100 μV) and those contaminated with other artifacts (peak-to-peak deflection exceeding ± 100 μV, or containing baseline drifts) were excluded from averaging. The analyzing epoch was time-locked to the onset of facial expression stimuli, and the length of the ERP epoch was 1200 ms with a prestimulus baseline of 200 ms. Data from nine participants were excluded because of poor quality EEG recording or too few artifact-free trials (less than 20 trials per condition). The final sample was composed of 25 participants (12 masculinity and 13 femininity).

ERP analyses focused on the peak amplitudes of N170 (electrode sites: P7, P8, PO7, PO8) and VPP (electrode sites: CZ, CPZ) during the time window 130 ms to 200 ms. All ERP data were assessed by three-way ANOVAs. The Greenhouse-Geisser correction was applied where sphericity was violated. When the main effect or an interaction was significant, pairwise comparisons were performed with the Bonferroni correction.

3 Results

3.1 Behavioral Results

For facial expression discrimination task, the interaction between emotional states and facial expressions was found, $F(2, 62) = 6.02$, $p = .00$, $\eta^2 = .16$ (Fig. 2). Compared with the neutral state, subjects reacted faster to the fearful expression in the fearful state ($p = .001$), but no differences were found on both happy and neutral expressions ($p > .1$). There were no differences on gender role between masculinity and femininity, $F(1, 31) = .20$, $p > .1$, $\eta^2 = .01$.

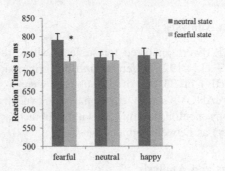

Fig. 2. Response times for the facial expression discrimination task illustrating an emotional states by facial expressions interaction (p = .001). Bars represent S.E.

3.2 ERP Results

N170 A significant main effect was observed for gender roles, $F(1,23) = 4.76$, $p = .04$, $\eta^2 = .08$, N170 was more negative for femininity than masculinity when discriminating facial expressions (Fig. 3. left column). The main effect of emotional states was also found, $F(1, 23) = 36.52$, $p = .00$, $\eta^2 = .61$, N170 was more negative in fearful state than neutral state (Fig. 3. right column).

A significant main effect was found for facial expressions, $F(2, 46) = 7.72$, $\varepsilon = .77$, $p = .003$, $\eta^2 = .25$. Post hoc tests showed, N170 was more negative to fearful and happy expressions than neutral expressions ($p < .01$), but there was no difference

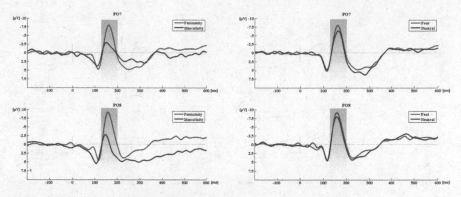

Fig. 3. Grand-average N170 elicited by subjects with different gender roles (left column) and in fearful and neutral states (right column) at PO7 and PO8 sites.

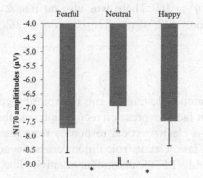

Fig. 4. The mean amplitudes of N170 for three facial expressions (fearful, neutral and happy). Bars represent S.E.

between happy and fearful expressions ($p > .1$), showing common effects for fear and happiness relative to neutral (Fig. 4).

No significant interaction was found between gender roles and facial expressions, $F(1, 23) = 2.6$, $p > .1$, $\eta^2 = .10$. The interaction between gender roles and emotional states was not found, $F(1, 23) = .00$, $p > .1$, $\eta^2 = .00$. There were also no interactions between emotional states and facial expressions, $F(2, 46) = .44$, $p > .1$, $\eta^2 = .02$.

VPP A significant main effect was found on gender roles, $F(1,23) = 8.41$, $p = .01$, $\eta^2 = .27$, VPP was more positive for femininity than masculinity (Fig. 5 left column). A main effect was found on facial expressions, $F(2,46) = 7.72$, $p = .00$, $\eta^2 = .25$, post hoc tests showed, VPP was more positive in response to fearful expressions than neutral and happy expressions ($p < .05$), no difference was found between happy and neutral expressions ($p > .1$) (Fig. 5 right column).

The main effect for emotional states was not significant, $F(1, 23) = 2.67$, $p > .1$, $\eta^2 = .10$. There was neither interaction between gender roles and facial expressions, $F(1, 23) = .23$, $p > .1$, $\eta^2 = .01$, nor between gender roles and emotional states,

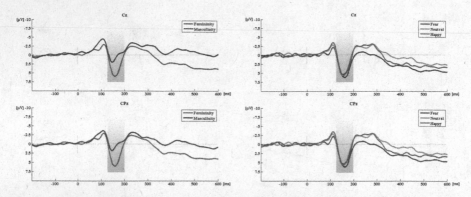

Fig. 5. Grand-average VPP elicited by subjects with different gender roles (left column) and for fearful, neutral and happy facial expressions (right column) at Cz and CPz sites.

$F(1, 23) = 1.84$, $p > .1$, $\eta^2 = .07$. There was also no interaction between emotional states and facial expressions, $F(2, 46) = .57$, $p > .1$, $\eta^2 = .02$.

4   Discussion

The current study was conducted with the aim of exploring gender role differences of female college students on facial expression recognition, and the influence of emotional state. Both behavioral and electrocortical responses were measured during the facial expression discrimination task. Gender role differences on facial expression recognition were observed on N170 and VPP, the emotional negativity bias was found on VPP, and behavioral results showed the effect of emotion congruency.

4.1   Gender Role Differences on Facial Expression Recognition: Evidence on Early ERP Components

Gender role differences were observed here. The femininities showed greater amplitudes on both N170 and VPP, indicating the sensitivity of femininities on facial expressions recognition. The results were consistent with previous studies concerned on sex differences, which also observed a more negative N170 in females than males on face coding task [4, 5]. Likewise, greater amplitude on VPP was found in femininities than masculinities. The similar effect on VPP might be explained by the relationship between N170 and VPP. Joyce et al. found that N170 and VPP were remarkably coincident in both physiological and functional properties and they were thought to be different manifestations of the same events [21].

Besides, considering the factor of biological and psychological gender, an interaction between biological sex and psychological gender identity was observed in a study examined sex difference in lateralization on the processing of facial expressions [22]. It is worth noting that biological sex and psychological gender may act as the same role or different [22]. In the current study, we have concerned females with different gender roles, thus, more research is required for a better understanding of the relationship between biological sex and psychological gender identity in facial expression recognition.

4.2   Emotional Negativity Bias: Evidence on VPP

Previous studies have found that the amplitude of several ERP components, such as N170, VPP, P300 and LPP, show enhancement to emotional negative faces than to neutral or positive faces [7, 8, 16], called the negative bias effects. According to this theory, emotional negative information is processed preferentially throughout the information processing stream due to important adaptive values. Here, the amplitude of VPP was larger for fearful faces than happy and neutral faces and showed a predominance of negative face processing, which was in line with previous studies. For N170, there was also significant main effect for emotional facial expressions, due to greater negativity to fearful and happy relative to neutral faces, which did not reflect the negativity bias. In one study investigated the precedence of potential threat (fearful expressions) over positive (happy expressions) and neutral signals, N170 showed similar results, with enhanced negativity for both fearful and happy relative to neutral faces [7]. According to these results, the positive component (VPP) may be more sensitive to negative valence, while the negative component (N170) may be modulated by the more general effects of emotional arousal [7]. Although emotional negativity bias was found for the female subjects here, no gender role difference was found, which was not consistent with the sex difference study. According to the study of Li et al., compared to men, women are sensitive to emotionally negative stimuli of lesser saliency [6], thus, further studies are welcomed to investigate whether the valence intensity has influence on gender role types or not.

4.3   Emotion Congruency: Evidence on Behavior

At the behavioral level, both femininity and masculinity reacted faster to fear facial expressions in a fearful state than neutral state. These results could be explained as emotion congruency that referred to emotional congruent facial expressions could be processing preferentially [23]. Evidence from previous studies supported this view [24]. Nevertheless, in the present study, we only examined the neutral state and fearful state for processing facial expressions, neglecting the happy state. For a better explanation on emotion congruency, which is not the primary goal of this study, further researches are needed to test face recognition among those three emotional states and make some comparisons with previous studies.

As the behavioral result, the results of the ERPs did not reflect the influence of emotional states on femininity and masculinity either, indicating the difference of biological sex and psychological gender identity. However, under the fearful state, N170 showed higher negativity than that under the neutral state, showing the effect of emotional states on facial expression recognition task. This result can be taken to reflect attention mechanisms similar to the work of Finucane et al. [25]: during a fear experience, participants are better able to inhibit irrelevant information than in neutral condition, resulting in faster response time to a target or like here, a higher activity in N170.

5   Conclusion

Our study gained insight into the difference of gender roles for facial expressions recognition of female college students under neutral and fearful states. Both the behavioral and physiological response (N170, VPP) were used as the index of measurement. Results showed that: (1) femininities were more sensitive than masculinities in the early stage of facial expression recognition. (2) subjects showed a predominance of negative face processing, but no gender role difference was found. (3) the fearful emotional state promoted individuals to recognize facial expressions, and there was no gender role difference. In conclusion, gender role differences exist in the early stage of facial expressions recognition, and femininities are more sensitive than masculinities. Our ERP results provide neuroscience evidence for differences in the early components of facial expression cognition process between the two gender roles of females.

Acknowledgments. We thank Dr. Jianping Cai for providing language help. This research was financially supported by Young Scientist Fund of National Natural Science Foundation of China (NSFC) (31300924), NSFC general program (61375116), the Fund of University of Jinan (XKY1508, XKY1408).

References

1. Groen, Y., et al.: Are there sex differences in ERPs related to processing empathy-evoking pictures? Neuropsychologia **51**(1), 142–155 (2013)
2. Meyers-Levy, J., Loken, B.: Revisiting gender differences: what we know and what lies ahead. J. Consum. Psychol. **25**(1), 129–149 (2015)
3. Stevens, J.S., Hamann, S.: Sex differences in brain activation to emotional stimuli: a meta-analysis of neuroimaging studies. Neuropsychologia **50**(7), 1578–1593 (2012)
4. Choi, D., et al.: Gender difference in N170 elicited under oddball task. J. Physiol. Anthropol. **34**, 7 (2015)
5. Sun, Y., Gao, X., Han, S.: Sex differences in face gender recognition: an event-related potential study. Brain Res. **1327**, 69–76 (2010)
6. Li, H., Yuan, J., Lin, C.: The neural mechanism underlying the female advantage in identifying negative emotions: an event-related potential study. NeuroImage **40**(4), 1921–1929 (2008)
7. Williams, L.M., et al.: The 'when' and 'where' of perceiving signals of threat versus non-threat. NeuroImage **31**(1), 458–467 (2006)
8. Batty, M., Taylor, M.J.: Early processing of the six basic facial emotional expressions. Cogn. Brain. Res. **17**(3), 613–620 (2003)
9. Foti, D., Hajcak, G., Dien, J.: Differentiating neural responses to emotional pictures: Evidence from temporal-spatial PCA. Psychophysiology **46**, 521–530 (2009)
10. Bem, S.L.: Gender schema theory: a cognitive account of sex typing. Psychol. Rev. **88**(4), 354–364 (1981)
11. Bem, S.L.: The Lenses Of Gender: Transforming The Debate On Sexual Inequality. Yale University Press, New Haven (1993)
12. Wood, W., et al.: Chapter two - biosocial construction of sex differences and similarities in behavior. In: Advances in Experimental Social Psychology, pp. 55–123. Academic Press (2012)

13. Ridgeway, C.L.: Framed before we know it: how gender shapes social relations. Gender Soc. **23**(2), 145–160 (2009)
14. Wieser, M.J., Keil, A.: Fearful faces heighten the cortical representation of contextual threat. NeuroImage **86**, 317–325 (2014)
15. Leleu, A., et al.: Contextual odors modulate the visual processing of emotional facial expressions: an ERP study. Neuropsychologia **77**, 366–379 (2015)
16. Righart, R., de Gelder, B.: Rapid influence of emotional scenes on encoding of facial expressions: an ERP study. In: Social Cognitive and Affective Neuroscience, pp. 270–278. Oxford University Press, United Kingdom (2008)
17. Liu, D., et al.: A new sex-role inventory (CSRI-50) indicates changes of sex role among chinese college students. Acta Psychol. Sinica **43**(6), 639–649 (2011)
18. Schaefer, A., et al.: Assessing the effectiveness of a large database of emotion-eliciting films: a new tool for emotion researchers. Cogn. Emot. **24**(7), 1153–1172 (2010)
19. Tottenham, N., et al.: The NimStim set of facial expressions: judgments from untrained research participants. Psychiatry Res. **168**(3), 242–249 (2009)
20. Smith, E., et al.: Electrocortical responses to NIMSTIM facial expressions of emotion. Int. J. Psychophysiol. **88**(1), 17–25 (2013)
21. Joyce, C., Rossion, B.: The face-sensitive N170 and VPP components manifest the same brain processes: the effect of reference electrode site. Clin. Neurophysiol. **116**(11), 2613–2631 (2005)
22. Bourne, V.J., Maxwell, A.M.: Examining the sex difference in lateralisation for processing facial emotion: does biological sex or psychological gender identity matter? Neuropsychologia **48**(5), 1289–1294 (2009)
23. Halberstadt, J.B., Niedenthal, P.M.: Emotional state and the use of stimulus dimensions in judgment. J. Pers. Soc. Psychol. **72**(5), 1017–1033 (1997)
24. Schmid, P.C., Schmid, M., Mast, Mood effects on emotion recognition. Motiv. Emot. **34**(3), 288–292 (2010)
25. Finucane, A.M., Power, M.J.: The effect of fear on attentional processing in a sample of healthy females. J. Anxiety Disord. **24**(1), 42–48 (2010)

Brain Big Data Analytics, Curation and Management

Overview of Acquisition Protocol in EEG Based Recognition System

Hui-Yen Yap[1(✉)], Yun-Huoy Choo[2], and Wee-How Khoh[3]

[1] Centre for Diploma Programme, Multimedia University (MMU),
Melaka, Malaysia
hyyap@mmu.edu.my
[2] Computational Intelligence and Technologies (CIT) Lab,
Faculty of Information and Communication Technology,
Universiti Teknikal Malaysia Melaka (UTeM), Melaka, Malaysia
huoy@utem.edu.my
[3] Faculty of Information Science and Technology,
Multimedia University (MMU), Melaka, Malaysia
whkhoh@mmu.edu.my

Abstract. Electroencephalogram (EEG) signals are unique neurons' electrical activity representation, which can support biometric recognition. This paper investigates the potential to identify an individual using brain signals and highlight the challenges of using EEG as a biometric modality in a recognition system. The understanding of designing an effective acquisition protocol is essential to the performance of the EEG-based biometric system. Different acquisition protocols of EEG based recognition i.e. relaxation, motor and non-motor imaginary, and evoked potentials were presented and discussed. Universality, permanence, uniqueness, and collectability are suggested as key requirements for constructing a viable biometric recognition system. Lastly, a summary of recent EEG biometrics studies was depicted before concluding on the findings. It is observed that both motor and non-motor imagery and event-related potential (ERP) outperformed the method of relaxation in acquisition protocol.

Keywords: EEG-based recognition · Brainwaves · Biometrics · Acquisition protocols · Electroencephalography

1 Introduction

Brain-computer interface (BCI) is a viable concept which provides a communication pathway between brain and external device. It can be viewed as a computer-based system that acquires brain signals, analyzes the signals obtained, and translates them into commands that are relayed to an output device in order to carry out the desired action. Initially, this technology has been studied with the main aim of helping patients with severe neuromuscular disorders. Today, BCI research has been further widened to include non-medical applications. For example, security and authentication studies in BCI utilize brain signals as a biometric modality. The motivation of brainwaves authentication lies in better privacy compliant than other biometric modalities. Brain-print is unique, thus it is more secure than other static physiological biometrics like

Y. Zeng et al. (Eds.): BI 2017, LNAI 10654, pp. 129–138, 2017.
https://doi.org/10.1007/978-3-319-70772-3_12

face, iris and fingerprints. Brainprint is almost impossible to be duplicated by an impostor, hence improves its resistance to spoofing attacks. Brain responses can be elicited through EEG recordings using specific protocols, ranging from resting state, imagined movements, to visual stimulation and so forth. To the best of our knowledge, only the resting state and visual evoke potentials (VEP) are commonly used in the neurophysiology area. Other protocols are not being discussed much for recognition purposes [1]. Hence, this paper aims to provide a comprehensive review of brainwaves elicitation process and its potential on various acquisition protocols, which is essential in EEG recognition system.

2   Signal Acquisition

The brain signals can be acquired using different approaches. It can be classified as invasive and noninvasive methods. The invasive method requires surgical intervention to implant electrodes under the scalp. Due to medical risks and the associated ethical concerns, researchers tend to avoid invasive approach. Common assessment methods of noninvasive approach are either contact or contactless with no implanting of external objects into subject's brain [2]. For example, functional magnetic resonance imaging (fMRI) and magnetoencephalography (MEG) are contactless neuroimaging techniques, while functional near infrared spectroscopy (fNIRS) and Electroencephalography (EEG) methods need to apply electrode sensors on subject's scalp during data acquisition. Among all, EEG is the most widely used method in many research because it is a direct, inexpensive and the simplest method to acquire brain signals.

2.1   The Noninvasive Electroencephalography Method

Electroencephalography (EEG) is an electrophysiological monitoring method to record electrical activities of the brain by measuring voltage fluctuations accompanying neurotransmission activity within the brain. With the advancement of bio-signal data acquisition technologies, EEG signals can be captured easily by placing the wired or wireless electrodes cap on the scalp during recording. The challenges of EEG signals lie in the low spatial resolution and poor signal-to-noise ratio. EEG signals are typically low in spatial resolution on the scalp, due to the physical dimension of the surface electrodes and dispersion of the signals, which are generated by the sources on the cortex, within the head structures before they reach the scalp [1]. The signal-to-noise ratio of EEG is very low, thus include high noise during acquisition. Hence, sophisticated data analysis and relatively large numbers of subjects are needed to extract useful information from EEG [3] signals. Despite these two primary limitations, EEG method presents a high temporal resolution, on the order of milliseconds rather than seconds, which allows dynamic studies to understand the underlying mechanisms by means of computational methods.

3   EEG Signal Based Recognition System

Good reasons to support the use of EEG brainwaves as biometrics for user recognition include unique and less likely to be synthetically generated [4]. Moreover, it is almost impossible to steal a person's as the brainprint as is sensitive to stress. Also, an intruder cannot force a person to reproduce his/her mental pass-phrase. Figure 1 illustrates an overall framework of user recognition using EEG signals.

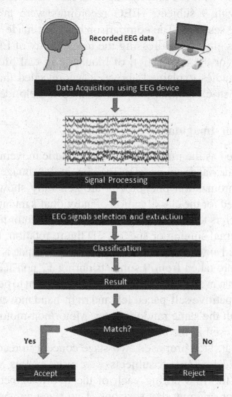

Fig. 1.   Framework of brainwave user recognition

The performance of an EEG based biometric system depends on the proper design of the acquisition protocol. The acquisition protocol is typically divided into three different categories, and details are discussed as follows:

3.1   Relaxation

Users are usually request seated in a comfortable chair with both arms resting and they are asked to perform few minutes of resting state with either eye closed (EC) or eyes open (EO). Occipital alpha waves during the periods of EC are the strongest EEG brain signals [5]. With this reason, it has motivated a number of researchers to implement relaxation protocol with EC during EEG recording. The idea of implementing EEG signals for biometric recognition was first proposed by [6, 7]. An EC in resting condition protocol was employed. Experiments were carried out with 4 subjects whose EEG recordings were taken. Both EEG recordings for these studies lasted for 3 continuous minutes in 1 session. Different classification methods were applied respectively. In the following years, several authors such as [8–11] also follow the track to apply the same protocol in their researches and achieved promising results.

In [12], 40 healthy subject's EEG recordings were recorded while performing the simple activity of resting with EO and resting with EC. Another study in [13] inherits same approach [12] with 9 subjects' EEG recordings were taken in two different sessions in two separate sessions. The comparison has been made between EO and EC. Longitudinal recordings allowed addressing the repeatability of EEG features which is a very important issue for the application of biometric in real life scenarios.

From findings of studies explained above, it has revealed that EEG as a physiological signal has biometric potential and can be used to help identify subjects.

3.2 Motor/Non-motor Imaginary

Motor imagery is defined as the imagination of kinesthetic movements of the left hand, right hand, foot, tongue, and so on. During the motor imagery, distinct mu/beta event-related (de)synchronization phenomena are generally shown around the motor cortex, which can be used for the classification of individual's intention [14]. In contrary, non-motor imaginary refers to mental imagery tasks except motor imagery tasks, such as mental calculation, internal singing or speech, 3D figure rotation, letter composing and etc. These protocols need extensive training and the performance is not promising [2]. In [4], EEG recordings were taken from 9 subjects during 12 non-feedback sessions over three days. The subject sat in a normal chair and requested them to perform several mental tasks: imagination of repetitive self-paced left and right-hand movements and generation of words beginning with the same random letter. Motor/non-motor imaginary protocol showed potential to be used in the biometric system.

In the following year, [15] proposed two-stage concept in recognizing individuals. The data were collected by [16]. 5 subjects were used in this study and their EEG signals were recorded for 10 s during each of the five imagined activities and each activity was repeated for different day sessions. Two-stage authentication was used to verify the claimed individuality of the subject using EEG features vectors. This study served as a pilot study and has good potential as a biometric as it is highly resistant to fraud.

In [17], two different datasets were used in this study; one was composed of imagined speech EEG data [18]. 6 subjects imagined speaking two syllables, /ba/or / ku/with no semantic meaning at different rhythms while their electrical recordings were recorded. The other dataset was obtained from a publicly available database [19]. For this set of EEG data, experiments were conducted in which subjects were presented with visual stimuli consisting of black and white pictures of objects (*The method of using stimuli will be discussed in next section*). A total of 122 subjects' EEG data were enrolled. The results have indicated that the potential of using imagined speech EEG data and for biometric identification due to its strong inter-subject variation.

Another study used four mental imagery tasks consisting baseline measurement, limb movement, visual counting activity and geometric figure rotation activity were performed by 5 subjects with using low-cost EEG device. The result of the study has shown that low-cost EEG authentications systems with mental task acquisition protocol may be viable [20].

The authors in [21] had collected EEG recordings from 15 subjects who sat in a quiet and closed room to undergo two 40 to 50 min sessions on separate days. The subjects performed several mental tasks such as breathing in EC mode, simulated finger movement, sports tasks, singing a song or reciting a passage, and pass-thought task. The study was able to maintain high-level authentication accuracy but unfortunately not user identification system.

3.3   Exposed to Stimuli (Evoked Potentials)

An evoked potential is an electrical response recorded from the nervous system after the presentation of stimuli. It can be divided into two categories: Event-Related Potential (ERP) and Steady State Evoked Potential (SSVP). ERP signals can be elicited using different stimulation paradigms such as sensory stimulus, a cognitive event or the execution of the motor response [2]. It is a stimulus-averaged signal time-locked to the presentation of some events of interest. When ERPs are created through averaging over many trials of a particular type of event, brain activity not related to the event of interest is reduced and brain activity related only to the event of interest is emphasized [22]. The typical employed ERP stimulation protocol during EEG acquisition of is the elicitation of Visual Evoked Potentials (VEP) which used to perform the brain perceives the processes visual inputs [1].

In [23], the authors accessed the feasibility of ERPs over a database 20 subjects. EEG recordings were recorded when subjects were exposed to stimuli, which consist of pictures of objects. These pictures are common black and white line drawings like an airplane, a banana, a ball, etc. The positive results obtained in this paper show the feasibility of ERPs to be applied in the system. The authors of [24] continue investigated the performance of EEG ERPs in a group of 102 subjects with using the similar protocol. Here a mental task consisting in recognizing and remembering shown objects was proposed.

In [25], the stability of the EEG signals over the course of time is explicitly investigated. While the EEG recordings were captured, 20 Subjects were presented with 75 acronyms and asked to read them silently. Three sessions have been conducted respectively in a period of 6 months. Their observations showed that the brain activities of an individual remain stable over a long period of time.

An ERP biometric protocol was proposed by [22] to elicit unique responses from subjects with multiple functional brain systems. A total of 56 subjects participated in this experiment. Subjects were exposed to 400 images: 100 sine gratings, 100 low-frequency words, 100 images of food, and 100 celebrity faces. 300 of the 400 were presented in black and white. The total duration of the experiment was approximately 1.5 h. The authors were able to achieve 100% identification accuracy.

On the other hand, SSEP is evoked by a stimulus modulated at a fixed frequency and occur as an increase in EEG activity at the stimulation frequency. The stimulation could be either visual as in Steady State Visually Evoked Potentials (SSVEP), auditory as in steady-state auditory evoked potentials (SSAEP) or even somatosensory as in Steady-state somatosensory evoked potential (SSSEP). The application of using SSVEP in BCI system has widely tested. To date, only one study has been proposed to use SSVEP to identify individual identity. EEG recordings were taken from 5 subjects who were instructed to look the stimuli displaying on an RGB LED stimuli device. The experiment achieved the True Acceptance Rate (TAR) of 60% to 100% revealing the potential of the proposed protocol.

4   Analysis and Discussions

Table 1 summarized the acquisition protocols implemented in the past studies as well as the database set up and system performances. Studies reported from 1–8 utilized relaxation of either EC or EO. It is found that EC could actually gain better results than single EO method. Studies of 9–13 used either motor imagery or non-motor imagery or both with EC while 14–18 utilized ERP as acquisition protocol. From the overall results, it is obvious that both motor and non-motor imagery and ERP outperformed than the method of relaxation in its acquisition protocol.

In order to be considered as a viable biometric, it must satisfy the following requirements:

- Universality – It refers to the fact that everyone should have that characteristic.
- Permanence – It refers to the characteristics of individual should be sufficiently invariant with respect to the matching criteria, over a period of time. Reproducibility of EEG biometrics in different acquisition sessions has raised a question that does the brain activities and responses stable for a different period of time? The issue of intra-individual variability has gained the concerns from scientific researchers in the past. However, this issue has not received much attention from researchers in the EEG-based biometric community. Few studies have attempted to investigate the long-term stability of human activities, Although the sample size of the studies was limited, the findings are valuable and show EEG signal fulfils the basic permanence requirement. In addition, circadian rhythm influences might affect the acquisition of EEG signal [1]; no studies have been done before. More exhaustive analysis with different acquisition protocols is still needed.
- Uniqueness – It means the characteristic should be unique for each individual and thus can be distinguishable among different people. Prior studies have extensively evaluated the use of EEG signals as a biometric with promising results.
- Collectability – It refers to a characteristic of an individual is measurable with some practical device. Traditionally, EEG signals are captured with clinical grade EEG device which consists of a number of electrodes. It is an inconvenience as it spent the time to set up. Thus, the minimization of the number of electrodes is a crucial issue need to be overcome to improve user's quality of experience.

EEG signals are normally analyzed in two different ways: frequency domain or time domain. The former one separates the oscillatory signals in different bands (i.e., alpha, beta, and gamma) while the latter averaging the signals at the onset of a

Table 1. Summarized of studies of EEG biometrics system.

No	Paper	Year	Acquisition protocol	Database	Session	Performance measure
1	Poulos et al. [6]	1999	EC	4	1	GAR = 80% − 100% CRR = 80% − 95%
2	Poulos et al. [7]	1999	EC	4	1	GAR = 72% − 84%
3	Riera et al. [8]	2008	EC	51	4	EER = 3.4%
4	Campisi et al. [9]	2011	EC	48	1	CRR = 96.98%
5	Su et al. [10]	2012	EC	40	2	CRR = 95%
6	La Rocca et al. [11]	2012	EC	45	1	CRR = 98.73%
7	Paranjape et al. [12]	2012	EO	40	1	GAR = 49% − 82%
8	La Rocca et al. [13]	2012	EC/EO	9	2	CRR = 100%
9	Marcel and Millan [4]	2012	Motor imagery	9	3	HTER = 8.1% − 12.3%
			Non-motor imagery			HTER = 12.1%
10	Palaniappan [15]	2008	EC	5	1	FRR = 1.5% − 0%
			Non-motor imagery			FAR = 0%
11	Brigham and Kumar [17]	2010	Non-motor imagery: Imagined speech	6	4	GAR = 99.76%
			ERP	120	4	GAR = 98.96%
12	Corey et al. [20]	2011	EC	5	1	FRR = 0.024 − 0.051
			Motor imagery			FAR = 0.007 − 0.011%
			Non-motor imagery			
13	Chuang et al. [21]	2013	EC	15	2	FRR = 0.280% FAR = 0%
			Non-motor imagery			FRR = 0.093% − 0.440% FAR = 0% − 0.120
14	Palaniappan et al. [23]	2003	ERP	20	1	Classification rate = 94.18%
15	Palaniappan et al. [24]	2007	ERP	102	1	Classification rate = 98.12%
16	Maria et al. [25]	2015	ERP	20	3	Classification rate = 84% − 99%
17	Maria et al. [22]	2016	ERP	56	1	Classification rate = 100%
18	Phothisonothai [26]	2015	SSVEP	5	1	TAR = 60% − 100%

particular event. Relaxation and mental/non-mental protocol belong to the first way as their recorded EEG data is an ongoing, continuous signal that can be collected in the absence of any particular stimulation. The implementation of these protocols is convenient but it raises a problem that the mental state of a user is uncontrollable when EEG data is enrolled during data collection in which each session is days or weeks apart. Besides, the EEG is not collected time-locked to any type of stimulation, which means it does not reflect the narrow, specific and cognitive process, making the classifier's task much more difficult. As a consequence, performance degrades over days and vary based on different mental tasks given. Therefore, incremental learning could be a solution to improve the performance of the system.

The protocol of using stimulation during EEG elicitation belongs to the second way. It allows the experimenter tightly control the cognitive state of the user being reflected in the resultant EP activity and make an analysis of the desired cognitive state of a user. ERP has been used as biometric measures in few biometric studies and the results reported are promising. Though it shows potential to allow better user recognition, however, due to the significantly small size of an ERP, it usually takes a large number of trials to accurately measure it correctly. The system seems to be impractical to put into real time as previous studies requested the users to undergo a lengthy acquisition period in order to get an accurate measure. The reduction of acquisition length needs to be considered by researchers to reduce or even eliminate the tiredness of users.

5   Conclusion

This paper provides an overview of acquisition protocols in the EEG-based recognition system. In summary, ERP is arguably that may provide more accurate biometric identification as its elicitation method allows experimenter control the cognitive state of the user during the acquisition period. In consideration of the difference in experiment settings for each study (i.e. device selection, electrodes configuration, data acquisition period and task involvement), more exhaustive analysis with various type of acquisition protocols is necessary. In recent year, the BCI researchers are focusing on a combination of different types of BCI systems in order to improve the user experience. In a hybrid BCI, two systems can be combined sequentially or simultaneously. In a simultaneous hybrid BCI, both systems are processed in parallel. Input signals used in simultaneous hybrid BCIs can be two different brain signals, one brain signal, or one brain signal and another input. In sequential hybrid BCIs, the output of one system is used as the input of the other system [26]. To the best of our knowledge, the concept of combining different types of acquisition protocols hasn't received any attention from the biometric community. Perhaps it can be served as research opportunities to improve the performance in user recognition.

Acknowledgments. The work was funded by UTeM Short Term High Impact grant (PJP/2016/FTMK/HI3/S01474).

References

1. Campisi, P., Rocca, D.L.: Brain waves for automatic biometric-based user recognition. IEEE Trans. Inf. Forensics Secur. **9**, 782–800 (2014)
2. Abdulkader, S.N., Atia, A., Mostafa, M.S.M.: Brain computer interfacing: applications and challenges. Egypt. Inf. J. **16**, 213–230 (2015)
3. Schlögl, A., Slater, M., Pfurtscheller, G.: Presence research and EEG. In: Proceedings of the 5th International Working Presence, pp. 154–160 (2002)
4. Marcel, S., del Millan, J.: R.: Person authentication using brainwaves (EEG) and maximum a posteriori model adaptation. IEEE Trans. Pattern Anal. Mach. Intell. **29**, 743–748 (2007)
5. Palva, S., Palva, J.M.: New vistas for α-frequency band oscillations. Trends Neurosci. **30**, 150–158 (2007)
6. Poulos, M., Rangoussi, M., Alexandris, N.. Neural network based person identification using EEG features. In: Proceedings of the 1999 IEEE International Conference on Acoustics, Speech, and Signal Processing. ICASSP 1999, pp. 1117–1120 (1999)
7. Poulos, M., Rangoussi, M., Chrissikopoulos, V., Evangelou, A.: Person identification based on parametric processing of the EEG. In: Proceedings 6th IEEE International Conference on Electronics, Circuits and Systems 1999. ICECS 1999, vol. 1, pp. 283–286 (1999)
8. Riera, A., Soria-Frisch, A., Caparrini, M., Grau, C., Ruffini, G.: Unobtrusive biometric system based on electroencephalogram analysis. EURASIP J. Adv. Signal Process. **2008**, 143728 (2008)
9. Campisi, P., Scarano, G., Babiloni, F., DeVico Fallani, F., Colonnese, S., Maiorana, E., Forastiere, L.: Brain waves based user recognition using the "eyes closed resting conditions" protocol. In: 2011 IEEE International Workshop on Information Forensics and Security, WIFS 2011 (2011)
10. Su, F., Zhou, H., Feng, Z., Ma, J.: A biometric-based covert warning system using EEG. In: Proceedings - 2012 5th IAPR International Conference on Biometrics, ICB 2012, pp. 342–347 (2012)
11. Rocca, D.L., Campisi, P., Scarano, G.: EEG biometrics for individual recognition in resting state with closed eyes. In: International Conference on Biometrics Special Interest Group, pp. 1–12 (2012)
12. Paranjape, R.B., Mahovsky, J., Benedicenti, L., Koles', Z.: The electroencephalogram as a biometric. In: Conference Proceedings of Canadian Conference on Electrical and Computer Engineering 2001. (Cat. No. 01TH8555), vol. 2, pp. 1363–1366 (2001)
13. Rocca, D.L., Campisi, P., Scarano, G.: On the repeatability of EEG features in a biometric recognition framework using a resting state protocol, pp. 419–428 (2013)
14. Hwang, H.J., Kim, S., Choi, S., Im, C.H.: EEG-based brain-computer interfaces: a thorough literature survey. Int. J. Hum. Comput. Interact. **29**, 814–826 (2013)
15. Palaniappan, R.: Two-stage biometric authentication method using thought activity brain waves. Int. J. Neural Syst. **18**, 59–66 (2008)
16. Keirn, Z.A., Aunon, J.I.: A new mode of communication between man and his surroundings. IEEE Trans. Biomed. Eng. **37**, 1209–1214 (1990)
17. Brigham, K., Kumar, B.V.K.V.: Subject identification from Electroencephalogram (EEG) signals during imagined speech. In: IEEE 4th International Conference on Biometrics: Theory, Applications and Systems, BTAS 2010 (2010)
18. D'Zmura, M., Deng, S., Lappas, T., Thorpe, S., Srinivasan, R.: Toward EEG sensing of imagined speech. In: Jacko, J.A. (ed.) HCI 2009. LNCS, vol. 5610, pp. 40–48. Springer, Heidelberg (2009). doi:10.1007/978-3-642-02574-7_5
19. EEG Database. http://kdd.ics.uci.edu/databases/eeg/eeg.html

20. Ashby, C., Bhatia, A., Tenore, F., Vogelstein, J.: Low-cost electroencephalogram (EEG) based authentication. In: 2011 5th International IEEE/EMBS Conference on Neural Engineering, NER 2011, pp. 442–445 (2011)
21. Chuang, J., Nguyen, H., Wang, C., Johnson, B.: I think, therefore i am: usability and security of authentication using brainwaves. In: Adams, A.A., Brenner, M., Smith, M. (eds.) FC 2013. LNCS, vol. 7862, pp. 1–16. Springer, Heidelberg (2013). doi:10.1007/978-3-642-41320-9_1
22. Ruiz-Blondet, M.V., Jin, Z., Laszlo, S.: CEREBRE: a novel method for very high accuracy event-related potential biometric identification. IEEE Trans. Inf. Forensics Secur. **11**, 1618–1629 (2016)
23. Palaniappan, R., Ravi, K.V.R.: A new method to identify individuals using signals from the brain. In: ICICS-PCM 2003 - Proceedings of the 2003 Joint Conference of the 4th International Conference on Information, Communications and Signal Processing and 4th Pacific-Rim Conference on Multimedia, pp. 1442–1445 (2003)
24. Palaniappan, R., Mandic, D.P.: Biometrics from brain electrical activity: a machine learning approach. IEEE Trans. Pattern Anal. Mach. Intell. **29**, 738–742 (2007)
25. Ruiz Blondet, M.V., Laszlo, S., Jin, Z.: Assessment of permanence of non-volitional EEG brainwaves as a biometric. In: EEE International Conference on Identity, Security and Behavior Analysis. ISBA 2015 (2015)
26. Pfurtscheller, G., Allison, B.Z., Brunner, C., Bauernfeind, G., Solis-Escalante, T., Scherer, R., Zander, T.O., Mueller-Putz, G., Neuper, C., Birbaumer, N.: The hybrid BCI. Front. Neurosci. **4**, 30 (2010)

A Study on Automatic Sleep Stage Classification Based on Clustering Algorithm

Xuexiao Shao[1,2], Bin Hu[1,2(✉)], and Xiangwei Zheng[1,2]

[1] School of Information Science and Engineering, Shandong Normal University,
Jinan 250014, China
binhu@sdnu.edu.cn
[2] Shandong Provincial Key Laboratory for Distributed Computer Software
Novel Technology, Jinan 250014, China

Abstract. Sleep episodes are generally classified according to EEG, EMG, ECG, EOG and other signals. Many experts at home and abroad put forward many automatic sleep staging classification methods, however the accuracy of most methods still remain to be improved. This paper firstly improves the initial center of clustering by combining the correlation coefficient and the correlation distance and uses the idea of piecewise function to update the clustering center. Based on the improvement of K-means clustering algorithm, an automatic sleep stage classification algorithm is proposed and is adopted after the wavelet denoising, EEG data feature extraction and spectrum analysis. The experimental results show that the classification accuracy is improved and the sleep automatic staging algorithm is effective by comparison between the experimental results with the artificial markers and the original algorithms.

Keywords: Clustering algorithm · Sleep staging · K-means · EEG

1 Introduction

Sleep has an extraordinary meaning for mankind, it is closely related to people's life and has an important role to the maintenance of human body, the body could be relaxed, at the same time, sleep can enhance the assimilation, and reduce the level of alienation. However, the disease caused by sleep has become another major medical challenge. The latest sleep survey published by China Sleep Research Institute, shows that the rate of sleep disorder disease in Chinese adults is about 38.2% [1] which is higher than that in western developed countries. In addition, insomnia and personal physical condition are closely related, people with disease are often more likely to suffer from insomnia. The study of sleep has become a hot topic in the field of biological information and neuroscience. Among them, the study of sleep staging can effectively assess the quality of sleep, for the treatment of the disease, and thus get wide attention.

Sleep episodes are classified according to EEG, EMG, ECG, EOG and other signals. The structure of sleep-wakefulness cycle is divided into two phases: non-rapid eye movement sleep (NREM) and rapid eye movement sleep (REM), the differences are whether there are eyeball paroxysmal fast movement and different brain waves.

Y. Zeng et al. (Eds.): BI 2017, LNAI 10654, pp. 139–148, 2017.
https://doi.org/10.1007/978-3-319-70772-3_13

EEG waveforms are waveforms with multiple intermediate frequency components, usually divided from high to low: β wave (13–40 Hz), α wave (8–13 Hz), θ wave (4–7 Hz), δ Wave (0–4 Hz) [2], and sleep can be divided into different stages according to brain frequency and data characteristics based on EEG signal. At present, many experts at home and abroad have also put forward many methods to achieve automatic sleep staging. Most of the methods for EEG information analysis to be improved accuracy, for some of the extraction of features is still inadequate. Therefore, we improved clustering algorithm, extracted EEG data characteristics, analyzed spectrum, improved sleep staging accuracy and achieved the automatic staging of sleep results.

In this paper, by improving the general clustering algorithm, it can automatically analyze its timing characteristics. Based on the EEG data set of CAP Sleep Database in PhysioBank, the EEG data is analyzed and the automatic sleep staging process is realized by using wavelet transform and improved K-means clustering algorithm to improve the initial clustering center and iterative method.

2   Related Work

The processing of sleep data is carried out by qualified medical practitioners or professionals, that is, artificial labeling. However, artificial labeling is often time-consuming and laborious. To solve this problem, domestic and foreign scholars carry out research on automatic sleep staging, and hope to achieve automatic sleep staging through related algorithms and techniques. Fell et al. [8] studied the frequency domain and nonlinear methods. The sleep process was divided into four stages: S1, S2, SS and REM. Extracted frequency domain features such as power, spectral edge, Dimension D2, maximum Lyapunov parameter L1 and approximate Kolmogorov entropy K2; Donglai Jiao et al. [10] studied the differences in the mean energy dissipation of the W and S1 in EEG, and analyzed the relative entropy of the two sleep stages and the multi-sample validation. In addition, SVM was used to classify the data, in order to achieve automatic sleep staging.

In the study, K-means algorithm is a typical clustering algorithm, which has the advantages of fast and simple algorithm. The K-means algorithm belongs to the numerical clustering algorithm, which requires the simultaneous extraction of N kinds of features. K-means is a distance-based iterative algorithm. Euclidean distances are often used in the EEG data processing. Therefore, Tapas Kanungo et al. [13] proposed a heuristic Lloyd's K-means clustering algorithm, which can be easily implemented by using the k-d tree. Alex Rodriguez et al. [14] proposed an algorithm that uses the fast search method to find the density extremes for clustering, which automatically discovers anomalous data and excludes clustering, which can identify clusters of arbitrary shapes and spatial dimensions. Overcoming the other clustering algorithms can only identify the shortcomings of spherical clusters. Koupparis et al. [15] combined with short-time Fourier transform and K-means clustering as semi-automatic sleep staging method, can effectively distinguish W, REM and S1 period. The improvement of K-means clustering algorithm can effectively improve the data analysis ability and realize the automatic staging of sleep.

3 Automatic Sleep Staging Classification Algorithm Based on K-Means Clustering

The overall system of sleep automatic staging in Fig. 1 is as follows:

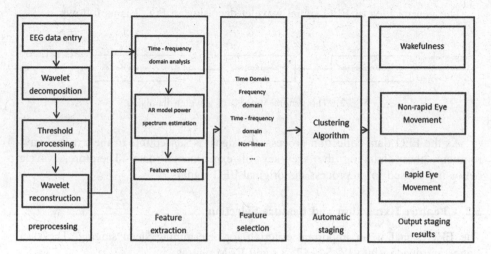

Fig. 1. The Sleep automatic staging system view

(1) The data input: The EEG dataset of the CAP Sleep Database in PhysioBank is used for sleep automatic staging.

(2) Preprocessing: The EEG data in the database is subjected to wavelet denoising.

(3) Feature extraction and Feature selection: Spectrum analysis of the denoised EEG signal and sleep - related feature vector extraction.

(4) Automatic staging: Selecting the different stages of the eigenvector as an input based on the improved K-means clustering algorithm.

(5) Output staging results: After clustering, the EEG signal is divided into 5 stages.

In this paper, a whole system of sleep automatic staging is designed (only the classification algorithm). Through the improvement of the traditional clustering algorithm, the automatic staging system based on the improved K-means algorithm is used, and the technique of wavelet denoising is used to realize the the study.

3.1 Denoising

We can use the method of wavelet transform to preprocess the original signal of sleep EEG. The fundamental wave, β wave, α wave, θ wave and δ wave of the brain wave signal are extracted by wavelet transform method.

For any function $f(t) \in L^2(R)$, the continuous wavelet transform is as follows:

$$W_f(a, b) = <f, \psi_{a,b}> = \frac{1}{\sqrt{|a|}} \int_R f(t)\psi(\frac{t-b}{a})dt \qquad (1)$$

Its inverse transformation is as follows:

$$f(t) = \frac{1}{C_\psi} \int_{R^+} \int_R \frac{1}{a^2} W_f(a,b) \psi(\frac{t-b}{a}) dadb \qquad (2)$$

The general idea of EEG signal wavelet denoising in Fig. 2 is as follows:

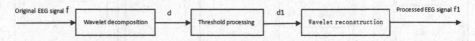

Fig. 2. The general process of wavelet denoising

As the EEG data collection process, the signal is susceptible to the external environment, thus changing the accuracy of experimental data. Therefore, wavelet denoising is used to preprocess the original EEG signals.

3.2   Feature Extraction and Feature Selection

The EEG signal is an important criterion for determining sleep staging. The sleep stages are divided into W, S1, S2, SS and REM phases.

The EEG feature reflects the potential information of the sleep signal. In this paper, four frequency domain features are extracted at EEG and EOG, and a time domain feature is extracted from the EMG. Calculating the ratio of δ, θ, and α to the total frequency of R_δ, R_θ, R_α.

$$R_\delta = \max\{E_\delta(C_i)/E_T(C_i)\} \qquad (3)$$

$$R_\theta = \max\{E_\theta(C_i)/E_T(C_i)\} \qquad (4)$$

$$R_\alpha = \max\{E_\alpha(C_i)/E_T(C_i)\} \qquad (5)$$

Among them, E_δ, E_θ, E_α, E_T are expressed as the energy of δ, θ, α and the whole band. The time series signal is transformed into the frequency domain by fast Fourier transform, the sum of squares of the bands is calculated in the frequency domain. The letter i represents the data point of the PSG signal. E_{LR}–The energy of the REM phase is calculated for EOG.

3.3   Automatic Sleep Stage Classification Based on Improving K-Means Algorithm

The K-means algorithm is the simplest clustering algorithm, the algorithm is fast and simple. However, it is also subject to some practical application and its own conditions. K is given in advance in K-means algorithm, it needs to be based on the initial clustering center to determine an initial division. In order to overcome the problem that the k-means algorithm converges to local problems, it is necessary to improve the K-means algorithm by improving the influence of K value on the clustering quality.

(1) Selecting the appropriate initial cluster center. In this paper, based on the discussion of correlation coefficient and correlation distance, the initial center of clustering is selected based on density idea. For EEG sample data, there is a certain correlation between the time series of EEG data. The correlation coefficients are defined as follows:

$$P_{x_i x_j} = \frac{cov(x_i, x_j)}{\sqrt{D(x_i)}\sqrt{D(x_j)}} = \frac{E((x_i - E_{x_i})(x_j - E_{x_j}))}{\sqrt{D(x_i)}\sqrt{D(x_j)}} \tag{6}$$

In the formula (6), $Cov(x_i, x_j)$ is the covariance of x_i, x_j. $D(x_i)$, $D(x_j)$ are the variance of x_i, x_j, respectively. $P_{x_i x_j}$ is called the correlation coefficient, used to measure the degree of correlation between random variables. $P_{x_i x_j} \in [-1, 1]$, the greater the correlation coefficient, the higher the correlation between the variables x_i, x_j. When $P_{x_i x_j}$ is 1 or -1, there is a definite linear correlation between x_i and x_j.

The relevant distance is calculated as follows:

$$d_{x_i, x_j} = 1 - P_{x_i x_j} \tag{7}$$

For these data relationships, the density is defined as a number of data points randomly distributed within a certain range. Now, setting $D = \{x_1, x_2, \cdots, x_n\}$, the density of x_i is as follows:

$$\rho_i = max\{d\lfloor x_i, x_j \rfloor\} \tag{8}$$

Among them, x_j belongs to the set of points which is closest to x_i. The smaller the ρ, the more dense the clustered lattice. In the clustering process, ρ minimum point is the first clustering center. When the next cluster center is determined, the set of clusters formed by the first ρ minimum is removed from the data set D. In the remaining sets, the smallest point is chosen to form a new clustering center, continuously, until k clustering centers are selected. K-means and other clustering algorithms, most of them choose Euclidean distance, but based on the temporal and spatial correlation of specific time series data and the enlightenment of hierarchical clustering algorithm. Euclidean distance neglects the correlation between data. Therefore, this paper uses the relevant distance as the distance metric of the algorithm, taking full account of the correlation between the time series data.

(2) Iterative updating of clustering centers. After selecting the initial center, the K-means clustering will iterate to update the prototype, calculate the average of all the points in the class as the new clustering center, the mean vector of the new clustering center is as follows:

$$u_i' = \frac{1}{|D_i|} \sum_{x \in D_i} x \tag{9}$$

For a certain set of data, the data distribution law conforms to a certain normal distribution principle. Thus, in conjunction with the idea of a piecewise function, the sample data that may have a normal distribution is discussed in detail so as to be accurate to the threshold of each segment. Now, normal distribution probability and data correlation coefficient equivalent:

$$P_{x_i x_j} = \emptyset \left(\frac{x - u}{\sigma} \right) \tag{10}$$

Define the degree of dissimilarity between data *dif*:

$$dif_{i,j} = dist_{i,j} = d_{i,j} \tag{11}$$

$$\omega_i = \arg \min d_{i,j} \tag{12}$$

Among them, u is the mean of the distance from the interior point to the center of the cluster, σ is the distance standard deviation from the interior point to the center of the cluster, and $dist_{i,j}$ represents the relevant distance between the two sample points. According to the principle of positive distribution, the different values of σ are divided into different thresholds, and the relevant distance is calculated by segmentation, which reduces the interference of outliers or noise, and thus gets more accurate clustering data samples.

Algorithm 1 Improved K-means Algorithm

1: After the data feature calculated, the selected EEG eigenvector is used as the input of the clustering algorithm staging system and the correlation coefficient $P_{x_i x_j}$ between the two characteristic data is calculated.
2: Calculating the correlation distance $d_{x_i x_j}$ between two data points, and further calculating the density ρ of each point. In all the obtained ρ, selecting one of the smallest, as the first cluster center, while obtaining a clustering set D_1.
3: Repeat (1) ~ (2) .Until C points are selected as the initial clustering center ($K = 2 * C$ in the algorithm).
4: The data remaining in the set D is allocated to the nearest class according to the distance from the nearest cluster center.
5: Calculating the distance from each point to the center in each class, calculating u, σ, according to different σ. Segment calculation, and we can obtain the smallest ω as the new clustering center.
6: Recalculated and assigned individual sample objects, according to step (S4).Until the function converges or the cluster center no longer changes.
7: The output of clustering results, the data which have different characteristic waveform characteristics will distinguish between different sleep staging.

---

4   Experimental Results and Analysis

The sleep data in this study are derived from the EEG data of CAP Sleep Database in PhysioBank. In this paper, the sleep EEG signal is used as the research object to study the sleep staging, and the program is debugged and edited by MATLAB.

4.1   Sleep Data Set

In this paper, the EEG signal of sleep staging start to be studied, we select the three samples, and select ins 1–3 data set for the data simulation experiments. A polysomnography (PSG), which includes a number of physiological parameters recorded throughout the nighttime sleep, is used to analyze the diagnostic methods of human sleep. The data format used in this article is stored in the *.edf data format, and using MATLAB for algorithm debugging and editing.

In each stage of sleep, EEG will show different characteristics of the waveform. Artificial labeling is often based on the characteristics of these waveforms to determine the stage, we use the clustering algorithm to extract these main feature vectors, and the system automatically performs the automatic staging process according to their characteristics. The Table 1 is the relationship between the sleep phase and the EEG signal.

Table 1. The association between sleep stages and EEG signals

EEG features	Frequency	Amplitude	Tense
Alpha	8–13 Hz	20–60 uV	W/S1/REM
Beta	13 Hz	2–20 uV	W
Theta	4–8 Hz	50–75 uV	S1/S2/SS
Delta	0–4 Hz	75 uV	SS
Sleep spindle wave	12–14 Hz	20–50 uV	S2

4.2   Evaluation Metrics

Sum of Squares for Error (SSE) is a commonly used criterion function for K-means clustering algorithm and can be used to evaluate the quality of clustering. The smaller the SSE, the smaller the error of samples and the center, the better the clustering effect. It is used as an evaluation algorithm running. The formula is defined as follows:

$$J(k,u) = \sum_{i=1}^{N} \left\| x^{(i)} - u_k(i) \right\|^2 \tag{13}$$

The iterative optimization is used to approximate the solution. N data sources, K clustering centers, u_k represents the clustering centers in the formula. The data in each class is different from each cluster center. And, J is the smallest, it means that the best effect of its segmentation.

4.3   Experimental Results and Discussion

Based on the improved K-means clustering algorithm, the sleep data showed the staging results as follows (Fig. 3):

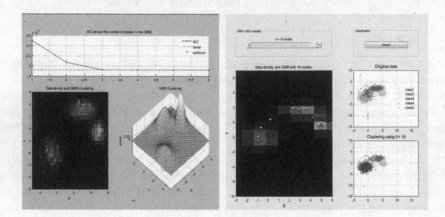

Fig. 3. Experimental results of automatic sleep staging

Clustering center C = 5, according to the formula, when case = two, K = 2 × C, so the figure K = 10. According to the clinical requirements of sleep staging, if the two stages of data occurred in the same band of data, then accounted for more than half of the time as this section of data on behalf of the sleep staging. The number of clustering centers is adjusted to observe the classification accuracy of sleep staging. First, according to the expert's judgment, the data segments of the three subjects who are mainly analyzed in this paper are counted. Then, the number of segments of the sleep phase of the automatic identification results and the results of artificial labeling should be calculated. At last, the number of segments matched is divided by the total number of segments, the classification accuracy of each stage is obtained, and the results are compared with the artificial markers. As shown in Table 2.

Table 2. Comparison between clustering algorithm and artificial marker

Subjects	Classification accuracy					
	W	S1	S2	SS	REM	Total
Test1						
K = 10	78%	15%	76%	96%	93%	75%
K = 12	75%	15%	76%	96%	90%	74%
Test2						
K = 10	77%	51%	79%	97%	91%	80%
K = 12	70%	21%	78%	95%	91%	76%
Test3						
K = 10	74%	45%	77%	94%	97%	81%
K = 12	72%	23%	77%	94%	95%	73%

Table 2 shows the accuracy of sleep staging, when the three subjects in the different K value of 10, 12 cases. At $K = 10(C = 5)$, we choose five clustering centers, the accuracy rate will be higher, which is due to the feature extraction and selection, we select the sleep staging feature vector, and then take into account the diversity of sleep states, we should be more objectively to select k.

Table 3. SSE after the first iteration

Method	SSE		
	Test1	Test2	Test3
Original K-means	1101.53	1097.45	1201.03
Improved K-means	1100.05	1095.76	1198.95

Table 3 shows the SSE values of the two algorithms after the first iteration. The improved algorithm has less SSE value than the original clustering algorithm, the smaller the SSE, the better clustering effect.

The improved clustering algorithm is compared with the original algorithm. The results are shown in the following table.

Table 4. Comparison of classification results between original clustering algorithm and improved clustering algorithm

Method	Classification accuracy		
	Test1	Test2	Test3
Original K-means	65%	57%	57%
Improved K-means	76%	66%	64%

From Table 4, we can see that the improved algorithm has improved significantly in the classification accuracy. For the data obtained by the three subjects, it can be seen that the classification effect is very good, which also confirms the improved algorithm by adjusting the clustering center selection and iterativing method.

5 Conclusion

Based on the influence of outliers on clustering effect, an improved K-means clustering algorithm is proposed and applied to automatic sleep staging. The sleep is divided into five stages: W, S1, S2, SS and REM. Combined with wavelet denoising characteristics, so as to deal with EEG data better and better.

In this paper, through the improvement of the relevant algorithms, the clustering data mining method can simplify the regularization of EEG data and extract the relevant features, so as to realize the purpose of data mining and realize the automatic staging of sleep.

In the course of the experiment, we also found that due to the diversity of sleep status and the impact of unknown factors, the accuracy of sleep staging is not high, the data processing speed is relatively slow and other shortcomings, which need to be further improved. In the next step, we will do further research on the stage of sleep, try to add two factors in the algorithm, give full consideration to the impact of sleep stage of the subjective and objective factors to further improve the accuracy of sleep staging, and try to improve the structure of the algorithm. Make it more robust, accurate and efficient in processing data.

References

1. Tang, Q.: Study on automatic sleep staging based on EEG. Guangdong University of Technology (2016)
2. Antunes, M., Oliveira, L.: Temporal data mining: an overview. In: KDD 2001 Workshop on Temporal Data Mining (2001)
3. Mörchen, F.: Time-series Knowledge Mining. Ph.D. thesis, Deptartment of Mathematics and Computer Science, University of Marburg, Germany (2006)
4. Keogh, E., Chakrabarti, K., Pazzani, M.J., Mehrotra, S.: Dimensionality reduction for fast similarity search in large time-series databases. Knowl. Inf. Syst. 3(3), 263–286 (2001)
5. Huang, S.: Data mining on time series data. J. Softw. 15(1), 1–7 (2004)
6. Du, Y.: Study and Application of Time Series Mining Related Algorithm. University of Science and Technology of China (2007)
7. Cao, H., Leung, V., Chow, C., Chan, H.: Enabling technologies for wireless body area networks: a survey and outlook. IEEE Commun. Mag. 47(12), 84–93 (2009)
8. Fell, J., Röschke, J., Mann, K., et al.: Discrimination of sleep stages: a comparison between spectral and nonlinear EEG measures. Electroencephalogr. Clin. Neurophysiol. 98(5), 401–410 (1996)
9. Ronzhina, M., Janoušek, O., Kolářová, J., et al.: Sleep scoring using artificial neural networks. Sleep Med. Rev. 16(3), 251–263 (2012)
10. Jiao, D.-l., Feng, H., Yao, F., et al.: Study on sleep staging based on mean energy dissipation. Beijing Biomed. Eng. 32(2), 134–138 (2013)
11. Peng, Z., Wei, M., Guo, J., et al.: Study of sleep staging based on singular value of the first principal component. Adv. Mod. Biomed. 14(7), 1368–1372 (2014)
12. Khalighi, S., Sousa, T., Santos, J.M., et al.: ISRUC-Sleep: a comprehensive public dataset for sleep researchers. Comput. Methods Programs Biomed. 124, 180–192 (2016)
13. Charbonnier, S., Zoubek, L., Lesecq, S., et al.: Self-evaluated automatic classifier as a decision-support tool for sleep/wake staging. Comput. Biol. Med. 41(6), 380–389 (2011)
14. Xiao, S.-y., Wang, B., Zhang, J., et al.: Study on automatic staging of sleep based on improved K-means clustering algorithm. Biomed. Eng. 33(5), 847–854 (2016)
15. Sakellariou, D., Koupparis, A.M., Kokkinos, V., et al.: Connectivity measures in EEG microstructural sleep elements. Front. Neuroinf. 10 (2016)

Speaker Verification Method Based on Two-Layer GMM-UBM Model in the Complex Environment

Qiang He[1,2,3], Zhijiang Wan[5], Haiyan Zhou[1,2,3], Jie Yang[4],
and Ning Zhong[1,2,3,5(✉)]

[1] International WIC Institute, Beijing University of Technology,
Beijing 100024, China
[2] Beijing International Collaboration Base on Brain Informatics
and Wisdom Services, Beijing, China
[3] Beijing Key Laboratory of MRI and Brain Informatics, Beijing, China
[4] Beijing Anding Hospital of Capital Medical University, Beijing, China
[5] Department of Life Science and Informatics, Maebashi Institute of Technology,
Maebashi-City 371-0816, Japan
zhong@maebashi-it.ac.jp

Abstract. In order to improve speaker verification accuracy in the complex environment, a two-layer Gaussian mixture model-universal background model (GMM-UBM) model based on speaker verification method is proposed. For different layer, a GMM-UBM model was trained by different combination of speech features. The voice data of 3 days (36 h) were recorded from the complex environment, and the collected data was manually segmented into four classes: quiet, noise, target speaker and other speaker. Not only the segment data can be used to train GMM-UBM model, but also it can provide a criterion to assess the effectiveness of the model. The results show that the highest recall for the second and third day were 0.75 and 0.74 respectively, and the corresponding specificity were 0.29 and 0.19, which indicates the proposed GMM-UBM model is viable to verify the target speaker in the complex environment.

Keywords: Complex environment · Speech feature · Speaker verification · GMM-UBM

1 Introduction

Speech emotion recognition is an important part of the intelligence service based on brain big data [1]. To determine the basis of emotional changes in patients with depression by using speech data [2], which can greatly improve the judgment of depression and the whole intelligence service system [3]. As a basic research work, this paper adopts the long time voice to confirm the speaker, which is used to extract the speech features of the specific speaker in the complex environment [4], and then as the basis of emotional analysis and quantification evaluation

© Springer International Publishing AG 2017
Y. Zeng et al. (Eds.): BI 2017, LNAI 10654, pp. 149–158, 2017.
https://doi.org/10.1007/978-3-319-70772-3_14

of depression [5]. However, most of the current researches use GMM model or GMM-UBM model combined with some of the speech features for speech recognition or verification in a specific environment, and also make good recognition results. But there are few studies about the speaker verification in the complex environment, and the recognition rate is not ideal [6].

With the rapid development of speech recognition technology in recent years, it has been widely used in many fields [7]. The speech emotion recognition plays an important role in judging a person's emotion [8], and the mood change is the most intuitive manifestation in depressive patients, the expression of speech intensity, speech rate, speech intonation and speech emotion features is the important index to measure emotion. This paper, as a basic research work, through the long-term speech feature extraction and speaker verification to further strengthens the basis for emotional analysis and quantification of depression [9]. Under certain conditions, the speaker recognition rate will reach a high level. However, the external environment of voice signal acquisition is complex and changeable [10], the acquisition of voice signals can be located at different background noise places, such as school, square and station, the equipment used in voice signals acquisition process will have a certain impact on voice signals [11]. Therefore, the key of the speaker verification method is the matching problem of model training and application environment, and how to overcome the influence of these external environment on the system [12].

In view of this, a two-layer GMM-UBM model based on speaker verification method in the complex environment is proposed [13]. The project team recorded long-term voice in the complex environment and used multiple features such as short-term zero rate, short-term average amplitude, short-term energy, pitch period, formant and static MFCC features and first and second order difference coefficients are trained in the GMM-UBM model and the speaker verification of the recorded voice [14]. On the one hand, it explored and improved the applicability of the GMM-UBM model to the study of speaker verification in the complex environment [15]. On the other hand, this research work can be used as a basic work to extract the characteristic indexes of the specific person's daily speech intensity, speech rate, intonation and emotional characteristics, and further apply it to the field of emotional evaluation for depression [16].

The paper is organized as follows. Section 2 focuses on the methods of speaker verification based on two-layer GMM-UBM model in complex environments. Section 3 is the description of the test and the results. Finally, Sect. 4 give a release discussion.

2   Methods

2.1   Voice Data Acquisition and Preprocessing

The voice data used in this paper is obtained by Sony Z3 smartphone and BOYA Lapel Microphone, and the sampling frequency is 8 kHz and 16 bit quantization. In the daily life environment, wear a microphone and recording continuous long-term voice data. Each wearing time is 12 h (8 o'clock in the morning, the end of 8 pm), and a total of 3 days, the voice data of 36 h. In order to facilitate the evaluation of this algorithm and need to preprocess the voice data. On the one hand, the voice segmentation software is developed in MATLAB, and voice data are divided into target speaker, other speaker, quiet and noise, four classes, and the voice data after segmentation can be used as a gold standard for evaluating the speaker's ability to identify model categorization. The number of samples after division is shown in Table 1. On the other hand, the four classes of voice data are divided into training samples and test samples according to the time relationship. For example, the first day of the four classes voice data are divided into train samples for training GMM-UBM model, The last two days of the four classes voice data as a testing samples, and to test the effectiveness of the speaker verification based on GMM-UBM model.

Table 1. Number of four types of voice samples after division

Serial number	Target speaker	Other speaker	Quiet	Noise
1	438	669	98	270
2	544	964	36	259
3	1374	2067	41	99

2.2 Feature Extraction

Speech by pre-emphasis, plus Hamming window and sub-frame processing, each sample speech is divided into several frames, frame length is 20 ms, frame shift is 10 ms. Extracting the short-term zero rate, short-term average amplitude, short-term energy, pitch period, 4-order formant and 12-order static MFCC features(excluding the 0th order) and its first and second order difference coefficients, total of 44-order speech features of each frames speech signal. The short-term zero rate, short-term average amplitude, short-term energy can be used to distinguish between the speech from the quiet environment and other environment. The pitch period, formant and MFCC features contain speaker vocal and channel information that can be used to distinguish between speaker and other speaker speech.

2.3 Speaker Verification Architecture Based on Two-Layer GMM-UBM Model

According to the idea of different features combined, In the two-layer GMM-UBM model, the first layer model is trained with short-term zero rate, short-term average amplitude, short-term energy and static MFCC and aims to remove the noise segment. The second layer model is trained with static MFCC and dynamic features, pitch period and formant, and aims to maximize the separation of speaker and other speaker speech.

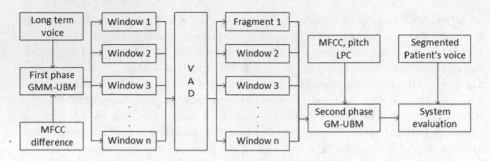

Fig. 1. Speaker verification architecture based on two-layer GMM-UBM model

3 Results

In this paper, two experiments are designed to evaluate the performance of speaker verification model. (1) In order to verify the validity of the model, the continuous long-term speech is segmented(fragmented) and classified, and the speakers speech is confirmed based on the GMM-UBM speaker verification model (2) Taking into account the automatic processing and verification of the speakers speech, the endpoint detection of continuous long-term speech is performed, and the speaker verification is based on the GMM-UBM speaker verification model after the endpoint detection of speech fragments.

3.1 Evaluation Criterion

In this paper, precision, recall rate, specificity and accuracy are used to evaluate the performance of the speaker verification model. In the calculation process, the speaker speech samples are defined as positive samples, the other three classes of speech samples are defined as negative samples. The process of the calculation will involve two indicators and some basic concepts: N, the total number of all samples; True Positives (TP), predicted by the model as positive samples and actually is the number of the positive samples, used to calculate how much of the actually positive sample is predicted correctly; False Positive (FP), predicted by the model as positive samples and actually is the number of negative samples, used to calculate the number of false positives samples of the model. True Negative (TN), predicted by the model as negative samples and actually is the number of negative samples, used to calculate the number of the actual negative sample is predicted to be negative; False Negatives (FN), predicted by the model as negative samples and actually is the number of positive samples, used to calculate the number of samples of the model omitted. The precision indicates the proportion of the actually positive sample is predicted to be the correct. The recall rate indicates the proportion of the negative samples predicted to be positive samples. The specificity indicates the proportion of the all negative samples predicted to be negative sample. The accuracy indicates the proportion of the sum of the positive samples and the negative samples being predicted to the correct number of samples. The calculation formula is as follows:

$$Precision = TP/(TP + FP)$$
$$Recall = TP/(TP + FN)$$
$$TNR = TN/(FP + TN)$$
$$Accuracy = (TP + TN)/N$$

(1)

3.2 GMM-UBM Speaker Verification Based on Segmented Voice Data

In order to prove the rationality of speaker verification using the GMM-UBM model for recorded voice in the complex environment, the speaker verification operation is first used for the segmented voice data by this model. Based on the segmentation of voice data, the GMM-UBM speaker verification model uses the first day of four classes samples as train sample, the remaining two days of four classes as test sample, directly using GMM-UBM model for speaker verification operation. As the GMM-UBM model needs to set the thresholds to confirm the speaker speech and other three kinds of sample speech, the ROC curve is used to represent the model classification effect under different thresholds. The ROC curve of the GMM-UBM speaker verification model under different thresholds is shown in the Fig. 2, As shown in the figure, for the second day and third day the segmentation of the voice data sample classification results, the different thresholds for the ROC curve value is close to the upper left of the coordinate axis. At the same time, compared with the coordinate diagonal, the ROC curve is far from the diagonal, which illustrates the rationality and feasibility of the speaker verification by using the GMM-UBM speaker verification model in the complex environment.

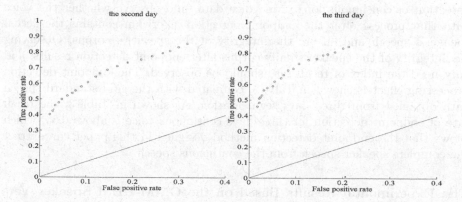

Fig. 2. ROC curve of GMM-UBM speaker verification model under different thresholds

In order to further visualize the classification effect of the model, when thresholds = 0, the classification results of the last two days samples are shown in Table 2. As shown in the table, when the model threshold is set to 0, the classification accuracy of the second day and third day segmentation speech samples is 0.804 and 0.815, respectively, which further illustrates the validity of the GMM-UBM speaker verification model.

Table 2. Classification results of two day speech segmentation samples when the model thresholds is set to 0

Serial number	Precision	Recall	TNR	Accuracy
1	0.715	0.643	0.12	0.804
2	0.867	0.661	0.073	0.815

3.3   GMM-UBM Speaker Verification Based on Continuous Long-Term Voice Data

Unlike speaker verification based on segmented voice data, the difficulty of the speaker verification for the continuous long-term voice data is: (1) The recording of voice in the complex environments contains noise, which make it difficult to locate the starting position and ending positions of the speaker's speech; (2) Due to the noise problem in the environment, it is necessary to design a speech feature with strong robustness and anti-noise capability to characterize the speaker's speech. In view of this, this paper uses endpoint detection and GMM-UBM model based on different features training strategy to process continuous voice data.

The Experiment Results of the Endpoint Detection. In order to facilitate the automatic processing and verification of the speaker speech, the endpoint detection of continuous long-term voice data, and observe whether the voice data that process after the endpoint detection algorithm contains the actual speaker's speech and ensure the integrity of the speaker's corpus. Observing the integrity of the speaker's daily corpus after endpoint detection results, and only need the index of recall rate should be observed. The endpoint detection processing effect is shown in Table 3, compared with the gold standard speech data obtained from three days segmentation. As shown in Table 3, the recall rate of endpoint detection for three days continuous voice is above 0.85, which shows that the endpoint detection method designed in this paper can extract the complete speaker speech from the continuous speech.

The Experimental Results Based on the GMM-UBM Speaker Verification Model of the First Layer. The first layer GMM-UBM model uses MFCC and its difference features, short-term zero rate, short-term average amplitude, short-term energy, four kinds of features to train, and aims to retain the speaker's speech and as much as possible to remove the noise segments. Based on the speech data after endpoint detection, the GMM-UBM model is

Table 3. Endpoint detection processing effect

Serial number	Precision	Recall	TNR	Accuracy
1	0.009	0.857	0.436	0.565
2	0.052	0.866	0.552	0.461
3	0.008	0.875	0.469	0.537

trained by the speech data of the first day, which is used to speaker verification. In order to better evaluate the effect of the long-term voice speaker verification, the confirmed speaker speech segment and the gold standard speech in accordance with the time alignment, statistical the length of speaker speech after verification, and then calculate the accuracy, recall rate and other indicators. At the same time, considering the threshold to the speaker verification effect of the GMM-UBM model, the ROC curve of the first layer GMM-UBM speaker verification model under different thresholds is shown in Fig. 3. Due to the first layer GMM-UBM speaker verification model is designed to retain the speaker's speech and as much as possible to remove the noise segment, the four indexes only need to focus on the recall rate and specificity of the two indicators, that is, to pursue of the highest recall rate and the lowest possible specificity. As shown in Fig. 3, the recall rate for speech data samples processed for the second and third days was up to 0.82 and 0.79, respectively, the corresponding specificity is 0.51 and 0.41, shown that the speaker verification results based on the first layer GMM-UBM model retain most of the actual speakers speech, but there are still more false samples in the results. In order to ensure that the speaker's speech is retained as much as possible, we choose to the thresholds for obtaining the highest recall rate to obtain the speaker verification result of the first layer GMM-UBM model and then enter into the second layer GMM-UBM model to form the final speaker verification result.

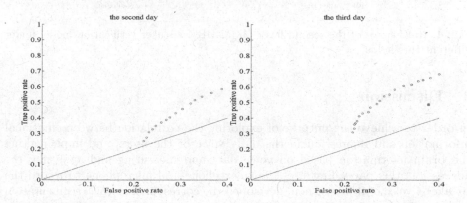

Fig. 3. ROC curve of the first layer GMM-UBM speaker verification model under different thresholds

The Experimental Results Based on the GMM-UBM Speaker Verification Model of the Second Layer. The second layer GMM-UBM model is trained by MFCC and its difference features, pitch period, formant and LPC coefficients, and aims to as much as possible distinguish speech samples such as target speakers speech, other speaker speech and noise. The training method differs from the first layer GMM-UBM model only in that the training features are inconsistent, and the other operations such as training sample, data alignment and index calculation are basically consistent. Considering the threshold to the speaker verification effect of the GMM-UBM model, the ROC curve of the second layer GMM-UBM speaker verification model under different thresholds is shown in Fig. 4. In the final speaker verification results, the recall rate for the continuous voice data samples processed for the second and third days was up to 0.75 and 0.74, respectively, the corresponding specificity is 0.29 and 0.19. Compared with the results of the first layer GMM-UBM speaker verification, although the recall rate has declined, the degree of specificity has fallen more than the recall rate declined, which is consistent with expected results, and further illustrate the validity of long-term speaker voice verification for two-layer GMM-UBM model based on different features of training strategy.

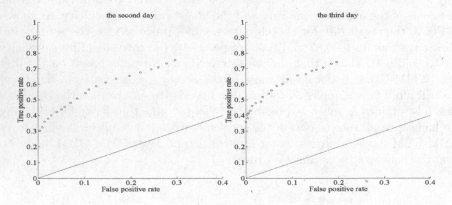

Fig. 4. ROC curve of the second layer GMM-UBM speaker verification model under different thresholds

4   Discussion

In order to achieve the purpose of exploring the correlation between emotional information and depression in the daily voice of the users, and implementing the brain information based on voice data and measuring and analyzing the depression, this paper first uses the smartphone and microphone to record the speaker's voice of 3 days. Then, developed the corresponding voice segmentation software, segment the long-term voice data and eventually classified as target speaker (subjects), other speaker, quiet and noise, not only can be used to train the model of speaker verification, but also can be used as a gold standard for assessing the validity of the model. According to the different phonetic features appearing in daily life, combined with the GMM-UBM training strategy based on different features, a two-layer structure GMM-UBM speaker verification model is proposed. Finally, in order to prove the validity of the GMM-UBM model in the speaker verification application, extracted the short-term zero rate, short-term average amplitude, short-term energy and static MFCC and dynamic features based on the speech samples of the first day segmentation, training the single layer GMM-UBM model was used to classify the speech samples segmented for the second and third day, and combining the four indexes precision, recall rate, specificity and accuracy to obtain experimental results. The furthermore, in order to extract the speakers speech from the continuous long-term voice in the complex environment, the voice data of the first day segmentation as a training data, and based on different features to train the two-layer GMM-UBM model, and the speaker verification operation based on the second and third days of continuous voice data. The experimental results show that the classification accuracy of the single layer GMM-UBM model for the second and third day were 0.804 and 0.815, respectively for the segmentation of speech samples, which shows that the effectiveness of the GMM-UBM model is used in the speaker verification. For the classification of continuous speech samples, calculate the first and second layer GMM-UBM speaker verification results. Based on the first layer GMM-UBM model speaker verification results show that the results retained most of the actual speaker voice, but there are still more false samples. Based on the second layer GMM-UBM model, the results show that the degree of specificity has fallen more than the recall rate declined, which is consistent with expected results, and further illustrate the validity of long-term speaker voice verification for two layer GMM-UBM model based on different features of training strategy.

Acknowledgments. This work is partially supported by the National Basic Research Program of China (No. 2014CB744600), National Natural Science Foundation of China (No. 61420106005), Beijing Natural Science Foundation (No. 4164080), and Beijing Outstanding Talent Training Foundation (No. 2014000020124G039).

References

1. Chen, J., Zhong, N.: Data-brain modeling for systematic brain informatics. In: Brain Informatics, International Conference, BI, Beijing, China, October 22–24 (2009)
2. Zhong, N., Motomura, S.: Agent-enriched data mining: a case study in brain informatics. IEEE Intell. Syst. **24**, 38–45 (2009)
3. Li, N., Mak, M.W.: SNR-invariant PLDA modeling in nonparametric subspace for robust speaker verification. IEEE/ACM Trans. Audio Speech/Language Process. **23**, 1648–1659 (2015)
4. Ding, I.J., Yen, C.T., Ou, D.C.: A method to integrate GMM, SVM and DTW for speaker recognition. Int. J. Eng. Technol. Innov. **4**, 38–47 (2014)

5. Wu, H., Wang, Y., Huang, J.: Blind detection of electronic disguised voice. In: IEEE International Conference on Acoustics, Speech and Signal Processing, pp. 3013–3017 (2013)
6. Wang, X., Yang, T., Yu, Y., Zhang, R., Guo, F.: Footstep-identification system based on walking interval. Intell. Syst. IEEE **30**, 46–52 (2015)
7. Turner, C., Joseph, A.: A wavelet packet and mel-frequency cepstral coefficients-based feature extraction method for speaker identification. Procedia Comput. Sci. **61**, 416–421 (2015)
8. Haris, B.C., Sinha, R.: Low-Complexity Speaker Verification with Decimated Supervector Representations. Elsevier Science Publishers B. V., Amsterdam (2015)
9. Wang, Y., Wu, H., Huang, J.: Verification of Hidden Speaker Behind Transformation Disguised Voices. Academic Press, Inc, Orlando (2015)
10. Kanagasundaram, A., Dean, D., Sridharan, S.: Improving PLDA speaker verification with limited development data. In: IEEE International Conference on Acoustics, Speech and Signal Processing, pp. 1665–1669 (2014)
11. Rajan, P., Afanasyev, A., Hautamki, V., Kinnunen, T.: From single to multiple enrollment i-vectors: practical PLDA scoring variants for speaker verification. Digit. Signal Proc. **31**, 93–101 (2014)
12. Xu, L., Yang, Z.: Speaker identification based on state space model. Int. J. Speech Technol. **19**, 1–8 (2016)
13. Rakhmanenko, I., Meshcheryakov, R.: Speech Features Evaluation for Small Set Automatic Speaker Verification Using GMM-UBM System (2016)
14. Sarkar, A.K., Tan, Z.H.: Text Dependent Speaker Verification Using un-supervised HMM-UBM and Temporal GMM-UBM. In: Interspeech (2016)
15. Dehak, N., Kenny, P.J., Dehak, R., Dumouchel, P., Ouellet, P.: Front-end factor analysis for speaker verification. IEEE Trans. Audio Speech Language Process. **19**, 788–798 (2011)
16. Jagtap, S.S., Bhalke, D.G.: Speaker verification using Gaussian mixture model. In: International Conference on Pervasive Computing, pp. 1–5 (2015)

Emotion Recognition from EEG Using Rhythm Synchronization Patterns with Joint Time-Frequency-Space Correlation

Hongzhi Kuai[1,2,3], Hongxia Xu[1,3], and Jianzhuo Yan[1,2,3]([✉])

[1] Faculty of Information Technology, Beijing University of Technology,
Beijing 100124, China
[2] Beijing Advanced Innovation Center for Future Internet Technology,
Beijing University of Technology, Beijing 100124, China
[3] Engineering Research Center of Digital Community,
Beijing University of Technology, Beijing 100124, China
kuaihongzhi@emails.bjut.edu.cn, {xhxccl,yanjianzhuo}@bjut.edu.cn

Abstract. Recently there has attracted wide attention in EEG-based emotion recognition (ER), which is one of the utilization of Brain Computer Interface (BCI). However, due to the ambiguity of human emotions and the complexity of EEG signals, the EEG-ER system which can recognize emotions with high accuracy is not easy to achieve. In this paper, by combining discrete wavelet transform, correlation analysis, and neural network methods, we propose an Emotional Recognition model based on rhythm synchronization patterns to distinguish the emotional stimulus responses to different emotional audio and video. In this model, the entire scalp conductance signal is analyzed from a joint time-frequency-space correlation, which is beneficial to the depth learning and expression of affective pattern, and then improve the accuracy of recognition. The accuracy of the proposed multi-layer EEG-ER system is compared with various feature extraction methods. For analysis results, average and maximum classification rates of 64% and 67.0% were obtained for arousal and 66.6% and 76.0% for valence.

Keywords: Affective computing · Emotion recognition · EEG · Rhythm synchronization patterns

1 Introduction

Emotion plays a vital role in our daily life as it influences our intelligence, behavior and social communication. Emotion recognition (ER) has recently attracted increasing attention in the fields of affective computing, brain-computer interface (BCI) and human-computer interface (HCI), many attempts have been undertaken to advance the different field of emotion recognition from visual (i.e., facial and bodily expression), audio, tactile (i.e., heart rate, skin conductivity, thermal signals etc.) and brain-wave (i.e., brain and scalp signals) modalities [1]. In the

Y. Zeng et al. (Eds.): BI 2017, LNAI 10654, pp. 159–168, 2017.
https://doi.org/10.1007/978-3-319-70772-3_15

existing ER study, as the EEG signal is reliable and not easy to camouflage and other characteristics, so gradually attracted people's attention. In particular, since Davidson and Fox [2] found that the electroencephalogram (EEG) signal generated from the central nervous system (CNS) could discriminate between positive and negative emotions in early 1980s, researchers started to look for the possibilities of using brain derived signals such as electroencephalogram (EEG) as the basis of emotional recognition. In recent years, with the rapid development of signal processing and machine learning technology and the improvement of computer data processing ability, the study of emotion recognition based on EEG signals has become a hot topic in the field of affective computing.

The steps for EEG-based emotion studies usually use some emotion elicitation materials to elicit certain emotions, record subjects brain activity or physiological signals, extract time domain, frequency domain or time-frequency domain features of these signals and classify the affective features through different classifiers. Koelstra et al. [3] presented methods by using time domain features based on statistical mothed. K-Nearest Neighbor (KNN), Bayesian Network (BN), Artificial Neural Network (ANN) and Support Vector Machine (SVM) classifier were used to classify two levels of valence states and two levels of arousal states. Murugappan et al. [4] presented methods by using frequency-domain feature, Fast Fourier Transform (FFT). KNN and Probabilistic Neural Network (PNN) classifier were used to classify five different emotions (happy, surprise, fear, disgust, neutral). Yohanes et al. [5] presented methods by using time-frequency domain feature, Discrete Wavelet Transform(DWT). Extreme Learning Machine (ELM) and SVM classifier were used to classify two emotions (happy and sad). In the field of neuroscience, emotion has been shown to have a strong correlation with rhythm, and this theory of relativity is also used in emotional recognition systems. Murugappan, Hadjidimitriou, Zheng et al. [6–8] extract the time domain or frequency domain features under different rhythms, and then study the classification effect of emotion under different rhythm.

Most of the above-mentioned emotional recognition system studies do not take into account the association between the brain regions, and the emotional features extracted often reflect the uniqueness of the electrodes corresponding to the brain regions. But some studies have shown that even if the cognitive task is very simple, its implementation also depends on multiple separate brain functional regions [9]. Specific functions require these separate brain functional areas to be combined, and through integration, collaboration can be completed. There is an indivisible relationship between cognition and emotion, and different emotional states often reflect the synergistic effect of different brain regions. Therefore, the information interaction between different brain regions in emotional recognition and the parallelism theory of brain function network should be paid enough attention [10]. Currently, some researchers use Magnitude Squared Coherence Estimate [11] and Asymmetry [12] to calculate features from Combinations of Electrodes, which simulates the information interactivity between different brain regions. However, these methods do not fully consider the effect of rhythm synchronization on affective recognition.

Based on the current research progress, this paper presents a heuristic multi-layer EEG emotion recognition model based on Rhythmic Synchronization Patterns, and studies the relationship between EEG signals and emotion. Wavelet decomposition and Magnitude Squared Coherence Estimate are integrated into the emotional model, which are used to construct emotional patterns of scalp EEG in different emotional states. RBF neural networks are used in the learn and classification of affective patterns. Since the alpha and beta rhythms have been shown to be more relevant to emotion, the model in this paper mainly identifies the rhythm patterns of the two bands [13]. Our current method is mainly compared with the traditional emotion recognition method based on time domain features and frequency domain features.

The rest of this paper is structured as follows. Section 2 describes the proposed Emotion Recognition Model using Rhythm Synchronization Patterns for discrimination of emotions based on a joint time-frequency-space correlation. Section 3 clarifies the experimental design. Section 4 presents the experimental results and discussion. Section 5 concludes this paper.

2 Architecture of Emotional Recognition Model Based on Rhythm Synchronization Patterns (RSP-ERM)

This section explains the emotional recognition model based on RSP from two aspects: functions of each layer and definition and evaluation of emotional state. The proposed emotional recognition model is shown in Fig. 1. The multichannel EEG data are the input and the valence/arousal level is the output. In the RSP-ERM, a joint time-frequency-space correlation pattern is calculated from combinations of electrodes, and a two-layer RBFNN is used to classify the emotional patterns.

2.1 Functions of Each Layer

The process of the emotional states discrimination in each layer of the proposed RSP-ERM is described as follows.

First Layer - Spectral Filtering: The EEG signal is very weak, so it is susceptible to the artifacts generated by other noise signals during the acquisition process. In the study of emotion recognition, the artifacts to be removed mainly include Electromyography (EMG), Electrooculogram (EOG), power frequency interference (50 HZ), electromagnetic interference and task-unrelated EEG. As the power frequency interference and electromagnetic interference are often presented in the high frequency band, it can be easily filtered out through the band-pass filter or low-pass filter. Therefore, the spectral filtering operation is defined as the first layer of the proposed model.

Second Layer - Rhythm extraction: Wavelet transform is a rapid development and more popular signal analysis method, which accurately reveals the signal's distribution characteristics in time domain and frequency domain, and

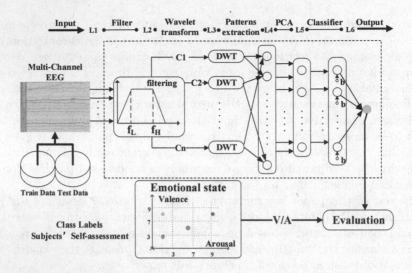

Fig. 1. The emotional recognition model based on rhythm synchronization patterns.

can analyze the signal by using a variety of scales. Therefore, the rhythm extraction operation is defined as the second layer of the proposed model. Based on the characteristics of wavelet multiresolution analysis, the wavelet coefficients over the frequency bands of alpha (8–16 Hz), beta (16–32 Hz) of EEG signals are extracted according to certain scale, and the information of each rhythm is obtained.

Third Layer - Function connection computation: Studies have shown that the cortical-subcortical interactions and interaction between different brain regions play an important role in the perception of emotional stimulus [15]. Therefore, brain connectivity features would be very informative to investigate the emotion patterns during the perception of emotional stimuli. In addition, the functional connectivity features can overcome the handedness issue because they are not reciprocal [16]. Based on this observation, the function connection computation is defined as the third layer of the proposed model. In this paper, the functional connectivity features of different rhythms in different channels are calculated by using the magnitude square coherence estimation (MSCE) [17], for constructing the functional connection matrix.

Fourth Layer - Principal feature learning: In general, the more features we get, the more conducive to the expression of emotional state, which is more conducive to the identification of emotional patterns. However, due to the functional leap of the basic patterns in the brain and the ambiguity of the emotion itself, there are many cross features between different emotional patterns. Some of these extracted features are irrelevant or redundant and have a negative effect on the accuracy of the classifier. In order to explore the main emotional patterns implied in the dynamic function connection matrix, the principal feature learning is defined as the fourth layer of the proposed model, which extract main features of the functional connection network matrix by using the principal component analysis (PCA). In addition, the filtered pattern feature will greatly improve the learning efficiency of the algorithm.

Fifth and sixth layers - Classification: Emotion recognition based on EEG signals is a pattern recognition process. While the reliability of the ground-truth, duration and intensity of emotions lead to the staggering between different emotional state patterns is particularly evident. Radial Basis Function Neural Networks takes full account of the fuzziness effects between category schema in pattern recognition applications, thus forming an effective interface and improving the accuracy of recognition. At the same time, because the model has the advantages of low computational complexity and fast convergence speed, it is widely used in the fields of pattern recognition and nonlinear function approximation. The fifth and sixth layers are a radial basis function neural networks for classifying different emotional states based on the chosen connectivity patterns.

2.2 Defining Emotional States - Class Label

In the study of emotional recognition, researchers usually use the following two perspectives to construct and understand the emotional space:

1. The discrete model considers that the emotional space is composed of a limited number of discrete basic emotions [24].
2. The dimensional model expresses the emotional space as the coordinate system of two-dimensional Valence - Arousal (Valence - Arousal, VA) [18] or three-dimensional Pleasure - Arousal - Dominance (Pleasure - Arousal - Dominance, PAD) [19].
   - Pleasure(Valence) refers to the individual feel the degree of pleasure, that is, positive and negative characteristics. The valence scale ranges from unhappy or sad to happy or joyful.
   - Arousal refers to the degree of psychological alertness and the degree of physiological activation of the individual. The arousal scale ranges from passive or bored to stimulated or active.
   - Dominance refers to whether the individual is in control or controlled [20]. The dominance scale ranges from submissive (or "without control") to dominant (or "in control, empowered").

This paper adopts the DEAP database, and the subjects were asked to rate the music videos according to the SAM [21] regarding emotional valence and arousal. This study mapped the scales (1-9) into two-dimensional valence-arousal model, which is similar to the one for emotion state classes [22]. The valence scale of 1-3 was mapped to "negative", 4-6 to "neutral" and 7-9 to "positive", respectively. The arousal scale of 1-3 was mapped to "passive", 4-6 to "neutral" and 7-9 to "active", respectively. The emotion state classes based on arousal-valence dimension is shown in Fig. 2.

Emotion is a process in which arousal and valence are integrated, and it is unreasonable to simply evaluate from one of its dimensions. Therefore, the emotional recognition model of this paper is mainly evaluated from two dimensions:

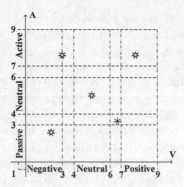

Fig. 2. The emotion state classes based on arousal-valence (A/V) dimension.

arousal and valence. In general, the classification problem based on the SAM scale evaluation is to use the intermediate value of the scale as the criterion for dividing the positive and negative samples. However, in reality, considering the ambiguity of emotion and the difference in sensitivity of the subjects to the stimulus, these trials of user's score near the middle of the SAM scale may not really represent the positive and negative state of emotion. Therefore, the negative and positive states from valence dimension and passive and active states from arousal dimension are investigated in classes $score \leq 3$ and $score \geq 7$. The output of our binary classifier Y is assigned to its class label using a hard threshold (step function). The binary classes are configured using:

$$\begin{cases} Y_a \begin{cases} = (0,1), if & ScoreA \leq 3, \\ = (1,0), if & ScoreA \geq 7. \end{cases} \\ Y_v \begin{cases} = (0,1), if & ScoreV \leq 3, \\ = (1,0), if & ScoreV \geq 7. \end{cases} \end{cases} \qquad (1)$$

where Y_a denotes the arousal group labels, Y_v denotes the valence group labels. Similarly, ScoreA indicates that the subjects rated the arousal score in the face of different audio and video stimuli, ScoreV indicates that the subjects rated the valence score in the face of different audio and video stimuli.

3   Experimental Design

3.1   Data Description

EEG data: The performance of the proposed emotional recognition model is investigated using DEAP Dataset. DEAP [23] is a multimodal dataset for analysis of human affective states. 32 Healthy participants (50% female), aged between 19 and 37 (mean age 26.9), participated in the experiment. 40 one-minute long excerpts of music videos were presented in 40 trials for each subject. There are 1280(32subjects × 40trials) emotional state samples. During the experiments,

EEG signals were recorded with 512 Hz sampling frequency, which were down-sampled to 128 Hz and filtered between 4.0 Hz and 45.0 Hz, and the EEG artifacts are removed.

Sample Distribution: Based on the above DEAP dataset, the proposed model is learned and tested for classifying the negative-positive states (ScoreV $\leq$ 3 or $\geq$ 7) and passive-active states (ScoreA $\leq$ 3 or $\geq$ 7), respectively. The sample size of negative state is 222; the sample size of positive state is 373; the sample size of passive state is 226; the sample size of active state is 297.

3.2 Learning and Testing Process

The learning process of Emotional Recognition Model based on Rhythm Synchronization Patterns, consists of the following three stages:

1. Computing the Function connection patterns of model in second, and third layers in an unsupervised manner (computing the rhythm's MSCE features).
2. Selecting of main connection patterns in fourth layer using PCA.
3. Computing the network parameters for fifth and sixth layers in a supervised manner (classification of labeled data).

In testing phase, stages 1 and 2 are repeated. The selected features are then classified using parameters calculated in learning phase.

3.3 Contrast Methods

In the contrast method, the statistical features of the time domain and the power spectral density feature in the frequency domain are calculated for 32-channels EEG signals in different emotional states. These features make up the vector as a target for classifier to learn as follows:

Time Domain Features (Statistics): In the different emotions, the 15 statistical features of each brain wave are calculated, and finally constitute the emotional state vector (15 $\times$ 32 channels). The statistical features include mean, root mean square, root amplitude, absolute mean, skew, kurtosis, variance, max, min, peak-to-peak, waveform index, peak index, pulse index, margin index, kurtosis index.

Frequency Domain Features(PSD): Power Spectral Density estimate via Welch's method. In different emotional states, the theta, alpha, beta and gamma rhythms of the EEG signals in each channel are extracted, and then their PSD are calculated and averaged to form the emotional state vector (4 $\times$ 32 channels).

The contrast method differs from this RSP-ERM in the model's second and third layers. The process of constructing the rhythm synchronization pattern in RSP-ERM's second and third layers was replaced by the process of time domain's and frequency domain's feature extraction.

4   Experimental Results and Discussion

All the methods are running under 10-flods Cross validation. Table 1 shows the classification results for each feature original after PCA by RSP-ERM. Spread of radial basis functions is default.

Table 1. The classification accuracy for EEG-based arousal and valence recognition.

Feature Selection	Classification Accuracy (%)	
	Arousal	Valence
Statistics	56.86	62.71
PSD	57.25	65.42
Alpha-rhythm pattern	63.92	65.25
Beta-rhythm pattern	61.37	66.61

By performing experiments on the DEAP dataset, the results show that the model based on rhythm synchronization patterns has higher recognition accuracy than other two contrast methods based on time domain or frequency domain features. Therefore, it is proved indirectly that the unique rhythm patterns of brain regions connectivity dose exist in the perception of individual emotional stimuli and it can be learned as a feature by the classifier. In addition, it can be seen from the experimental results that the recognition rate of valence is higher than the recognition rate of arousal in all emotion recognition methods. Arousal is the physiological and psychological state of being awoken or of sense organs stimulated to a point of perception. And the intensity of the above-mentioned state has a great relationship with the sensitivity of the subject to the emotional stimulation. The experimental results show that the difference in the sensitivity of the individual to the emotional stimulus will have a greater impact on the EEG signal, and this change will bring great challenges to the emotional recognition system. Although this approach has greater advantages over other methods in Arousal's identification, deep learning of sensitivity is still the focus of the next step, which can improve the effectiveness of the entire emotional recognition system.

5   Conclusions

The study of the association between EEG rhythm and emotional state has been a hotspot in the field of neuroscience, and it has a very strong theoretical basis. In addition, the existing studies have shown that there is a synchronization phenomenon in the brain-wave rhythm, which characterizes the functional integration and collaboration between different brain regions. Inspired by the above biological theory, this paper presents a multi-layer emotional recognition model based on rhythm synchronization patterns of multi-channel EEG signals. In this model, the original signal is mapped to a specific rhythm space by time-frequency analysis. And the correlation analysis method is used to simulate the synchronization pattern of brain-wave rhythm for different emotional states, thus realizing the integration of multi-channel signals in time-frequency-space dimension. From the experimental results, it can be found that the proposed method is useful to detect the emotion information from EEG recordings in a good accuracy.

Acknowledgements. The work is supported by Science and Technology Project of Beijing Municipal Commission of Education (No. KM201710005026), National Basic Research Program of China (No. 2014CB744600), Beijing Key Laboratory of MRI and Brain Informatics.

References

1. Gunes, H.: Automatic, dimensional and continuous emotion recognition (2010)
2. Davidson, R.J., Fox, N.A.: Asymmetrical brain activity discriminates between positive and negative affective stimuli in human infants. Science **218**(4578), 1235–1237 (1982)
3. Sohaib, A.T., Qureshi, S., Hagelbäck, J., Hilborn, O., Jerčić, P.: Evaluating classifiers for emotion recognition using EEG. In: Schmorrow, D.D., Fidopiastis, C.M. (eds.) AC 2013. LNCS (LNAI), vol. 8027, pp. 492–501. Springer, Heidelberg (2013). doi:10.1007/978-3-642-39454-6_53
4. Murugappan, M., Murugappan, S.: Human emotion recognition through short time Electroencephalogram (EEG) signals using Fast Fourier Transform (FFT). In: 9th IEEE International Colloquium on Signal Processing and its Applications (CSPA), pp. 289–294. IEEE Press, Kuala Lumpur (2013)
5. Yohanes, R.E., Ser, W., Huang, G.B.: Discrete Wavelet Transform coefficients for emotion recognition from EEG signals. In: 2012 Annual International Conference of the Engineering in Medicine and Biology Society (EMBC), pp. 2251–2254. IEEE Press, San Diego, CA, USA, August 2012
6. Murugappan, M., Juhari, M.R.B.M., Nagarajan, R., Yaacob, S.: An investigation on visual and audiovisual stimulus based emotion recognition using EEG. Int. J. Med. Eng. Inf. **1**(3), 342–356 (2009)
7. Hadjidimitriou, S.K., Hadjileontiadis, L.J.: Toward an EEG-based recognition of music liking using time-frequency analysis. IEEE Trans. Biomed. Eng. **59**(12), 3498–3510 (2012)
8. Zheng, W.L., Dong, B.N., Lu, B.L.: Multimodal emotion recognition using EEG and eye tracking data. In: 2014 36th Annual International Conference of the Engineering in Medicine and Biology Society (EMBC), pp. 5040–5043. IEEE Press, Chicago, August 2014
9. Herreras, E.B.: Cognitive neuroscience: the biology of the mind. Cuadernos de Neuropsicología **4**(1), 87–90 (2010)
10. De Bie, T., Cristianini, N., Rosipal, R.: Eigenproblems in pattern recognition. In: Handbook of Geometric, Computing, pp. 129–167. Springer, Berlin (2005). doi:10.1007/3-540-28247-5_5

11. Khosrowabadi, R., Heijnen, M., Wahab, A., Quek, H.C.: The dynamic emotion recognition system based on functional connectivity of brain regions. In: Intelligent Vehicles Symposium (IV), pp. 377–381 IEEE Press, San Diego, June 2010
12. Petrantonakis, P.C., Hadjileontiadis, L.: A novel emotion elicitation index using frontal brain asymmetry for enhanced EEG-based emotion recognition. IEEE Trans. Inf. Technol. Biomed. 15(5), 737–746 (2011)
13. Rowland, N., Meile, M.J., Nicolaidis, S.: EEG alpha activity reflects attentional demands, and beta activity reflects emotional and cognitive processes. Science 228(4700), 750–752 (1985)
14. Yijun, W.: Brain-Computer Interface System Based on Rhythm Modulation - From Offline to Online. Doctoral dissertation, Tsinghua University, Beijing (2007)
15. Kober, H., Barrett, L.F., Joseph, J., Bliss-Moreau, E., Lindquist, K., Wager, T.: Functional grouping and cortical–subcortical interactions in emotion: a meta-analysis of neuroimaging studies. Neuroimage 42(2), 998–1031 (2008)
16. Sporns, O.: Brain connectivity. Scholarpedia 2(10), 4695 (2007)
17. Kay, S.M., Estimation, S.M.: Theory and Application. Prentce-Hall, Englewood Cliffs (1988)
18. Feldman Barrett, L., Russell, J.A.: Independence and bipolarity in the structure of current affect. J. Pers. Soc. Psychol. 74(4), 967 (1998)
19. Izard, C.: Special section: on defining emotion. Emot. Rev. 2(4), 363–385 (2010)
20. Liu, Y., Fu, Q., Fu, X.: The interaction between cognition and emotion. Chin. Sci. Bull. 54(22), 4102–4116 (2009)
21. Morris, J.D.: Observations: SAM: the self-assessment manikin; an efficient cross-cultural measurement of emotional response. J. Advertising Res. 35(6), 63–68 (1995)
22. Jirayucharoensak, S., Pan-Ngum, S., Israsena, P.: EEG-based emotion recognition using deep learning network with principal component based covariate shift adaptation. Sci. World J. (2014)
23. Koelstra, S., Muhl, C., Soleymani, M., Lee, J.S., Yazdani, A., Ebrahimi, T., Pun, T., Nijholt, A., Patras, I.: Deap: a database for emotion analysis; using physiological signals. IEEE Trans. Affect. Comput. 3(1), 18–31 (2012)
24. Eerola, T., Vuoskoski, J.K.: A comparison of the discrete and dimensional models of emotion in music. Psychol. Music 39(1), 18–49 (2011)

Informatics Paradigms for Brain and Mental Health

Patients with Major Depressive Disorder Alters Dorsal Medial Prefrontal Cortex Response to Anticipation with Different Saliences

Bin Zhang$^{(\boxtimes)}$ and Jijun Wang

Shanghai Key Laboratory of Psychotic Disorders,
Shanghai Mental Health Center, Shanghai Jiao Tong University
School of Medicine, Shanghai, China
zhang.bin845@foxmail.com

Abstract. Patients with major depressive disorder (MDD) show an impaired ability to modulate emotional states and deficits in processing emotional information. Many studies in healthy individuals showed a major involvement of the medial prefrontal cortex (MPFC) in the modulation of emotional processing. Recently, we used emotional expected pictures with high or low salience in a functional MRI (fMRI) paradigm to study altered modulation of MPFC in MDD patients and to explore the neural correlates of pathological cognitive bias, and investigated the effect of symptom severity on this functional impairment. Data were obtained from 18 healthy subjects and 18 MDD patients, diagnosed according to the ICD-10 criteria. Subjects lay in a 3 T Siemens Trio scanner while viewing emotional pictures, which were randomly preceded by an expectancy cue in 50% of the cases. Our study showed a lower effect of salience on dorsal MPFC (DMPFC) activation during anticipatory preparation in depressed subjects. Differential effects for high salient versus low salient pictures viewing were also found in DMPFC in depressed and healthy subjects, and this effect was significantly higher for depressed subjects and correlated positively with Hamilton rating scale for depression (HAMD) scores. Comparing the anticipation effect on subsequent picture viewing between high and low salient pictures, differential effects were higher for depressed subjects during high salient picture viewing and lower during low salient picture viewing. Therefore, we could shed light on DMPFC functioning in depressive disorder by separating cognitive and consumatory effects in this region.

Keywords: Major depressive disorder · Salience expectancy · Functional MRI · Medial prefrontal cortex

1 Introduction

Major depressive disorder (MDD) is a highly prevalent psychiatric disorder, typically characterized by a lack of interest in one-pleasurable activities [1]. MDD patients show an impaired ability to modulate emotional states and deficits in processing positive emotional information [2].

© Springer International Publishing AG 2017
Y. Zeng et al. (Eds.): BI 2017, LNAI 10654, pp. 171–180, 2017.
https://doi.org/10.1007/978-3-319-70772-3_16

For healthy individuals, attention is generally biased towards positive stimuli [3], while the patients with MDD deficit in shifting the focus of their attention to positive stimuli [2]. Kellough et al. found that depressive individuals decreased attention for positive stimuli and increased for negative stimuli [4], suggesting that depressive patients has inefficient attentional disengagement from negative stimuli. And, the study of Shafritz et al. showed that depressive individuals require greater cognitive effort to divert attention away from negative stimuli, while the healthy individuals are for positive stimuli, and this inhibitory processing had been associated with the activity in the pregenual anterior cingulate cortex (pgACC) [5]. This biased attention in depression has been suggested to rely on an easier access to negative memories or cognitions [6].

Recently, brain imaging studies has revealed that many of the brain regions responsible for normally regulating mood show disrupted function in depression. And, functional MRI (fMRI) has been applied to examine differential activations in certain brain regions between MDD patients and healthy controls. Resting-state functional connectivity studies with fMRI data have reported abnormal signal fluctuations in anterior cingulate cortex (ACC) [7], amygdala [8] and medial prefrontal cortex (MPFC) [9] in MDD individuals. Several studies investigated expectancy induced modulation of emotional pictures with emotional expectancy paradigm [10–12], and found that MPFC regulated attentional modulation of emotion processing. Bermpohl et al. found altered of anticipation and picture viewing interaction in MPFC of MDD patients [11]. Zhang et al. found MDD patients had greater activity in MPFC as a function of positive anticipation [10].

Previous study showed self-directedness is considered an indicator for many mental disorders, especially depression [13]. People with low self-directedness are influenced by salient attention [14]. Recently, we used emotional expected pictures with different salient scores in an fMRI paradigm to study altered modulation of MPFC in MDD patients and to explore the neural correlates of pathological cognitive bias. This pivotal analysis suggested abnormal anticipatory modulation of emotional pictures in a small sample, however on an uncorrected level and with potential age effects. In the current study, we aimed to investigate the effect of symptom severity on this functional impairment.

2   Methods

2.1   Subjects

Eighteen MDD patients that currently experiencing depressive episodes and 18 age and gender matched HC participants performed the fMRI experiment. Patients were clinically diagnosed according to the ICD-10 criteria and severity was assessed using the 24-items Hamilton rating scale for depression (HAMD) (Table 1). Exclusion criteria were major non-psychiatric medical illness, history of seizures, prior electroconvulsive therapy treatments, illicit substance use or substance use disorders, and pregnancy. The exact number of previous depressive episodes was not available for all patients. The length of the current episode was between one and twelve months. All patients were medicated according to clinical standards with a selective serotonin reuptake inhibitor

Table 1. Demographic characteristics of study participants

	HC (n = 18)	MDD (n = 18)	Group effect P value
Gender (male/females)	9/9	9/9	0.522
Age (years)	39.28 ± 13.02	43.77 ± 10.83	0.203
HAMD	–	18.05 ± 8.04	–

(SSRI), anti-psychotic medication, or a selective noradrenalin reuptake inhibitor (SNRI), tetracyclic antidepressant (TCA), or mood stabilizers.

2.2 Task Design

The experiment was designed and presented by Presentation (Neurobehavioral Systems, http://www.neurobs.com) software. The emotional expectancy paradigm used visual cues and pictures with different arousals. A total of 40 pictures were selected from the International Affective Picture System (IAPS), and pictures were divided into high salient (n = 20) and low salient (n = 20) pictures according to the ratings of perceived arousal.

Each picture was presented for 4 s. Half of the trials were preceded by an expectancy task that predicts 100% validity of the types of the upcoming stimulus; e.g., an upwards-pointing arrow indicated that a high salient picture will follow; a downward pointing arrow indicated that a low salient picture will follow. Half of the pictures were preceded without any cue (unexpected pictures), meaning these pictures were followed directly by after fixation. In the paradigm, we used the cue-picture mismatch events in order to bring new condition to the participants, therefore to increase their attention towards pictures: 10 visual cues were followed directly by fixation but not pictures. Each expectancy cue lasted 3–5 s; the baseline period with a fixation cross lasted 8.5–10.5 s (Fig. 1).

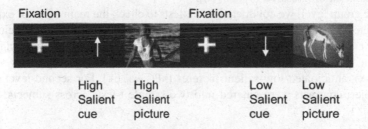

Fixation   Fixation

High   High   Low   Low
Salient   Salient   Salient   Salient
cue   picture   cue   picture

Fig. 1. Paradigm for emotional expectancy task

2.3 Data Acquisition and Analysis

The fMRI was performed using a 3-Tesla Siemens MAGNETOM Trio MR system (Siemens, Erlangen, Germany) with an echo-planar imaging (EPI) sequence: TR/TE = 1250/25 ms; FA = 70°, FOV = 220 mm × 220 mm; matrix = 44 × 44; slice thickness = 5 mm (no slice gap); 26 interleaved transverse slices; voxel size $5 \times 5 \times 5$ mm^3, and the task scan contained 488 image volumes.

Structural images were acquired using magnetization-prepared rapid gradient echo (MPRAGE) sequence with the following scan parameters: repetition time (TR) = 8.2 ms; echo time (TE) = 3.2 ms; inversion time (TI) = 450 ms; flip angle (FA) = 12°; field of view (FOV) = 256 mm × 256 mm; matrix = 256 × 256; slice thickness = 1 mm, no gap; and 188 sagittal slices.

Data were preprocessed using Statistical Parametric Mapping Software (SPM 8, http://www.fil.ion.ucl.ac.uk/spm). The 488 volumes were corrected for time delay between different slices and realigned to the first volume. Head motion parameters were computed by estimating translation in each direction and the angular rotation on each axis for each volume. Each subject had a maximum displacement of less than 3 mm in any cardinal direction (x, y, z), and a maximum spin (x, y, z) less than 3°. Individual structural images were linearly coregistered to the mean functional image; then the transformed structural images were segmented into grey matter (GM), white matter, and cerebrospinal fluid. The GM maps were linearly coregistered to the tissue probability maps in the MNI space. The motion-corrected functional volumes were spatially normalized to the individual's structural image using the parameters estimated during linear coregistration. The functional images were resampled into $3 \times 3 \times 3$ mm^3 voxels. Finally, all datasets were smoothed with a Gaussian kernel of $8 \times 8 \times 8$ mm^3 FWHM.

At the single subject level, we modeled five regressors of interest and convolved with the canonical hemodynamic response (CHR) function on the base of general linear model (GLM). The first regressor indicated fixation. Two regressors indicated the effect of expectancy during cuing trails irrelevant of whether the picture was followed or not (expectancy of high salient pictures [Xhs] and expectancy of low salient pictures [Xls]). Two regressors indicated the effect of expected arousal during picture display session (expected high salient pictures [xPhs] and expected low salient pictures [xPls]). The voxel time series were high-pass filtered at 1/128 Hz to account for non-physiological slow drifts in the measured signal and modeled for temporal autocorrelation across scans with an autoregressive model [10].

On the group level, we conducted four t-tests to check the main effect of expectancy and expected picture viewing: anticipation of high-salient pictures > anticipation of low-salient pictures (Xhs > Xls), anticipated high-salient pictures > anticipated low-salient pictures (xPhs vs. xPls), anticipated high-salient pictures > fixation (xPhs > fix), anticipated low-salient pictures (xPls vs. fix). For second-level analysis, single subject contrasts were entered into two-sample t-tests across subjects.

3   Results

3.1   Anticipation Period Findings

Effect of Salience on Anticipation period. With one sample t test, salience during anticipation period of the contrast (Xhs > Xls) revealed a specific effect in the dorsal MPFC (DMPFC) and the ACC (p < 0.001, uncorrected, Fig. 2) in healthy controls (HC).

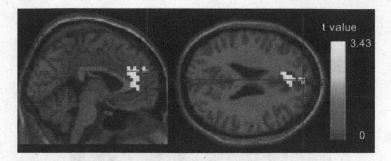

Fig. 2. During Xhs > Xls, health controls showed significantly increased activity in the dorsal MPFC and ACC (warm color, p < 0.001 uncorrected).

Differences in Anticipation between MDD and Healthy Subjects. With two sample t test, activation in DMPFC and dorsolateral prefrontal cortex (DLPFC) during high-salient compared to low-salient anticipation (Xhs > Xls) was higher in healthy subjects compared to depressed patients (p < 0.001, uncorrected, Fig. 3).

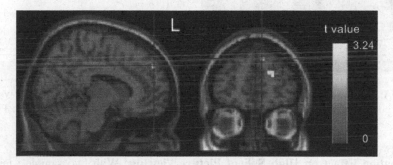

Fig. 3. Group difference of high salient vs. low salient expectancy (Xhs > Xls), health controls showed significantly increased activity in the dorsal MPFC and DLPFC (warm color, p < 0.001 uncorrected).

3.2   The Findings of Anticipation Effect on Picture Viewing

Anticipation Effect on High and Low Salient Picture Viewing. With two sample t test, depressed patients showed increased activation in DMPFC for the contrast of xPhs vs. xPls, compared to controls (p < 0.001 uncorrected, Fig. 4). Moreover, DMPFC activation for this contrast was positively correlated with HAMD score in depressed subjects (p < 0.05, Fig. 5).

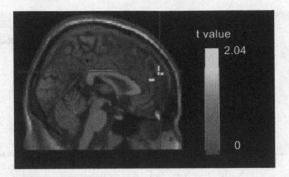

Fig. 4. During the contrast of xPhs > xPls, MDD patients showed significantly increased activity in the DMPFC (warm color, p < 0.001 uncorrected).

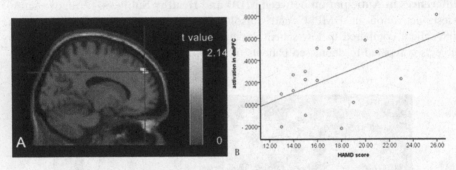

Fig. 5. A: Regression analysis of HAMD with effect of expected picture viewing, revealed a positive correlation of HAMD and activation in DMPFC (p < 0.05, uncorrected). **B**: Positive correlation between HAMD score and the mean betas of DMPFC in patients.

Anticipation Effect on High Salient Picture Viewing. With two sample t test, the anticipation effect on high-salient picture viewing was significantly higher for depressed subjects in DMPFC (p < 0.001, uncorrected, Fig. 6).

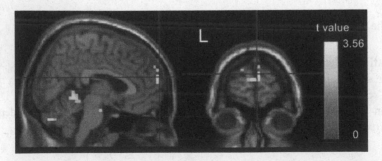

Fig. 6. During the contrast of xPhs > fix, MDD patients showed significantly increased activity in the DMPFC (warm color, p < 0.001 uncorrected).

Anticipation Effect on Low Salient Picture Viewing. With two sample t test, the anticipation effect on low-salient picture viewing was significantly decreased in depressed subjects in DMPFC. The analysis revealed effects in the DMPFC (p < 0.001, uncorrected, Fig. 7).

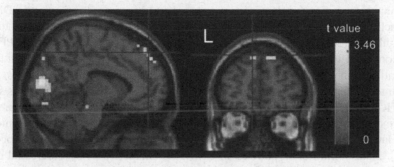

Fig. 7. During the contrast of xPls > fix, MDD patients showed significantly increased activity in the DMPFC (warm color, *p* < 0.001 uncorrected).

4 Discussion

The current study found that DMPFC, one core region of anterior DMN which previous associated with altered attention proceeding in depressed patients, was related to emotional anticipant and picture viewing with different kinds of saliences. To our knowledge, this is the first study to investigate altered neural activation during salient effect on anticipant and pictures viewing in MDD patients.

DMPFC is a code region in emotional regulation, and specifically in attentional modulation of emotion processing [12, 15]. Several studies have found that subjects showed higher activation in DMPFC, when they directed their attention to the emotional content of stimuli compared to implicit tasks [16, 17]. And, previous clinical studies reported MDD patients showed abnormal attentional modulation of emotion processing. The study of Bermpohl and colleagues showed that health participants increased activation in the DMPFC during the expected emotional vs. unexpected emotional picture perception, but this effect was absent among MDD patients [11]. Zhang and colleagues found that MDD patients had greater activation in DMPFC than HCs during positive vs. neutral emotion expectancy condition [10]. Altered activations in DMPFC of MDD patients for picture perception and expectancy were observed in the current study, which is consistent with the previous studies. And, we firstly observed that altered activation on salient effect in MDD patients.

Our study showed, during the contrast of Xhs > Xls, specific effect in the DMPFC in health subjects, and a lower effect of salience on DMPFC activation during anticipatory preparation in depressed subjects compared to HCs. DMPFC is one core region of the anterior default mode network (DMN), identified to be associated with attention proceeding deficit of emotional pictures perception in depressed patients [10, 11]. Previous study found insufficient deactivation of anterior DMN in MDD patients occurred in response to positive emotion anticipation, and decreased decoupling from anterior towards posterior DMN during positive vs. neutral anticipation in MDD patients [10]. So, from the previous and current study, MDD patients alters anterior DMN response to anticipation: for high salient one, MDD patients could not reach to normal activation in anterior DMN; for emotional one, MDD patients could not deactivated in anterior DMN.

In our study, we found depressed patients showed increased activation in DMPFC as a function of high salient vs. low salient expected pictures. In the previous study by Bermpohl et al. (2009) [11], they found that, during, HC increased the left DMPFC and PCC, compared to MDD patients. What is more, they found there was insufficient deactivation of the left DMPFC for positive picture conditions in MDD patients, and in the remitted patients, the activation of the left DMPFC went back to the level is similar as the HC. So the patients have abnormal attentional modulation of positive emotion. Furthermore, DMPFC is one important region of the default mode network, which has been related to rumination, autobiographical memory and self-referential processing [18–20]. And, Grimm et al. showed that MDD is characterized by differential deactivation in the anterior DMN [21].

This effect was negatively correlated with HAMD scores and could be shown to affect the subsequent picture condition. Differential effects for high- versus low-salient picture viewing were found in DMPFC in depressed and healthy subjects, this effect was significantly higher for depressed subjects and correlated positively with HAMD scores. Comparing the anticipation-effect on subsequent picture viewing between high and low-salient pictures, differential effects were higher for depressed subjects during high-salient picture viewing and lower during low-salient picture viewing.

These observations should however be seen within some relevant limitations. We have only 18 subjects each group which is the small sample, so it may be the reason that there were the trend, but there were no significant results. In our future study, we will enlarge the sample size.

Therefore, our study exceeds previous work by Bernpohl et al. and Zhang et al. [10, 11], relating differential effects during picture viewing to alterations during preparatory anticipation. Therefore, we could shed light on DMPFC functioning in depressive disorder which is recently highly discussed in the literatures [22, 23] by separating cognitive and consumatory effects in this region.

Acknowledgment. This work was funded by the Qihang Foundation of Shanghai Mental Health Center (2016-QH-03), Shanghai Jiao Tong University Foundation (YG2016MS37) and the International Communication Foundation Science and Technology Commission of Shanghai Municipality (16410722500) awarded to Dr. Bin Zhang. This work was also funded by Science and Technology Commission of Shanghai Municipality (14411961400) awarded to Dr. Jijun Wang.

References

1. Belmaker, R.H., Agam, G.: Major depressive disorder. New Engl. J. Med. **358**(1), 55–68 (2008). doi:10.1056/NEJMra073096
2. Disner, S.G., Beevers, C.G., Haigh, E.A., Beck, A.T.: Neural mechanisms of the cognitive model of depression. Nat. Rev. Neurosci. **12**(8), 467–477 (2011). doi:10.1038/nrn3027

3. Gotlib, I.H., Kasch, K.L., Traill, S., Joormann, J., Arnow, B.A., Johnson, S.L.: Coherence and specificity of information-processing biases in depression and social phobia. J. Abnorm. Psychol. **113**(3), 386–398 (2004). doi:10.1037/0021-843X.113.3.386

4. Kellough, J.L., Beevers, C.G., Ellis, A.J., Wells, T.T.: Time course of selective attention in clinically depressed young adults: an eye tracking study. Behav. Res. Ther. **46**(11), 1238–1243 (2008). doi:10.1016/j.brat.2008.07.004

5. Shafritz, K.M., Collins, S.H., Blumberg, H.P.: The interaction of emotional and cognitive neural systems in emotionally guided response inhibition. NeuroImage **31**(1), 468–475 (2006). doi:10.1016/j.neuroimage.2005.11.053

6. MacLeod, C., Rutherford, E., Campbell, L., Ebsworthy, G., Holker, L.: Selective attention and emotional vulnerability: assessing the causal basis of their association through the experimental manipulation of attentional bias. J. Abnorm. Psychol. **111**(1), 107–123 (2002)

7. Zhang, B., Li, M., Qin, W., Demenescu, L.R., Metzger, C.D., Bogerts, B., Yu, C., Walter, M.: Altered functional connectivity density in major depressive disorder at rest. Eur. Arch. Psychiatry Clin. Neurosci. **266**(3), 239–248 (2016). doi:10.1007/s00406-015-0614-0

8. Cullen, K.R., Westlund, M.K., Klimes-Dougan, B., Mueller, B.A., Houri, A., Eberly, L.E., Lim, K.O.: Abnormal amygdala resting-state functional connectivity in adolescent depression. JAMA Psychiatry **71**(10), 1138–1147 (2014). doi:10.1001/jamapsychiatry.2014.1087

9. Drevets, W.C., Thase, M.E., Moses-Kolko, E.L., Price, J., Frank, E., Kupfer, D.J., Mathis, C.: Serotonin-1A receptor imaging in recurrent depression: replication and literature review. Nucl. Med. Biol. **34**(7), 865–877 (2007). doi:10.1016/j.nucmedbio.2007.06.008

10. Zhang, B., Li, S., Zhuo, C., Li, M., Safron, A., Genz, A., Qin, W., Yu, C., Walter, M.: Altered task-specific deactivation in the default mode network depends on valence in patients with major depressive disorder. J. Affect. Disord. **207**, 377–383 (2017). doi:10.1016/j.jad.2016.08.042

11. Bermpohl, F., Walter, M., Sajonz, B., Lucke, C., Hagele, C., Sterzer, P., Adli, M., Heinz, A., Northoff, G.: Attentional modulation of emotional stimulus processing in patients with major depression–alterations in prefrontal cortical regions. Neurosci. Lett. **463**(2), 108–113 (2009). doi:10.1016/j.neulet.2009.07.061

12. Bermpohl, F., Pascual-Leone, A., Amedi, A., Merabet, L.B., Fregni, F., Gaab, N., Alsop, D., Schlaug, G., Northoff, G.: Attentional modulation of emotional stimulus processing: an fMRI study using emotional expectancy. Hum. Brain Mapp. **27**(8), 662–677 (2006). doi:10.1002/hbm.20209

13. Cloninger, C.R., Svrakic, D.M., Przybeck, T.R.: Can personality assessment predict future depression? A twelve-month follow-up of 631 subjects. J. Affect. Disord. **92**(1), 35–44 (2006). doi:10.1016/j.jad.2005.12.034

14. Anderson, B.A.: A value-driven mechanism of attentional selection. J. Vis. **13**(3) (2013). doi:10.1167/13.3.7

15. Northoff, G., Heinzel, A., Bermpohl, F., Niese, R., Pfennig, A., Pascual-Leone, A., Schlaug, G.: Reciprocal modulation and attenuation in the prefrontal cortex: an fMRI study on emotional-cognitive interaction. Hum. Brain Mapp. **21**(3), 202–212 (2004). doi:10.1002/hbm.20002

16. Fichtenholtz, H.M., Dean, H.L., Dillon, D.G., Yamasaki, H., McCarthy, G., LaBar, K.S.: Emotion-attention network interactions during a visual oddball task. Brain Res. Cogn. Brain Res. **20**(1), 67–80 (2004). doi:10.1016/j.cogbrainres.2004.01.006

17. Winston, J.S., O'Doherty, J., Dolan, R.J.: Common and distinct neural responses during direct and incidental processing of multiple facial emotions. NeuroImage **20**(1), 84–97 (2003)

18. Zhu, X., Wang, X., Xiao, J., Liao, J., Zhong, M., Wang, W., Yao, S.: Evidence of a dissociation pattern in resting-state default mode network connectivity in first-episode, treatment-naive major depression patients. Biol. Psychiat. **71**(7), 611–617 (2012). doi:10. 1016/j.biopsych.2011.10.035

19. Whitfield-Gabrieli, S., Ford, J.M.: Default mode network activity and connectivity in psychopathology. Annu. Rev. Clin. Psychol. **8**, 49–76 (2012). doi:10.1146/annurev-clinpsy-032511-143049

20. Raichle, M.E., MacLeod, A.M., Snyder, A.Z., Powers, W.J., Gusnard, D.A., Shulman, G.L.: A default mode of brain function. Proc. Natl. Acad. Sci. U.S.A. **98**(2), 676–682 (2001). doi:10.1073/pnas.98.2.676

21. Grimm, S., Ernst, J., Boesiger, P., Schuepbach, D., Boeker, H., Northoff, G.: Reduced negative BOLD responses in the default-mode network and increased self-focus in depression. World J. Biol. Psychiatry **12**(8), 627–637 (2011). doi:10.3109/15622975.2010. 545145. The official journal of the World Federation of Societies of Biological Psychiatry

22. Satterthwaite, T.D., Cook, P.A., Bruce, S.E., Conway, C., Mikkelsen, E., Satchell, E., Vandekar, S.N., Durbin, T., Shinohara, R.T., Sheline, Y.I.: Dimensional depression severity in women with major depression and post-traumatic stress disorder correlates with fronto-amygdalar hypoconnectivty. Mol. Psychiatry **21**(7), 894–902 (2016). doi:10.1038/ mp.2015.149

23. Sheline, Y.I., Price, J.L., Yan, Z., Mintun, M.A.: Resting-state functional MRI in depression unmasks increased connectivity between networks via the dorsal nexus. Proc. Natl. Acad. Sci. U.S.A. **107**(24), 11020–11025 (2010). doi:10.1073/pnas.1000446107

Abnormal Brain Activity in ADHD: A Study of Resting-State fMRI

Chao Tang[1], Yuqing Wei[1], Jiajia Zhao[1], Xin Zhang[2],
and Jingxin Nie[1(✉)]

[1] Center for Studies of Psychological Application, Institute of Cognitive
Neuroscience, South China Normal University, Guangzhou 510631, China
niejingxin@gmail.com
[2] School of Electronics and Information Engineering,
South China University of Technology, Guangzhou 510000, China

Abstract. The prevalence rate of ADHD varies from age to age. To better understand the development of ADHD from childhood to adolescence, different age groups of ADHD from large dataset are needed to explore the development pattern of brain activities. In this study, amplitude of low frequency fluctuation (ALFF), fractional amplitude of low frequency fluctuation (fALFF) and regional homogeneity (ReHo) were extracted from resting-state functional magnetic resonance imaging (rs-fMRI) of both ADHD subjects and typical developing (TD) subjects from 7 to 16 years old. The result showed that the different areas mainly appear at the bilateral superior frontal cortex, anterior cingulate cortex (ACC), precentral gyrus, right superior occipital lobe, cerebellum and parts of basal ganglia between all ADHD subjects and all TD subjects. Besides, compared with TD, there were different brain activity patterns at different ages in ADHD, which appear at the left ACC and left occipital lobe. The result can inspire more studies on comparisons between functional connectivity methods.

Keywords: Rs-fMRI · ADHD · Brain development

1 Introduction

Attention deficit hyperactivity disorder (ADHD) is generally considered to be a neurodevelopmental disorder with high incidence in childhood [1, 2]. It is mainly characterized by lack of attention, excessive activity (restless in adult), or difficulty controlling behavior which is not appropriate for a person's age [3]. It was estimated that the prevalence of ADHD in pre-school children (3–6 years old) in Europe is 1.8–1.9% [4]. Although there is no global consensus, meta-regression analyses have estimated the worldwide ADHD prevalence at between 5.29% and 7.1% in children and adolescents, and at 3.4% in adults [5–7]. In addition, about 30–50% of people diagnosed with ADHD in childhood continue to have symptoms into adulthood and about 2–5% of adults also have the symptoms [8]. Since the incidence of ADHD varies between ages, in order to understand the development of ADHD from childhood to adolescence, it is necessary to explore the difference patterns of regional brain activities between ADHD subjects and typical developing (TD) subjects from different age groups.

© Springer International Publishing AG 2017
Y. Zeng et al. (Eds.): BI 2017, LNAI 10654, pp. 181–189, 2017.
https://doi.org/10.1007/978-3-319-70772-3_17

Functional magnetic resonance imaging (fMRI) has been widely used to measure brain activities in-vivo. Compared to the task-related fMRI, resting-state fMRI (rs-fMRI) does not require subject to perform any task, which greatly simplifies the fMRI procedure for patients with difficulty to accomplish certain tasks. Brain activity could be characterized by different measurements, such as amplitude of low frequency fluctuation (ALFF), fractional amplitude of low frequency fluctuation (fALFF) and regional homogeneity (ReHo). Since Zang et al. [9] applied ALFF to probe the abnormal spontaneous neuronal activities of ADHD patients, ALFF has been widely used in the studies of various mental diseases, such as schizophrenia [10, 11], autism spectrum disorder [12], attention deficit hyperactivity disorder [13]. However, some researchers found that ALFF of ADHD patients increased abnormally in some brain areas but the energy consumption of these regions did not increase correspondingly, which was likely to be caused by noise. Therefore, Zou et al. [14] proposed fALFF to reduce the abnormal value in ALFF. At the same time, Zang et al. [15, 16] firstly proposed the regional homogeneity approach and explored the functional abnormalities of Parkinson's patients using ReHo. Subsequently, many studies have validated the feasibility of ReHo in the analysis of fMRI data from multiple aspects [17, 18].

The different brain activities between ADHD and TD have been identified in previous studies. For example, comparing the value of ALFF between 17 ADHD boys (7:51 ± 1.96 years old) and 17 matched controls (9.73 ± 1.57 years old), Yang et al. [13] found that ADHD showed higher ALFF in the left superior frontal gyrus and sensorimotor cortex (SMC) as well as lower ALFF in the bilateral anterior, middle cingulate and the right middle frontal gyrus (MFG). Using ALFF and ReHo on a smaller sample, in contrast to 12 controls (12.5 ± 14.1 years old), the 12 ADHD (11 ± 14.8 years old) patients exhibited significant resting-state brain activities in the bilateral VI/VII (BA 17/18/19), left SI (BA 3), left AII (BA 22), bilateral thalamus, left dorsal brainstem and midbrain [19]. For 29 boys with ADHD (11.00 ± 16.50 years old) and 27 matched controls (11.25 ± 14.92 years old), Cao et al. [20] indicated that ReHo of ADHD patients decreased in the frontal–striatal–cerebellar circuits, but increased in the occipital cortex. However, study of ADHD with small sample size is difficult to cover the brain activity patterns of ADHD which vary with age.

In this study, a large rs-fMRI dataset with 266 ADHD subjects and 719 TD subjects from 7 to 16 years old were adopted. ALFF, fALFF and ReHo of each subject were calculated and compared to study the abnormal brain activity of ADHD.

2 Method

2.1 Dataset

Data were acquired from the database–1000 Functional Connectomes Project (1000-FCP) [21]. It is a neuroimaging database that collects resting-state fMRI data from multiple sites. For the TD participants in our study, inclusion criteria included: age from 7 to 16 years old, with no mental disease, image at least cover 95% of brain. Especially, the ADHD data are acquired from the ADHD-200 dataset, which is a sub set of 1000-FCP. It contains resting-state fMRI and anatomical MRI images aggregated across 8 independent imaging sites, which are obtained from children and adolescents with ADHD (ages: 7–21 years old). Finally we got 266 ADHD subjects and 719 TD subjects with age from 7 to 16 years. To further explore the developmental changes of regional brain activities in ADHD, We divided the TD participants and the ADHD participants into two groups respectively, one for childhood from 7 to 11 years old (TD: 407; ADHD: 169) and the other for adolescence from 12 to 16 years of old (TD: 312; ADHD: 97).

2.2   Image Processing

Resting state fMRI data were preprocessed with DPARSF [22]. The first ten time point was removed to avoid magnetization instability. All images were corrected for slice timing to minimize the difference during image acquisition and realigned to the middle volume to avoid excessive head motion. Then, these images were spatially normalized to a standard template (Montreal Neurological Institute) and resampled to 3 mm × 3 mm × 3 mm voxel resolution. Spatial smoothing was performed with a Gaussian kernel of 4 mm full-width at half-maximum (FWHM) to improve the SNR (signal-to-noise ratio). Besides, the mean signal of white matter and cerebrospinal fluid were removed as covariates. After that, linear trend removal as well as band-pass filtering (0.01–0.1 Hz) were also performed. The brain is divided into 90 ROIs by AAL (automated anatomical labeling) atlas [23] in order to further localize the local variation. Finally, ALFF, fALFF and ReHo of each voxel across participants were be calculated with REST [24].

3   Statistics Analysis

All statistics analysis were performed with SPM12 [25].The brain activity measurements (ALFF, fALFF and ReHo) between ADHD and TD are compared by two-sample t-test on each voxel, taking a significant threshold of $P < 0.01$, with age as covariate, and corrected for multiple comparisons with false discovery rates (FDR). Voxels with $P < 0.01$ and cluster size > 270 mm^3 were regarded to show a significant group difference. We also performed two-sample t-test on ALFF, fALFF and ReHo for childhood and adolescence groups with the method above.

4   Result

4.1   The Comparison of ALFF, fALFF and ReHo Between Two Groups

Compared with TD, ADHD showed significant divergence of ALFF, fALFF and ReHo in extensive regions. The main group differences between ADHD and TD on three measures are shown on Table 1 and Fig. 1.

4.2   The Comparison of ALFF, fALFF and ReHo of Two Age Groups

As we can see, there were some subtle but important differences of the three measurements between two groups in childhood and adolescence. The main age groups differences between ADHD and TD on three measures are shown on Table 2 and Fig. 2.

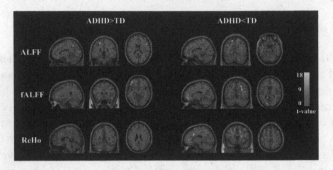

Fig. 1. The bright areas mean that the major differences between all ADHD subjects and all TD subjects. (ADHD: Attention deficit hyperactivity disorder group, TD: Typically developing group)

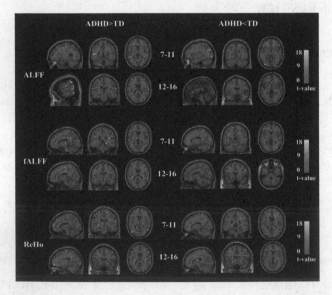

Fig. 2. The bright areas mean that the major differences between ADHD and TD in different age groups. (ADHD: Attention deficit hyperactivity disorder group, TD: Typically developing group)

Table 1. Regions showing significant differences in ALFF, fALFF and ReHo between the ADHD and TD

Regions	p-value	Regions	p-value
ALFF: ADHD > TD		**ALFF: ADHD < TD**	
Left caudate nucleus	0.01	Right superior occipital lobe	<0.01
Left medial superior frontal cortex	0.002	Left superior temporal gyrus	<0.001
Bilateral postcentral gyrus	0.002	Bilateral precuneus	<0.001
Right pallidum	<0.001	Bilateral cerebellum	0.001
Bilateral thalamus	0.004		
Bilateral putamen	0.004		
Vermis	0.001		
fALFF: ADHD > TD		**fALFF: ADHD < TD**	
Bilateral superior frontal cortex	<0.001	Right anterior cingulate cortex	0.002
Right supplementary motor area	0.014	Bilateral superior parietal lobe	0.001
Bilateral pallidum	0.009	Bilateral superior occipital lobe	0.002
Bilateral putamen	0.001	Right precuneus	0.007
Bilateral thalamus	<0.001	Bilateral cerebellum	0.001
Right caudate nucleus	0.011		
Vermis	0.001		
ReHo: ADHD > TD		**ReHo: ADHD < TD**	
Bilateral cerebellum	0.002	Right anterior cingulate cortex	<0.001
Bilateral precentral gyrus	0.02	Bilateral inferior occipital lobe	<0.001
Bilateral supplementary motor area	<0.001	Bilateral superior frontal cortex	0.002
Right postcentral gyrus	<0.001	Bilateral cerebellum	0.003
Left thalamus	0.001		
Right caudate nucleus	0.001		
Right inferior parietal lobe	0.002		

Table 2. Regions showing significant differences in ALFF, fALFF and ReHo between the ADHD and TD in two age groups

Regions	p-value	Regions	p-value
ALFF: ADHD > TD (childhood)		**ALFF: ADHD < TD (childhood)**	
Left insula	<0.001	Bilateral superior parietal lobe	0.004
Right superior frontal cortex	0.001	Bilateral inferior occipital lobe	0.001
Bilateral precentral gyrus	<0.001	Bilateral cerebellum	0.001
Left thalamus	<0.001		
Left precuneus	<0.001		
Left anterior cingulate cortex	<0.001		
ALFF: ADHD > TD (adolescence)		**ALFF: ADHD < TD (adolescence)**	
Right postcentral gyrus	0.001	left angular gyrus	0.002
Right caudate nucleus	0.001	Right medial superior frontal cortex	0.001
Left cerebellum	0.005	Left occipital lobe	0.001
Right inferior frontal cortex	0.001	Bilateral cerebellum	<0.001
fALFF: ADHD > TD (childhood)		**fALFF: ADHD < TD (childhood)**	
Bilateral precuneus	0.005	Right anterior cingulate cortex	0.002
Left supplementary motor area	0.007	Bilateral superior parietal lobe	0.001

(*continued*)

Table 2. (*continued*)

Regions	p-value	Regions	p-value
Right paracentral lobule	0.002	Bilateral superior occipital lobe	0.002
Right postcentral gyrus	0.013	Right precuneus	0.007
Bilateral thalamus	0.006	right cerebellum	0.005
Left medial superior frontal cortex	0.002		
fALFF: ADHD > TD (adolescence)		**fALFF: ADHD < TD (adolescence)**	
Bilateral postcentral gyrus	0.002	Right cerebellum	<0.001
Right supplementary motor area	0.007	Right middle frontal cortex	0.013
Right middle cingulate cortex	0.005		
Left insula	0.008		
Left middle orbital frontal cortex	0.004		
ReHo: ADHD > TD (childhood)		**ReHo: ADHD < TD (childhood)**	
Left putamen	0.008	bilateral superior parietal lobe	<0.001
Left superior medial frontal cortex	0.008	Left inferior occipital lobe	0.003
Left caudate nucleus	<0.001	Right superior frontal cortex	<0.001
Right postcentral gyrus	<0.001	Bilateral cerebellum	<0.001
left anterior cingulate cortex	<0.001		
ReHo: ADHD > TD (adolescence)		**ReHo: ADHD < TD (adolescence)**	
Bilateral pallidum	<0.001	Right precuneus	0.009
Right superior frontal cortex	0.006	Right posterior cingulate cortex	0.002
Left putamen	<0.001	Left thalamus	0.009
Left paracentral lobule	<0.001	Right middle frontal cortex	0.003
Bilateral insula	<0.001		
Right supplementary motor area	<0.001		
Verimis	0.001		

5   Discussion

In this study, three measurements were used to describe the voxel-based local changes in brain activities from different aspects. Firstly, ALFF reflects the level of spontaneous activity of single voxel according to the level of oxygen content, which can directly observe the changes of regional brain activities [9]. Similar to ALFF, fALFF is the ALFF of a given frequency band expressed as a fraction of the sum of amplitudes across the entire frequency range in a given signal, which represents the relative contribution of specific low frequency oscillations to the whole detectable frequency range [26]. Finally, ReHo depicts the coherence of neural activity of a specific brain region with its neighboring or adjacent brain regions [27]. Using the three measurements, we found, after removing the confounding factor age, all ADHD subjects showed abnormal regional activities in the movement pathway and cognitive control circuits, which was in line with previous studies [28, 29]. For all of three measurements, on the one hands, ADHD showed stronger activation than TD in the parts of basal ganglia (caudate nucleus, putamen and pallidum), supplementary motor area, precentral gyrus, cerebellum and thalamus which was related to dysfunction of movement control and execution functions [30, 31]. On the other hands, ADHD exhibited the lower activation than TD in superior occipital lobe in three measurements, which may be associated with attention lapse [32]. Besides, there were its unique activation areas for each index which was mainly concentrated in brain areas involved in advanced cognitive control. For example, the decreased ALFF, fALFF and ReHo appeared in the left superior temporal gyrus, right anterior cingulate cortex and superior frontal cortex respectively.

From the developmental point of view, compared to TD group, the ALFF of left anterior cingulate cortex (ACC) in ADHD group was significantly increased in childhood, which may be connected to compensation mechanism of inattention in ADHD [10]. But in adolescence, there was no significant difference. The result may demonstrate that the degree of ADHD symptoms has improved with age, which could be associated with drug treatment and other factors [33]. As for fALFF, the difference between the two age groups was not very obvious, which was caused by the insensitivity of fALFF to age. Finally, in childhood, the ReHo value of ACC in ADHD was higher than TD, but the differences of ReHo between ADHD and TD disappeared in adolescence. This result suggests that the complex cognitive ability of patients with ADHD has improved with age, specially, in the aspects of regulating stimulus selection and response selection in the attention process [13]. In contrast, the decreased ReHo of ADHD appeared at left inferior occipital lobe in childhood and disappears in adolescence, which could be related to pay attention to multiple irrelevant visual stimuli from the environment simultaneously. Thus, the abnormal activities of ACC and left inferior occipital lobe may jointly result in inattention problems with ADHD [9, 34].

In conclusion, our study shows the abnormal brain activities of ADHD in rs-fMRI which are different from TD using three measurements (ALFF, fALFF and ReHo) from the perspective of development. Moreover, it facilitates our understanding for the mechanism of ADHD and examines the effectiveness of these three measurements. However, there are some limitations in this study. First, the gender factor is not considered, which may affect the result. Second, the division standard of age groups is arbitrary to a certain extent and we still need to explore more suitable criterion. Finally, the source of data is acquired from different institutes, which causes some bias of parameter setting that may influence the further analysis.

Acknowledgements. This paper was supported by NFSC (National Natural Science Foundation of China) (Grant No. 61403148).

References

1. Sroubek, A., Kelly, M., Li, X.: Inattentiveness in attention-deficit/hyperactivity disorder. Neurosci. Bull. **29**(1), 103–110 (2013). doi:10.1007/s12264-012-1295-6
2. Clauss-Ehlers, C.S. (ed.): Encyclopedia of Cross-Cultural School Psychology. Springer, Boston (2010). doi:10.1007/978-0-387-71799-9
3. American Psychiatric Association: Diagnostic and Statistical Manual of Mental Disorders, 5th edn, pp. 59–65. American Psychiatric Publishing, Arlington (2013)
4. Wichstrøm, L., Berg-Nielsen, T.S., Angold, A., Egger, H.L., Solheim, E., Sveen, T.H.: Prevalence of psychiatric disorders in preschoolers. J. Child Psychol. Psychiatry **53**(6), 695–705 (2012). doi:10.1111/j.1469-7610.2011.02514.x

5. Polanczyk, G., de Lima, M.S., Horta, B.L., Biederman, J., Rohde, L.A.: The worldwide prevalence of ADHD: a systematic review and metaregression analysis. Am. J. Psychiatry **164**(6), 942 (2007). doi:10.1176/ajp.2007.164.6.942

6. Fayyad, J., De, G.R., Kessler, R., Alonso, J., Angermeyer, M., Demyttenaere, K., et al.: Cross-national prevalence and correlates of adult attention-deficit hyperactivity disorder. Br. J. Psychiatry **190**(5), 402–409 (2007). doi:10.1192/bjp.bp.106.034389

7. Willcutt, E.G.: The prevalence of DSM-IV attention-deficit/hyperactivity disorder: a meta-analytic review. Neurother. J. Am. Soc. Exp. Neurother. **9**(3), 490–499 (2012). doi:10.1007/s13311-012-0135-8

8. Ginsberg, Y., Quintero, J., Anand, E., Casillas, M., Upadhyaya, H.P.: Underdiagnosis of attention-deficit/hyperactivity disorder in adult patients: a review of the literature. Primary Care Companion J. Clin. Psychiatry **16**(3), 470–472 (2014). doi:10.1007/s13311-012-0135-8

9. Zang, Y.F., He, Y., Zhu, C.Z., Cao, Q.J., Sui, M.Q., Liang, M., et al.: Altered baseline brain activity in children with ADHD revealed by resting-state functional MRI. Brain Dev. **29**(2), 83–91 (2007). doi:10.1016/j.braindev.2006.07.002

10. Welsh, R.C., Chen, A.C., Taylor, S.F.: Low-frequency BOLD fluctuations demonstrate altered thalamocortical connectivity in schizophrenia. Schizophr. Bull. **36**(4), 713 (2010). doi:10.1093/schbul/sbn145

11. Yu, R., Chien, Y.L., Wang, H.L., Liu, C.M., Liu, C.C., Hwang, T.J., et al.: Frequency-specific alternations in the amplitude of low-frequency fluctuations in schizophrenia. Hum. Brain Mapp. **35**(2), 627–637 (2014). doi:10.1002/hbm.22203

12. Itahashi, T., Yamada, T., Watanabe, H., Nakamura, M., Ohta, H., Kanai, C., et al.: Alterations of local spontaneous brain activity and connectivity in adults with high-functioning autism spectrum disorder. Mol. Autism **6**(1), 30 (2015). doi:10.1186/s13229-015-0026-z

13. Yang, H., Wu, Q.Z., Guo, L.T., Li, Q.Q., Long, X.Y., Huang, X.Q., et al.: Abnormal spontaneous brain activity in medication-naive ADHD children: a resting state fMRI study. Neurosci. Lett. **502**(2), 89–93 (2011). doi:10.1016/j.neulet.2011.07.028

14. Zou, Q.H., Zhu, C.Z., Yang, Y., Zuo, X.N., Long, X.Y., Cao, Q.J., et al.: An improved approach to detection of amplitude of low-frequency fluctuation (ALFF) for resting-state fMRI: fractional ALFF. J. Neurosci. Methods **172**(1), 137–141 (2008). doi:10.1016/j.jneumeth.2008.04.012

15. Zang, Y., Jiang, T., Lu, Y., He, Y., Tian, L.: Regional homogeneity approach to fMRI data analysis. Neuroimage **22**(1), 394–400 (2004). doi:10.1016/j.neuroimage.2003.12.030

16. Wu, T., Long, X., Zang, Y., Wang, L., Hallett, M., Li, K., et al.: Regional homogeneity changes in patients with Parkinson's disease. Hum. Brain Mapp. **30**(5), 1502 (2009). doi:10.1002/hbm.20622

17. Jiang, L., Xu, T., He, Y., Hou, X.H., Wang, J., Cao, X.Y., et al.: Toward neurobiological characterization of functional homogeneity in the human cortex: regional variation, morphological association and functional covariance network organization. Brain Struct. Funct. **220**(5), 2485–2507 (2015). doi:10.1007/s00429-014-0795-8

18. Liu, Y., Wang, K., Yu, C., He, Y., Zhou, Y., Liang, M., et al.: Regional homogeneity, functional connectivity and imaging markers of Alzheimer's disease: a review of resting-state fMRI studies. Neuropsychologia **46**(6), 1648–1656 (2008). doi:10.1016/j.neuropsychologia.2008.01.027

19. Tian, L., Jiang, T., Liang, M., Zang, Y., He, Y., Sui, M., et al.: Enhanced resting-state brain activities in ADHD patients: a fMRI study. Brain Develop. **30**(5), 342–348 (2008). doi:10.1016/j.braindev.2007.10.005

20. Cao, Q., Zang, Y., Sun, L., Sui, M., Long, X., Zou, Q., et al.: Abnormal neural activity in children with attention deficit hyperactivity disorder: a resting-state functional magnetic resonance imaging study. NeuroReport **17**(17), 1033–1036 (2006). doi:10.1097/01.wnr. 0000224769.92454.5d

21. 1000 Functional Connectomes Project. http://fcon_1000.projects.nitrc.org/

22. Data Processing Assistant for Resting-State fMRI. http://www.restfmri.net/forum/DPARSF

23. Tzouriomazoyer, N., Landeau, B., Papathanassiou, D., Crivello, F., Etard, O., Delcroix, N., et al.: Automated anatomical labeling of activations in SPM using a macroscopic anatomical parcellation of the MNI MRI single-subject brain. Neuroimage **15**(1), 273 (2002). doi:10. 1006/nimg.2001.0978

24. Resting-State fMRI Data Analysis Toolkit. http://www.restfmri.net/forum/REST

25. Statistical Parametric Mapping. http://www.fil.ion.ucl.ac.uk/spm/

26. Wei, C., Ji, X., Jie, Z., Feng, J.: Individual classification of ADHD patients by integrating multiscale neuroimaging markers and advanced pattern recognition techniques. Front. Syst. Neurosci. **6**, 58 (2012). doi:10.3389/fnsys.2012.00058

27. An, L., Cao, Q.J., Sui, M.Q., Sun, L., Zou, Q.H., Zang, Y.F., et al.: Local synchronization and amplitude of the fluctuation of spontaneous brain activity in attention-deficit/hyperactivity disorder: a resting-state fMRI study. Neurosci. Bull. **29**(5), 603–613 (2013). doi:10.1007/ s12264-013-1353-8

28. Posner, J., Park, C., Wang, Z.: Connecting the dots: a review of resting connectivity MRI studies in attention-deficit/hyperactivity disorder. Neuropsychol. Rev. **24**(1), 3–15 (2014). doi:10.1007/s11065-014-9251-z

29. Tibbetts, P.E.: Cognitive neuroscience: the biology of the mind. Q. Rev. Biol. (2009). doi:10.1086/603482

30. Conn, P.J., Battaglia, G., Marino, M.J., Nicoletti, F.: Metabotropic glutamate receptors in the basal ganglia motor circuit. Nat. Rev. Neurosci. **6**(10), 787–798 (2005). doi:10.1038/ nrn1763

31. Devinsky, O., Morrell, M.J., Vogt, B.A.: Contributions of anterior cingulate cortex to behaviour. Brain **118**(1), 279 (1995). doi:10.1093/brain/118.1.279

32. Posner, M.I., Petersen, S.E.: The attention system of the human brain. Annu. Rev. Neurosci. **13**(1), 25 (1990). doi:10.1146/annurev.ne.13.030190.000325

33. Shang, C.Y., Yan, C.G., Lin, H.Y., Tseng, W.Y., Castellanos, F.X., Gau, S.S.: Differential effects of methylphenidate and atomoxetine on intrinsic brain activity in children with attention deficit hyperactivity disorder. Psychol. Med. **46**(15), 3173 (2016). doi:10.1017/ S0033291716001938

34. Cubillo, A., Halari, R., Ecker, C., Giampietro, V., Taylor, E., Rubia, K.: Reduced activation and inter-regional functional connectivity of fronto-striatal networks in adults with childhood attention-deficit hyperactivity disorder (ADHD) and persisting symptoms during tasks of motor inhibition and cognitive switching. J. Psychiatr. Res. **44**(10), 629–639 (2010). doi:10. 1016/j.jpsychires.2009.11.016

Wearable EEG-Based Real-Time System for Depression Monitoring

Shengjie Zhao, Qinglin Zhao, Xiaowei Zhang$^{(\boxtimes)}$, Hong Peng, Zhijun Yao,
Jian Shen, Yuan Yao, Hua Jiang, and Bin Hu

School of Information Science and Engineering,
Lanzhou University, Lanzhou, China
{zhaoshj15,qlzhao,zhangxw,pengh,yaozj,
yaoy2015,hjiang15,bh}@lzu.edu.cn

Abstract. It has been reported that depression can be detected by electrophysiological signals. However, few studies investigate how to daily
monitor patient's electrophysiological signals through a more convenient
way for a doctor, especially on the monitoring of electroencephalogram
(EEG) signals for depression diagnosis. Since a person's mental state and
physiological state are changing over time, the most insured diagnosis of
depression requires doctors to collect and analyze subject's EEG signals every day until two weeks for the clinical practice. In this work, we
designed a real-time depression monitoring system to capture the user's
EEG data by a wearable device and to perform real-time signal filtering,
artifacts removal and power spectrum visualization, which could be combined with psychological test scales as an auxiliary diagnosis. In addition
to collecting the resting EEG signals for real-time analysis or diagnosis
of depression, we also introduced an external audio stimulus paradigm
to further make a detection of depression. Through the machine learning
method, system can give a credible probability of depression under each
stimulus as a user's self-rating score from continuous EEG data. EEG
signals collected from 81 early-onset patients and 89 normal controls are
used to build the final classification model and to verify the practical
performance.

Keywords: Depression monitoring · Wearable device · Real-time signal
processing · Auxiliary diagnosis

1 Introduction

Depression is a disorder of the representation and regulation of mood and emotion [1], which is more than a low mood. People with depression may experience lack of interest in daily activities, poor concentration, low energy, feeling
of worthlessness, and at its worst, could lead to suicide [2]. The exact cause of
depression is not known. Many researchers believe it is caused by chemical imbal-
ances in the brain, which may be hereditary or caused by events in a person's
life.

© Springer International Publishing AG 2017
Y. Zeng et al. (Eds.): BI 2017, LNAI 10654, pp. 190–201, 2017.
https://doi.org/10.1007/978-3-319-70772-3_18

The clinical detection and diagnosis of depression by doctors is mainly based on questionnaire or interview [3] and there are no objective evaluation criteria of depression in existing clinical practice. However, experienced doctors are lacking and at the same time the diagnosis often takes long time, which limit the diagnosis of depression more or less. Based upon this information, we aim to develop pervasive supporting technology and service. Trends like "self-diagnosis" already show the potential of collect personal sensor data for health improvement, which can be found in many mobile phone apps [4] and wearable device monitoring your sleep quality, blood pressure or heart rate [5]. It is easy to collect personal electrophysiological data. Challenge is how to make sense of these data. Among the human physiological signals, electroencephalogram (EEG) signals, which reflect the working status of human brain, can be considered as the most commonly used biopotential and have been widely studied for their clinical importance, especially in depression [6]. Although EEG recording will not always change the patient's planned treatment, it has been reported that it will help to speed up the diagnosis and the patient's access to the appropriate treatment [7], that is, to present such an auxiliary diagnosis to detect depression indirectly. In order to collect this complex and nonstationary random signal, multiple electrodes are placed on the scalp, usually using the electrode cap as a collection device. When the electrodes are placed, it is required to inject conductive gel or saline to reduce the resistance. However, traditional electrodes have the disadvantage of uncomfortable and it is challenging to maintain high quality EEG signals because conductive gel or saline has the defect of evaporation [8], which makes the contact resistance between electrode and scalp a significant increase, resulting in signal distortion.

This study focuses on how to acquire EEG signals in a comfortable and convenient manner and to ensure the reliability of the EEG data(raw EEG signals, denoised EEG signals and features), thus simplifying and benefiting the detection of depression. Recently, wearable and wireless Brain Computer Interface (BCI) has been used in the field of gaming control [9], drowsiness detection [10] and health monitoring [11,12]. Similarly, we can achieve the first step in simplification by using a wearable and wireless BCI device. The following steps of simplification depends on the characteristics of depression itself. Depression has its own unique essence that abnormalities in activation of prefrontal regions have been reported more frequently than for any other brain region, mostly in the direction of decreased bilateral or predominantly left-sided activation [13,14]. G Rajkowska [15] found morphometric evidence for neuronal and glial prefrontal cell pathology in major depression. In [16], both major depression disorder and post-traumatic stress disorder displayed more left than right-frontal activity. Obviously, forehead is an ideal location for attaching electrodes so we can expect to use the frontal asymmetry of EEG for depression detection.

Except for the above, another advantage of forehead is that it is not covered in hair, which reduce the preparation time for EEG placement. Furthermore, in the aspect of feature engineering, non-hairy regions of the forehead can be used to extract rich information that are associated with many cognitive abilities and dysfunctions. Therefore, Fp1, Fp2 and Fpz signals are chosen as raw EEG signals, which are subsequently used to extract depression related EEG features to construct a sample space. Real-time measures are also used to upgrade our system from depression detection to depression monitoring, which is the foundation of the system with convenience and intuitive. It will show how unobtrusive and easily available our real-time system can be used in home or other personal space privately, to monitor depression.

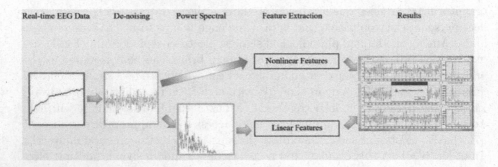

Fig. 1. The flow chart of our method

2   Related Work

Recently, EEG-based depression detection using machine learning techniques has been well studied. Some features are proved have significantly related to depression. In [17], highest classification accuracy of 83.3% is obtained by correlation dimension among other nonlinear features. In [18], signal entropy: approximate entropy, sample entropy, renyi entropy and bi-spectral phase entropy are extracted for depression detection. Both linear and nonlinear features of EEG are applied in these studies.

Of course, these methods are based on collected EEG signals. However, monitoring the state of the user's brain functioning while collecting EEG can complete online tasks and visualize features. Recently, real-time epileptic seizure detection [19], real-time concentration level detection [20] and neurofeedback technology [21] have been proposed. On the other hand, power spectrum is an important visual cue for doctor that most of the patients with depression have a strong power spectrum in low frequency band, so system requires real-time similarly. Our work embed the real-time system in depression detection, which is used as a method of auxiliary diagnosis or assisted rehabilitation and also facilitate the diagnosis of depression and long-term monitoring of the patient's condition during the onset. In addition, classification models can be constructed using different paradigms(resting EEG, audio/vedio stimulus), which can be combined with the data being collected to present the real-time probability of depression in an intuitive way. Furthermore, the real-time extracted features can also be served as the input of neurofeedback which give user the visual/audio feedback to show her/his mental state.

3 Methodology

The process of depression monitoring system consists of several steps: paradigm design, signals collection, data processing and result analysis. According to the standard flow of EEG signal acquisition, the paradigm of our experiment is mainly composed of two parts: resting EEG test and audio stimulation EEG test. As the flow chart shown in Fig. 1, during the testing process, EEG signals are recorded accordingly, and then artifacts that contaminate EEG signals are removed in real time. After that, relevant features are extracted and EEG signals are analyzed.

4 Making Sense of the Raw Data

4.1 Hardware

EEG signals are transmitted to the computer through a small equipment called three channel wearable EEG-collector [22]. Its sample rate is 250 Hz. Figure 2 presents the main instructions of hardware device. The hardware part of the wearing device will first encapsulate the data into frames and send it to the communication thread in the listening state, which is mainly used to resolve data via a specific protocol, and then remove the header, frame tail and tag bits, save valid information and send it to the computing thread for signal processing. Taking the acceptability of the delay, integrity of the information and computing costs into account, our system processes two seconds of data (500 sample points) every time.

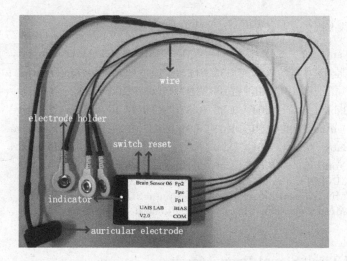

Fig. 2. Instructions of three channel wearable EEG-collector

4.2   Resting EEG

Before applying stimulus, participants need to do some preparations. After wearing EEG recording sensor, they were told to try to minimize facial movements and eyeball movements during 90 s resting EEG recording. The main intention of resting EEG test are not only available to doctors for real-time monitoring, but also provides them with daily, non-hospitalized and private EEG signals tracking through online diagnosis.

4.3   Stimulus

External stimulus test is a common way to induce emotional susceptibility to subjects, often using pictures, film clips, music or voice for emotional classification or scores for several specific emotional dimensions. It is an good idea that we can use the same approach to give predictions of depression risk. For pictures or movies, the test results are susceptible to electromyography (EMG) signals such as blink, eyeball movements and so on. In order to reduce the impact of EMG signals, it is better to use the method of audio stimulation to induce the prefrontal lobe activity for depression monitoring.

Audio stimulation can be roughly divided into three categories according to its property: negative, neutral, positive. Experiment in this paper selected two positive, two negative and two neutral totally six audio from IADS-2 [23], each of which lasts 6 s (Table1). After sorting the 6-segment audio in sequence and added an extra 6 s resting test at each interval, the experiment will eventually record 72 s of EEG data, and then the data can constitute 12 independent samples in accordance with 6-second time division.

Table 1. Audio stimulation profile

Name	PMN	PSD	AMN	ASD	DMN	DSD	Property
Cattle	5.01	1.85	6.04	1.85	4.56	1.75	Neutral
Painting	4.96	1.68	5.37	1.68	5.06	1.82	Neutral
Babies Cry	2.04	1.39	6.87	1.39	3.46	2.31	Negative
Dentist Drill	2.89	1.67	6.91	1.67	2.92	2.03	Negative
Baby	7.61	2.10	6.03	2.10	6.14	1.98	Positive
Croud	7.65	1.58	7.12	1.58	6.09	2.18	Positive

PMN: Pleasure Mean   PSD:Pleasure Standard Deviation
AMN:Arousal Mean   ASD:Arousal Standard Deviation
DMN:Dimensions Mean   DSD:Standard Deviation

4.4   Real-Time Signal Preprocessing

The raw EEG signals are noisy and difficult to analyze. Before signals can be visualized in our system, it had to be preprocessed. The noise present in the

EEG signals could be denoised using Finite Impulse Response filter. Unlike other methods that preprocess the entire EEG, it requires that the signals should be processed in time slices to achieve real-time purposes. As the above-mentioned basis for every two seconds to deal with, data points that need to be preprocessed is 500. In this system, there are four steps for EEG signal preprocessing.

Firstly, a mid-filter [24] is used to eliminate EEG signal drifting. After signals entry, we first perform mid-filtering on the raw EEG signals to remove the baseline drift. Since the band-pass filter frequency used is 1 to 40 Hz, the low frequency component in the band will still maintain a high level. The removal of baseline drift can effectively control the signal amplitude within a normal range, from which reflect the real EEG signal.

Secondly, EEG signal was filtered using a 1–40 Hz FIR filter to notch out power line noise at 50 and 60 Hz [25]. Since the length of FIR filter is 724, each calculated result will have an additional 224 points, which will be superimposed with the next 2 s signal.

Thirdly, through the welch method, the EEG signal is transferred from the time domain to the frequency domain. Our system can perform real-time frequency domain analysis.

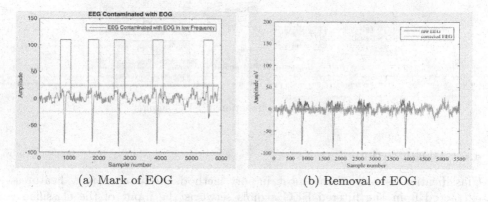

(a) Mark of EOG   (b) Removal of EOG

Fig. 3. Kalman filtered real EEG signal

Forthly, using the wavelet transform and Kalman filter [26] to eliminate the EOG signal. At present the mainstream removal of artifacts method is Independent Component Analysis [27], but for real-time system should adopt an approximate method. When the original EEG signal is convoluted by the wavelet operator, it can mark the blink interval, and then the signal of the interval will use the Kalman filter to fit the real EEG signal. As mentioned above, each audio stimulus lasts 6 s, which requires us to conduct a real-time removal of artifacts according to this time slice. The marked EOG area(left) and Kalman filtered real EEG signal(right) are shown in Fig. 3.

Of course, the influence of EMG cannot be completely disentangled, we added a data removal mechanism. Once a time slice of data is too much noise, it will be deleted. Since we are using multiple time slices to give a probability of depression (See in Sect. 5), the loss of some sampling points can be ignored. The worst case is that the subject will be asked to re-experiment if he has a more than 33% low quality EEG signals.

4.5   Feature Extraction

The EEG-based system for depression monitoring has integrated a number of algorithms for feature extraction. The applied algorithms can support online signal feature extraction in EEG. Corresponding to the filtered EEG signal, the system mainly selected two types of features as evaluations for classification model: linear and nonlinear features(Fig. 4). For linear features, we utilize the features of lower computational complexity due to real-time purposes. Max frequency, mean frequency and center frequency are calculated from power spectral. The results of feature selection in nonlinear features show that C0 complexity [28], permutation entropy [29] and Lempel-Ziv complexity (LZC) [30] are related to depression.

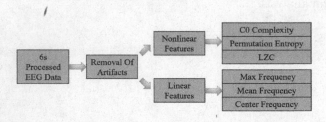

Fig. 4. Feature extraction of EEG signals

4.6   Classification

Classification is one of the most import methods in data mining. Features extracted from the filtered EEG signals serve as the input of the classifiers. We consider the probability of depression from the perspective of the frequency domain and the time domain in the EEG signals, so we remove influence of age and gender to increase the universality of our system, even with these two features the accuracy can achieve higher.

4.7   Visualization

As presented in Fig. 5, to visualize the monitoring results and better reflect the real-time advantage, we designed a visual interface, respectively, corresponding to Fp1, Fp2 and Fpz, which can clearly observe the denoised continuous EEG and power spectral every two seconds. For patients with depression, their power spectral of resting EEG will be more likely to focus on the low frequency band, which can be used as a reference for the doctors and researchers. Another important use of visualization is to determine the start time of the experiment. Due to emotional instability, frequent blink or eye movement, causing the EEG signal to be too noisy, such test results are inaccurate. It requires us to stabilize the amplitude of the EEG signal in a certain range (50 μV or less), and then start the experiment.

The above-mentioned algorithms of feature extraction, artifacts removal and classification require data to be fed in a bunch of 3000 samples at a time for one channel. Therefore we use the queue to buffer the data from wear device to the algorithms. The queue is refreshed once the buffer that receives the EEG signals is accumulated to 6 s (3000 sample points). In this way, through different machine learning algorithms, the system can display the recognition results in real time, which can be presented in multimedia such as audio, picture or video. This enhances interaction with the users and has good scalability, especially in the rehabilitation of mental illness. The source code of the system is mainly implemented by C++.

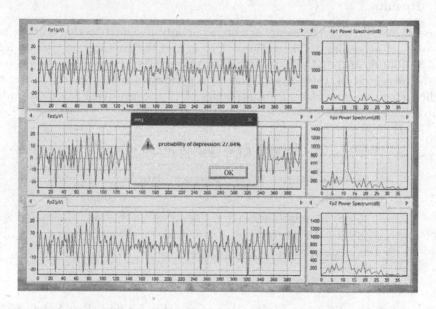

Fig. 5. Graphical interfaces and real-time classification results output

5 Experiment

To distinguish the user's real-time mental state, we introduced a probability of depression. In the ideal case (less blink and EMG), the experiment will eventually get a 12-segment sample (Including data recorded between two stimuli). Since participant received different types of audio stimulation (from negative to positive), the final results need to build 12 different models for classification, thus classification results can vote on the probability of depression. Based on the presented related work we expect that nearest neighbors, trees and SVM perform well.

5.1   Participants

In order to make the classification model can better respond to diagnosis of early-onset depression, we ruled out the subjects who had eaten psychiatric drugs and had suicidal behavior from psychology instruments. In addition, we also made psychometrical scale investigation of the subjects. In this study, MINI [31], PHQ-9 [32] which have got reliable results are adopted to further filtrate the subjects. Only patients with early-onset depression whose scale score is greater than the threshold can be chosen as a subject. Finally, a total of 170 people who met the condition agreed and signed informed consent before participants in the experiment. 81 patients with early-onset depression are recruited from Beijing ANDing hospital affiliated to capital University of medical sciences. 89 age and educational background matching subjects were selected as control group.

5.2   Results

To avoid over-fitting, 170 samples were randomly assigned to the training set and test set according to the ratio of 4 to 1, and the training set uses 10-fold cross validation to build the classification model. Classification results on test set are presented in Table 2. In General, a performance of about 77% can be reached that averages classification results of 12 consecutive EEG data under audio stimulation.

Table 2. Classification results

Average accuracy of the 12-segment data	
Classifier	Accuracy
Local classification(KNN + Naive Bayes)	78.40%
SVM (RBF Kernel)	77.80%
Xgboost(Gbtree + Logistic Regression)	75.80%

Regarding the different classification approaches we find that: Xgboost [33] or other classification trees seem work better when dealing with discrete features. Local classification [34] can reach a higher accuracy but only the category labels can be output. SVM [35] is better for its probability output mode, which can give the risk of depression that users could make comparison more accurately.

The system will eventually give a self-rating score for the risk of depression(Fig. 5), which full score is 100%. We tested our system on normal people, high-risk groups (like surgeon doctor) and patients with early-onset depression to verify its practical performance. Their self-rating scores show an increasing trend.

6   Conclusions and Future Work

The diagnosis of depression has always been a difficult problem. In this research, we propose to use the wearable equipment for real-time depression monitoring as an auxiliary diagnosis. Clean EEG signals and the power spectrum are displayed through real-time denoising and time-frequency conversion. Considering the extensibility of the system, we also introduced an audio stimulation test to assess the risk of depression. In our experiment, different classification models are constructed by pre-collected training data under different audio stimulations. And then we verify the reliability of the model and the stability of the system.

The disadvantage of the article is that the system does not design a good experimental paradigm. But the discovery mentioned above helps the experimental paradigm for the choice of audio stimulation, our future work is to verify whether the choice of high frequency stimulation will be more effective on detection of depression.

In general, this paper indicates that our system can be adopted as a health monitoring means for doctors and users in depression evaluation. Especially in psychological screening of schools, companies or hospitals. Data visualization and self-rating scores are helpful for the diagnosis of depression. Users only need some simple guidances can be self-test at home and the data can be automatically saved and uploaded to the server, which can facilitate doctors to track their health condition.

Acknowledgments. This work was supported by the National Basic Research Program of China (973 Program) (No. 2014CB744600) and the National Natural Science Foundation of China (grant No. 61210010, No. 61632014, No. 61402211).

References

1. Davidson, R.J., Pizzagalli, D., Nitschke, J.B., Putnam, K.: Depression: perspectives from affective neuroscience. Annu. Rev. Psychol. **53**(1), 545–574 (2002)
2. Kaplan, H.I., Sadock, B.J.: Comprehensive Textbook of Psychiatry. Williams & Wilkins, Baltimore (1980)
3. Bjelland, I., Dahl, A.A., Haug, T.T., Neckelmann, D.: The validity of the hospital anxiety and depression scale. an updated literature review. J. Psychosom. Res. **52**(2), 69–77 (2002)
4. Lupton, D., Jutel, A., Kawachi, I., Subramanian, S.V.: 'It's like having a physician in your pocket!' a critical analysis of self-diagnosis smartphone apps. Soc. Sci. Med. **133**, 128–135 (2015)
5. Isais, R., Nguyen, K., Perez, G., Rubio, R.: A low-cost microcontroller-based wireless ECG-blood pressure telemonitor for home care. In: Proceedings of the International Conference of the IEEE Engineering in Medicine and Biology Society, vol. 4, pp. 3157–3160 (2003)
6. Thibodeau, R., Jorgensen, R.S., Kim, S.: Depression, anxiety, and resting frontal EEG asymmetry: a meta-analytic review. J. Abnorm. Psychol. **115**(4), 715–729 (2006)

7. Ziai, W.C., Dan, S., Llinas, R., Venkatesha, S., Truesdale, M., Schevchenko, A., Kaplan, P.W.: Emergent EEG in the emergency department in patients with altered mental states. Clin. Neurophysiol. **123**(5), 910–917 (2012). Official Journal of the International Federation of Clinical Neurophysiology

8. Ferree, T.C., Luu, P., Russell, G.S., Tucker, D.M.: Scalp electrode impedance, infection risk, and EEG data quality. Clin. Neurophysiol. **112**(3), 536–44 (2001). Official Journal of the International Federation of Clinical Neurophysiology

9. Liao, L.D., Chen, C.Y., Wang, I., Chen, S.F., Li, S.Y., Chen, B.W., Chang, J.Y., Lin, C.T.: Gaming control using a wearable and wireless EEG-based brain-computer interface device with novel dry foam-based sensors. J. Neuroeng. Rehabil. **9**(1), 1–12 (2012)

10. Lin, C.T., Chang, C.J., Lin, B.S., Hung, S.H.: A real-time wireless brain computer interface system for drowsiness detection. IEEE Trans. Biomed. Circ. Syst. **4**(4), 214–222 (2010)

11. Pantelopoulos, A., Bourbakis, N.G.: A survey on wearable sensor-based systems for health monitoring and prognosis. IEEE Trans. Syst. Man Cybern. Part C Appl. Rev. **40**(1), 1–12 (2010)

12. Lin, C.T., Chuang, C.H., Cao, Z., Singh, A., Huang, C.S., Yu, Y.H., Nascimben, M., Liu, Y.T., King, J.T., Su, T.P.: Forehead EEG in support of future feasible personal healthcare solutions: Sleep management, headache prevention, and depression treatment. IEEE Access **PP**(99), 1 (2017)

13. Davidson, R.J., Abercrombie, H., Nitschke, J.B., Putnam, K.: Regional brain function, emotion and disorders of emotion. Curr. Opin. Neurobiol. **9**(2), 228–234 (1999)

14. George, M.S., Ketter, T.A., Post, R.M.: Prefrontal cortex dysfunction in clinical depression. Depression **2**(2), 59–72 (2010)

15. Rajkowska, G., Miguelhidalgo, J.J.: Gliogenesis and glial pathology in depression. CNS Neurol. Disord. Drug Targets **6**(3), 219 (2007)

16. Kemp, A.H., Griffiths, K., Felmingham, K.L., Shankman, S.A., Drinkenburg, W., Arns, M., Clark, C.R., Bryant, R.A.: Disorder specificity despite comorbidity: resting eeg alpha asymmetry in major depressive disorder and post-traumatic stress disorder. Biol. Psychol. **85**(2), 350–354 (2010)

17. Hosseinifard, B., Moradi, M.H., Rostami, R.: Classifying depression patients and normal subjects using machine learning techniques and nonlinear features from EEG signal. Comput. Methods Programs Biomed. **109**(3), 339 (2013)

18. Bairy, G.M., Bhat, S., Eugene, L.W.J., Niranjan, U.C., Puthankattil, S.D., Joseph, P.K.: Automated classification of depression electroencephalographic signals using discrete cosine transform and nonlinear dynamics. J. Med. Imaging Health Inform. **5**(3), 635–640 (2015)

19. Zandi, A.S., Javidan, M., Dumont, G.A., Tafreshi, R.: Automated real-time epileptic seizure detection in scalp eeg recordings using an algorithm based on wavelet packet transform. IEEE Trans. Biomed. Eng. **57**(7), 1639–51 (2010)

20. Sourina, O., Wang, Q., Liu, Y., Nguyen, M.K.: A real-time fractal-based brain state recognition from EEG and its applications. In: Biosignals 2011 - Proceedings of the International Conference on Bio-Inspired Systems and Signal Processing, Rome, Italy, 26–29 January, pp. 82–90 (2011)

21. Zotev, V., Phillips, R., Han, Y., Misaki, M., Bodurka, J.: Self-regulation of human brain activity using simultaneous real-time fMRI and EEG neurofeedback. Neuroimage **85**(2), 985–995 (2014)

22. Hu, B., Majoe, D., Ratcliffe, M., Qi, Y.: EEG-based cognitive interfaces for ubiquitous applications: developments and challenges. IEEE Intell. Syst. **26**(5), 46–53 (2011)
23. Bradley, M.M., Lang, P.J.: The international affective digitized sounds affective ratings of sounds and instruction manual. University of Florida (2007)
24. Pitas, I.: Nonlinear digital filters: principles and applications. Kluwer International (1990)
25. Wickert, M.: Modern digital signal processing. IEEE Trans. Educ. (2003). Special Issue on Circuits & Systems
26. Chen, Y., Zhao, Q., Hu, B., Li, J., Jiang, H., Lin, W., Li, Y., Zhou, S., Peng, H.: A method of removing ocular artifacts from EEG using discrete wavelet transform and Kalman filtering. In: IEEE International Conference on Bioinformatics and Biomedicine, pp. 1485–1492 (2016)
27. Jung, T.P., Humphries, C., Lee, T.W., Makeig, S., Mckeown, M.J., Iragui, V., Sejnowski, T.J.: Extended ICA removes artifacts from electroencephalographic recordings **10**, 894–900 (1998)
28. Fang, C., Gu, F., Xu, J., Liu, Z., Ren, L.: A new measurement of complexity for studying EEG mutual information. In: Proceedings of the International Conference on Neural Information Processing, Iconip'r98, Kitakyushu, Japan, 21–23 October 1998, pp. 435–437 (1998)
29. Bandt, C., Pompe, B.: Permutation entropy: a natural complexity measure for time series. Phys. Rev. Lett. **88**(17), 174102 (2002)
30. Lempel, A., Ziv, J.: On the complexity of finite sequences. IEEE Trans. Inf. Theory **22**(1), 75–81 (1976)
31. Bech, P., Kastrup, M., Rafaelsen, O.J.: Mini-compendium of rating scales for states of anxiety, depression, mania, schizophrenia with corresponding DSM-III syndromes. Acta Psychiatrica Scandinavica Supplementum **326**, 1–37 (1985)
32. Kroenke, K., Spitzer, R.L.: The PHQ-9: a new depression diagnostic and severity measure. Psychiatr. Ann. **32**(9), 509–521 (2002)
33. Chen, T., Guestrin, C.: XGBoost: a scalable tree boosting system, pp. 785–794 (2016)
34. Mao, C., Hu, B., Wang, M., Moore, P.: Learning from neighborhood for classification with local distribution characteristics. In: International Joint Conference on Neural Networks, pp. 1–8 (2015)
35. Cortes, C., Vapnik, V.: Support-vector networks. Mach. Learn. **20**(3), 273–297 (1995)

Group Guided Sparse Group Lasso Multi-task Learning for Cognitive Performance Prediction of Alzheimer's Disease

Xiaoli Liu[1,2], Peng Cao[1,2(✉)], Jinzhu yang[1,2], Dazhe Zhao[2], and Osmar Zaiane[3]

[1] College of Computer Science and Engineering, Northeastern University, Shenyang, China
`caopeng@mail.neu.edu.cn`
[2] Key Laboratory of Medical Image Computing of Ministry of Education, Northeastern University, Shenyang, China
[3] Computing Science, University of Alberta, Edmonton, AB, Canada

Abstract. Alzheimer's disease (AD), the most common form of dementia, causes progressive impairment of cognitive functions of patients. There is thus an urgent need to (1) accurately predict the cognitive performance of the disease, and (2) identify potential MRI (Magnetic Resonance Imaging)-related biomarkers most predictive of the estimation of cognitive outcomes. The main objective of this work is to build a multi-task learning based on MRI in the presence of structure in the features. In this paper, we simultaneously exploit the interrelated structures within the MRI features and among the tasks and present a novel Group guided Sparse group lasso (GSGL) regularized multi-task learning approach, to effectively incorporate both the relatedness among multiple cognitive score prediction tasks and useful inherent group structure in features. An Alternating Direction Method of Multipliers (ADMM) based optimization is developed to efficiently solve the non-smooth formulation. We demonstrate the performance of the proposed method using the Alzheimer's Disease Neuroimaging Initiative (ADNI) datasets and show that our proposed methods achieve not only clearly improved prediction performance for cognitive measurements, but also finds a compact set of highly suggestive biomarkers relevant to AD.

Keywords: Alzheimer's disease · Regression · Sparse learning · Multi-task learning

1 Introduction

Alzheimer's disease (AD) is a gradually progressive syndrome that mainly affects memory function, ultimately culminating in a dementia state. It has been proved that brain atrophy detected by MRI is correlated with neuropsychological deficits. Many clinical/cognitive measures have been designed to evaluate the cognitive status of the patients and used as important criteria for clinical

© Springer International Publishing AG 2017
Y. Zeng et al. (Eds.): BI 2017, LNAI 10654, pp. 202–212, 2017.
https://doi.org/10.1007/978-3-319-70772-3_19

diagnosis of probable AD. Many cognitive measures including Mini Mental State Examination (MMSE) and Alzheimer's Disease Assessment Scale cognitive sub-scale (ADAS-Cog) have been designed to evaluate the cognitive status of the patients and used as important criteria for clinical diagnosis of probable AD. It is known that there exist inherent correlations among multiple clinical vari-ables of a subject, and a joint analysis of data from multiple cognitive tasks is expected to improve the performance [1–3]. The assumption of the commonly used Multi-task learning (MTL) is that all tasks share the same data represen-tation with $\ell_{2,1}$ regularization, since a given imaging marker can affect multiple cognitive scores and only a subset of the imaging features (brain region) are relevant. This assumption of $\ell_{2,1}$ regularization is restrictive since it encourages all the tasks to share the same data representation. Sparse group Lasso (SGL) [4] allows the simultaneous selection of a common set of biomarkers for all the tasks and the selection of a specific set of biomarkers for different tasks. However, these methods ignore the useful structure information among the MRI features. Since brain structures tend to work together to achieve a certain function, brain imaging measures are often correlated with each other. The prior group informa-tion should be incorporated into the MTL models and guide the learning process (Fig. 1).

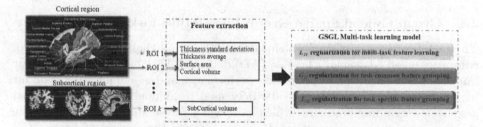

Fig. 1. Flow chart of the proposed GSGL-MTL method.

Many previous works extract only the volume or thickness measures of cor-tical regions of interest (ROIs) as the features [3,5]. To avoid manual measure bias caused by the single feature in this study, multiple features are extracted to measure the atrophy information of each ROI involving cortical thickness, sur-face area and volume from gray matter and white matter. The multiple shape measures from the same region provide a comprehensively quantitative evalu-ation of cortical atrophy, and tend to be selected together as joint predictors. It is hypothesized that not only a subset of MRI features, but also a subset of ROIs are relevant to each assessment. Therefore, we use this prior knowl-edge of interrelated structure to group relevant shape features together in the same region to guide the learning process. Based on this intuitive motivation, we simultaneously exploit the interrelated structures within features as well as among the tasks, and present a novel multi-task learning method to effectively incorporate both the relatedness among multiple cognitive score prediction tasks and useful inherent group structure in features. Inspired by the recent success of the group lasso regularization [6] as well as the term "bi-level analysis" [7], we propose a unified "bi-level" learning framework to jointly perform both individual feature-level and ROI-level analysis by group lasso regularization with the "grouping" effect such that it helps reduce the variances in the estimation of coefficient and improves the stability of biomarkers selection. Specially, we develop a novel multi-task learning formulation based on a group guided SGL. The regularizer consists of three components including an $\ell_{2,1}$ penalty, which ensures that a small subset of features will be selected for the regression models, and a $G_{2,1}$ penalty, which encourages the task-common ROI across multi-task. To relax the restrictive assumption of shared ROI imposed in the correlation among the cognitive tasks, a task-specific ROI based $\ell_{2,1}$-norm for each task is incorporated. The proposed formulation is challenging to solve due to the use of multiple non-smooth penalties. We present an Alternating Direction Method of Multipliers (ADMM)-type algorithm for solving the proposed non-smooth optimization problems efficiently. We conducted extensive experiments using data from the ADNI dataset to demonstrate our methods along various dimensions including prediction performance and biomarkers identification.

2 Proposed Method

2.1 Group Guided Sparse Group Lasso Multi-task Learning

The high feature-dimension problem is one of the major challenges in the study of computer aided Alzheimer's Disease (AD) diagnosis. Variable selection is of great importance to improve the prediction performance and model interpretation for high-dimensional data. Lasso is a widely used technique for high-dimensional association mapping problems, which can yield a sparse and easily interpretable solution via an ℓ_1 regularization. However, despite the success of Lasso, it is limited to considering each task separately and ignores the inherent structure of features. However, Lasso fails to capture the correlation information among the pairwise of group features. The pairwise correlations among group of features are very high, Lasso tends to select only one of the pairwise correlated features, resulting in ignoring the group effect.

Group regularizers like group lasso [6] via an $\ell_{2,1}$ regularization assumes covarying variables in groups, and have been extensively studied in the multi-task feature learning. The key assumption behind the group lasso regularizer is that if a few features in a group are important, then most of the features in the same group should also be important.

Multi-Task Learning (MTL) is a statistical learning framework which seeks at learning several models in a joint manner. It has been commonly used to obtain better generalization performance than learning each task individually [8,9]. The critical issues in MTL is to identify how the tasks are related and build learning models to capture such task relatedness. Consider a multi-task learning (MTL) setting with k tasks. Let $X \in \mathbb{R}^{n \times p}$ denote the matrix of covariates, $Y \in \mathbb{R}^{n \times k}$ be the matrix of responses with each row corresponding to a sample, and $\Theta \in \mathbb{R}^{p \times k}$

denote the parameter matrix, with column $\theta_{.h} \in \mathbb{R}^p$ corresponding to task h, $h = 1, \ldots, k$, and row $\theta_{i.} \in \mathbb{R}^k$ corresponding to feature i, $i = 1, \ldots, p$. The MRI measure features in the same brain region belong to a group. We assume the p features to be divided into q disjoint groups $\mathcal{G}_l, l = 1, \ldots, q$, with each group having m_l features respectively. The MTL problem can be set-up as one of estimating the parameters based on suitable regularized loss function:

$$\min_{\Theta} \quad L(Y, X, \Theta) + \lambda R(\Theta) \, , \tag{1}$$

where $L(\cdot)$ denotes the loss function and $R(\cdot)$ is the regularizer. In the current context, we assume the loss to be square loss, i.e.,

$$L(Y, X, \Theta) = \|Y - X\Theta\|_F^2 = \sum_{i=1}^{n} \|\mathbf{y}_i - \mathbf{x}_i\Theta\|_2^2 \, , \tag{2}$$

where $\mathbf{y}_i \in \mathbb{R}^{1 \times k}, \mathbf{x}_i \in \mathbb{R}^{1 \times p}$ are the i-th rows of Y, X, respectively corresponding to the multi-task response and covariates for the i-th sample. We note that the MTL framework can be easily extended to other loss functions. Base on some prior knowledge, we then add penalty $R(\Theta)$ to encode the relatedness among tasks.

Group Lasso regularized multi-task learning (GL-MTL) aims to obtain better generalization performance by exploiting the shared features among different tasks [9]. In our case, given that one imaging marker can affect multiple cognitive scores, the coefficients of the coefficient matrix of the same row is largely correlated. It has been successfully applied to capture biomarkers having affects across most or all responses in the application of AD prediction [10,11]. The GL-MTL model via the $\ell_{2,1}$-norm regularization considers

$$R(\Theta) = \|\Theta\|_{2,1} = \sum_{i=1}^{p} \|\theta_{i.}\|_2 \, , \tag{3}$$

and is suitable for simultaneously enforcing sparsity over features for all tasks.

The key point of Eq. (3) is the use of ℓ_2-norm for $\theta_{i.}$, which forces the weights corresponding to the i-th feature across multiple tasks to be grouped together and tends to select features based on the strength of k tasks jointly. There is a correlation in multiple cognitive measures, and the associated imaging predictors usually have more or less effect on all of these scores, which leads to a correlation between regression coefficients. By employing GL-MTL, the correlation information among different tasks can be incorporated into the model to build a more appropriate predictive model and identify a subset of the features.

One appealing property of the group lasso regularization in GL-MTL is that it encourages multiple predictors from related tasks to share a subset of features. However, the $\ell_{2,1}$-norm regularization only consider the shared representation from the features, neglecting the potentially grouping information among multiple neuroimaging measures. In order to address it, we consider prior information group information in features and multi-task learning simultaneously in one single framework. Specifically, We propose a Group guided Sparse Group Lasso regularized multi-task learning (GSGL-MTL) algorithm exploiting both the group structure of features and the multi-task correlation, to unify feature-level and ROI-level analysis in an unified multi-task learning framework. The GSGL-MTL formulation focuses on the following regularized loss function:

$$\min_{\Theta \in \mathbb{R}^{p \times k}} \frac{1}{2} \|Y - X\Theta\|_F^2 + \lambda_1 \|\Theta\|_{2,1} + \lambda_2 \|\Theta\|_{G_{2,1}} + \lambda_3 \|\text{vec}(\Theta)\|_{2,1} \ . \qquad (4)$$

where $\|\Theta\|_{G_{2,1}} = \sum_{l=1}^{q} w_l \sqrt{\sum_{j \in \mathcal{G}_l} \|\theta_{j\cdot}\|_2}$, $\|\text{vec}(\Theta)\|_{2,1} = \sum_{h=1}^{k} \sum_{l=1}^{q} w_l \|\theta_{\mathcal{G}_l h}\|_2$, and $w_l = \sqrt{m_l}$ is the weight for each group. The second and third norms are called Group guided Sparse Group Lasso norm (GSGL), where $\|\Theta\|_{G_{2,1}}$ encourages the task-common ROIs to induce the same group sparsity patterns across different tasks (coupling all tasks) and $\|\text{vec}(\Theta)\|_{2,1}$ encourages the task-specific ROIs to induce the different group sparsity patterns across different tasks (decoupled for each task), as illustrated in Fig. 2. Although we only consider the least squares loss function for regression here, the above formulation can be easily generalized to other convex loss functions for classification, such as hinge loss or logistic function.

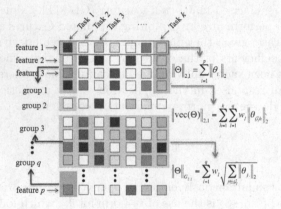

Fig. 2. The illustration of the GSGL-MTL method

2.2 Optimization

In this section, we present a novel solver for the problem in Eq. (1) based on the ADMM. The proposed formulation is, however, challenging to solve due to the use of three non-smooth penalties. It is easy to show that the objective function of the GSGL-MTL method is convex. To efficiently handle the two non-smooth constraints, we propose an optimization method which employs ADMM algorithm [12] to solve the proposed multi-task learning problem by decomposing a large global problem into a series of smaller local subproblems and coordinates the local solutions to identify the globally optimal solution [12].

Assume $R^{\lambda_1}_{\lambda_2,\lambda_3}(\Theta) = \lambda_1\|\Theta\|_{2,1} + \lambda_2\|\Theta\|_{G_{2,1}} + \lambda_3\|\text{vec}(\Theta)\|_{2,1}$, then Eq. (4) is equivalent to the following constrained optimization problem:

$$\min_{\Theta \in \mathbb{R}^{p \times k}} \frac{1}{2}\|Y - X\Theta\|_F^2 + R^{\lambda_1}_{\lambda_2,\lambda_3}(Q) \quad \text{subject to } \Theta - Q = 0 \,. \tag{5}$$

where Q is slack variables. Then Eq. (5) can be solved by ADMM. The augmented Lagrangian is $L_\rho(\Theta, Q, U) = \frac{1}{2}\|Y - X\Theta\|_F^2 + R^{\lambda_1}_{\lambda_2,\lambda_3}(Q) + \text{Tr}(U^T(\Theta - Q)) + \frac{\rho}{2}\|\Theta - Q\|^2$, where U is augmented Lagrangian multiplier.

Update Θ^{t+1}: In the $(t+1)$-th iteration, Θ^{t+1} can be updated by minimizing L_ρ with Q, U fixed: $\Theta^{t+1} = \text{argmin}_\Theta \frac{1}{2}\|Y - X\Theta\|_F^2 + \text{Tr}((U^t)^T(\Theta - Q^t)) + \frac{\rho}{2}\|\Theta - Q^t\|^2$. The optimization problem is quadratic. The optimal solution is given by $\Theta^{t+1} = F^{-1}B^t$, where $F = X^TX + \rho I$ and $B^t = X^TY - U^t + \rho Q^t$.

Update Q: The update for Q effectively needs to solve the following problem: $Q^{t+1} = \text{argmin}_Q \frac{\rho}{2}\|Q - \Theta^{t+1}\|^2 + R^{\lambda_1}_{\lambda_2,\lambda_3}(Q) - \text{Tr}((U^t)^TQ)$, which is equivalent to computing the proximal operator for $R^{\lambda_1}_{\lambda_2,\lambda_3}(\cdot)$. In particular, we need to solve

$$\Psi^{\lambda_1/\rho}_{\lambda_2/\rho,\lambda_3/\rho}(O^{t+1}) = \text{argmin}_Q \left\{ R^{\lambda_1/\rho}_{\lambda_2/\rho,\lambda_3/\rho}(Q) + \frac{1}{2}\|Q - O^{t+1}\|^2 \right\}, \tag{6}$$

where $O^{t+1} = \Theta^{t+1} + \frac{1}{\rho}U^t$.

The goal is to be able to compute $Q^{t+1} = \Psi^{\lambda_1/\rho}_{\lambda_2/\rho,\lambda_3/\rho}(O^{t+1})$ efficiently. It can be shown [13] that the proximal operator for the composite regularizer can be computed efficiently in three steps, and all of these steps can be executed efficiently using suitable extensions of soft-thresholding.

$$\Pi^{t+1} = \Psi^{\lambda_1/\rho}_{0,0}(O^{t+1}) = \text{argmin}_\Pi \left\{ \frac{\lambda_1}{\rho}\|\Pi\|_{2,1} + \frac{1}{2}\|\Pi - O^{t+1}\| \right\} \tag{7a}$$

$$\Gamma^{t+1} = \Psi^0_{\lambda_2/\rho,0}(\Pi^{t+1}) = \Psi^{\lambda_1/\rho}_{\lambda_2/\rho,0}(O^{t+1})$$

$$= \text{argmin}_\Gamma \left\{ \frac{\lambda_2}{\rho}\|\Gamma\|_{G_{2,1}} + \frac{1}{2}\|\Gamma - \Pi^{t+1}\| \right\} \tag{7b}$$

$$Q^{t+1} = \Psi^0_{0,\lambda_3/\rho}(\Gamma^{t+1}) = \Psi^0_{\lambda_2/\rho,\lambda_3/\rho}(\Pi^{t+1}) = \Psi^{\lambda_1/\rho}_{\lambda_2/\rho,\lambda_3/\rho}(O^{t+1})$$

$$= \text{argmin}_Q \left\{ \frac{\lambda_3}{\rho}\|\text{vec}(Q)\|_{2,1} + \frac{1}{2}\|Q - \Gamma^{t+1}\| \right\} \tag{7c}$$

The row-wise updates of (7a)–(7c) can be done by soft-thresholding as:

$$\pi_{i\cdot} = \frac{\max\left\{\|o_{i\cdot}\|_2 - \frac{\lambda_1}{\rho}, 0\right\}}{\|o_{i\cdot}\|_2}o_{i\cdot} \,, \tag{8a}$$

$$\gamma_{j\cdot} = \frac{\max\left\{\sqrt{\sum_{j \in \mathcal{G}_l}\|\pi_{j\cdot}\|_2} - \frac{\lambda_2 w_l}{\rho}, 0\right\}}{\sqrt{\sum_{j \in \mathcal{G}_l}\|\pi_{j\cdot}\|_2}}\pi_{j\cdot} \,, \tag{8b}$$

$$q_{\mathcal{G}_l h} = \frac{\max\left\{\|\gamma_{\mathcal{G}_l h}\|_2 - \frac{\lambda_3 w_l}{\rho}, 0\right\}}{\|\gamma_{\mathcal{G}_l h}\|_2}\gamma_{\mathcal{G}_l h} \,, \tag{8c}$$

where $\pi_{i\cdot}$, $o_{i\cdot}$, $\gamma_{j\cdot}$ are the i-th row of Π^{t+1}, O^{t+1}, Γ^{t+1}, $q_{\mathcal{G}_l h}$, $\gamma_{\mathcal{G}_l h}$ are rows in group $\mathcal{G}_l$ for task h of Q^{t+1} and Γ^{t+1}, respectively.

Dual Update for U: Following standard ADMM dual update, the update for the dual variable for our setting is as follows: $U^{t+1} = U^t + \rho(\Theta^{t+1} - Q^{t+1})$.

3   Experimental Results

3.1   Data and Experimental Setting

In this work, only ADNI subjects with no missing features or cognitive scores are included. This yields a total of $n = 816$ subjects, who are categorized into 3 baseline diagnostic groups: Cognitively Normal (CN, $n_1 = 228$), Mild Cognitive Impairment (MCI, $n_2 = 399$), and Alzheimer's Disease (AD, $n_3 = 189$). The dataset has been processed by a team from UCSF (University of California at San Francisco), who performed cortical reconstruction and volumetric segmentations with the FreeSurfer image analysis suite. There were $p = 319$ MRI features in total, including the cortical thickness average (TA), standard deviation of thickness (TS), surface area (SA), cortical volume (CV) and subcortical volume (SV) for a variety of ROIs. In order to sufficiently investigate the comparison, we further evaluate the performance on all the cognitive assessments (e.g. ADAS, MMSE and RAVLT, totally $k = 20$ tasks). To our best knowledge, no previous work uses all the cognitive scores for training and evaluation.

We use 10-fold cross valuation to evaluate our model and conduct the comparison. In each of twenty trials, a 5-fold nested cross validation procedure for all the comparable methods in our experiments is employed to tune the regularization parameters. Data was z-scored before applying regression methods. To have a fair comparison, we validate the regularization parameters of all the methods in the same search space (from 10^{-1} to 10^3) on a subset of the training set, and use the optimal parameters to train the final models. We evaluate all the algorithms in terms of both root mean squared error (rMSE), normalized mean squared error (nMSE) and the weighted R-value (wR) which are commonly used in multi-task learning problem.

3.2   The Results of Comparing with the Comparable Methods

In this section, we conduct empirical evaluation for the proposed methods by comparing with three single task learning methods: Ridge and Group Lasso, both of which are applied independently on each task. To verify the effect of individual components in our framework and show the contribution of individual components, we evaluate the three components of our approach: GL-MTL ($\lambda_2 = \lambda_3 = 0$), GSGL-MTL-s ($\lambda_2 = 0$) with promoting task-specific ROI and GSGL-MTL-c ($\lambda_3 = 0$) with promoting task-common ROI. Moreover, to illustrate how well our GSGL-MTL works, we comprehensively compare our proposed methods with several popular state-of-the-art MTL methods: SGL-MTL and

Table 1. Performance comparison of various methods on twenty cognitive prediction tasks. The best results are bolded, and superscript symbol * indicate that GSGL-MTL significantly outperformed that method on that score (Student's t-test at a level of 0.05 was used).

	Ridge	Group Lasso	GL-MTL	GSGL-MT-s	GSGL-MT-c	SGL-MTL	SRMTL	GSGL-MTL
ADAS	7.44±0.36	6.76±0.39	6.66±0.41	6.65±0.42	**6.63±0.45**	6.65±0.42	6.92±0.46	6.64±0.46
MMSE	2.56±0.14	2.21±0.07	2.19±0.10	2.18±0.09	**2.17±0.08**	2.19±0.09	2.40±0.31	**2.17±0.08**
TOTAL	11.16±0.73	9.96±0.87	9.65±0.69	9.64±0.75	**9.59±0.76**	9.64±0.75	10.39±0.81	9.60±0.77
TOT6	3.90±0.36	3.36±0.28	3.32±0.25	3.31±0.27	**3.30±0.26**	3.31±0.27	4.06±0.87	3.30±0.25
TOTB	1.98±0.12	1.66±0.15	1.67±0.14	1.66±0.15	**1.65±0.15**	1.66±0.14	3.02±1.78	**1.65±0.15**
T30	4.06±0.28	3.46±0.23	3.44±0.23	3.43±0.24	**3.42±0.25**	3.43±0.24	4.23±0.94	3.42±0.26
RECOG	4.31±0.42	3.98±0.21	3.62±0.27	3.62±0.24	3.61±0.22	3.62±0.24	4.11±0.73	**3.61±0.21**
ANIM	6.30±0.55	5.51±0.69	5.26±0.44	5.25±0.49	5.24±0.49	5.25±0.49	5.43±0.49	**5.23±0.49**
VEG	4.27±0.38	3.71±0.17	3.67±0.18	3.67±0.20	**3.66±0.20**	3.67±0.20	4.24±0.86	3.66±0.19
TRAILS-A	26.18±3.76	23.19±4.19	23.01±3.49	22.99±3.56	22.88±3.65	22.99±3.56	24.06±3.79	**22.87±3.66**
TRAILS-B	80.01±8.10	71.15±6.03	69.88±5.28	69.32±5.17	69.17±4.70	69.82±5.18	75.16±8.20	**69.13±4.65**
IMM	4.69±0.36	4.20±0.30	4.14±0.30	4.14±0.32	**4.12±0.32**	4.142±0.327	4.89±1.04	4.12±0.32
DEL	5.27±0.50	4.63±0.46	4.58±0.43	4.53±0.45	**4.56±0.46**	4.58±0.45	5.19±0.62	4.56±0.46
DRAW	1.15±0.10	0.97±0.10	0.96±0.11	0.95±0.11	0.96±0.11	0.96±0.11	2.69±2.06	**0.95±0.11**
COPY	0.77±0.06	0.67±0.07	0.64±0.10	0.64±0.09	**0.64±0.09**	**0.64±0.09**	3.35±3.07	**0.64±0.09**
BOSNAM	4.56±0.53	4.02±0.38	**3.95±0.47**	3.95±0.42	3.96±0.44	3.96±0.42	4.04±0.51	3.95±0.44
ANART	11.23±0.75	9.76±0.89	9.61±0.72	9.53±0.74	9.53±0.71	9.58±0.74	10.07±0.85	**9.52±0.71**
FOR	2.57±0.26	1.99±0.15	2.00±0.12	2.00±0.13	2.00±0.12	1.99±0.13	3.63±2.03	**1.99±0.13**
BAC	2.55±0.19	2.14±0.19	2.16±0.17	2.14±0.18	2.14±0.18	2.14±0.18	3.13±1.28	**2.13±0.18**
DIGIT	12.76±1.27	11.73±1.33	11.21±1.22	11.23±1.26	**11.16±1.23**	11.23±1.26	12.14±1.57	11.18±1.20
nMSE	10.35±1.08*	8.07±0.68*	7.76±0.63*	7.74±0.61*	7.64±0.55	7.74±0.61*	11.68±4.22*	**7.63±0.54**
wR.	0.29±0.04*	0.39±0.04*	0.40±0.05*	0.40±0.05*	0.41±0.05*	0.40±0.054*	0.39±0.04*	**0.41±0.04**

Sparse regularized multi-task learning formulation (SRMTL) [14]. The experimental results are shown in Table 1.

As can be seen from the Table 1, GSGL-MTL significantly outperformed the single task learning methods (Ridge and Group Lasso), and the recent state-of-the-art algorithms proposed in terms of nMSE and wR, which indicates that the interrelated structures within features and the correlation among the tasks are effectively captured by the GSGL norm.

3.3   Identification of MRI Biomarkers

Finally, we examined the biomarkers identified by different methods. The proposed GSGL-MTL is a group guided model which is able to identify a compact set of relevant neuroimaging biomarkers from the region level due to the group lasso on the features, which would provide us with better interpretability of the brain region.

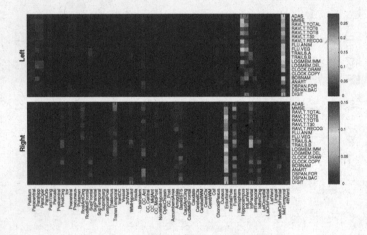

Fig. 3. Baseline matrix sparsity features.

Figure 3 is the heat maps of the regression weights of all ROIs in each hemisphere for each cognitive score at the baseline time calculated by GSGL-MTL through 10-fold cross validation experiments. Each item (i, j) indicates the weight of the i-th ROI for the j-th task, and is calculated by $w_i \sqrt{\sum_{q \in \mathcal{G}_i} \|\theta_{qi}\|_2}$, where q is the q-th MRI feature in the i-th ROI. The larger the absolute value of a coefficient, the more important its corresponding brain region is in predicting the corresponding cognitive score. The figure illustrates that the proposed GSGL-MTL clearly presented a sparsity across all the cortical measures from the level of ROI, which indicates a small portion of the brain regions is relevant to the cognitive outcome. We found that the imaging biomarkers identified by GSGL-MTL yielded promising patterns that are expected from prior knowledge on neuroimaging and cognition. Some important brain regions are selected, such as R. Middle Temporal, L. Hippocampus and R. Entorhinal, which are highly relevant to the cognitive impairment.

4 Conclusions

In this paper, we propose a Group guided Sparse group lasso (GSGL) regularized multi-task learning to learn the relationship between images and corresponding clinical scores from feature level and ROI level with taking the inherent group structure of the features into account. The experiments on the ADNI dataset have verified the effectiveness of GSGL-MTL, which offers consistently better performance than the baseline single task learning and several state-of-the-art multi-task learning algorithms. These promising results justify that by inducing both sparsity of feature and ROI level, GSGL-MTL captures useful information about AD. In the current work, only apriori group information is incorporated into multi-task predictive model, we are interested in the investigation of other structures in features, such as graph structure, which can help gain additional insights to understand and interpret data in future work.

Acknowledgment. This research was supported by NFSC (No. 61502091) and Fundamental Research Funds for the Central Universities (No. N161604001 and No. N150408001).

References

1. Zhang, D., Shen, D., Alzheimer's Disease Neuroimaging Initiative, et al.: Multimodal multi-task learning for joint prediction of multiple regression and classification variables in Alzheimer's disease. Neuroimage **59**(2), 895–907 (2012)
2. Yan, J., Huang, H., Risacher, S.L., Kim, S., Inlow, M., Moore, J.H., Saykin, A.J., Shen, L.: Network-guided sparse learning for predicting cognitive outcomes from MRI measures. In: Shen, L., Liu, T., Yap, P.-T., Huang, H., Shen, D., Westin, C.-F. (eds.) MBIA 2013. LNCS, vol. 8159, pp. 202–210. Springer, Cham (2013). doi:10.1007/978-3-319-02126-3_20
3. Wan, J., Zhang, Z., Yan, J., Li, T., Rao, B.D., Fang, S., Kim, S., Risacher, S.L., Saykin, A.J., Shen, L.: Sparse bayesian multi-task learning for predicting cognitive outcomes from neuroimaging measures in alzheimer's disease. In: IEEE Conference on Computer Vision and Pattern Recognition (CVPR), pp. 940–947 (2012)
4. Wang, J., Ye, J.: Two-layer feature reduction for sparse-group lasso via decomposition of convex sets. In: Advances in Neural Information Processing Systems, pp. 2132–2140 (2014)
5. Zhu, X., Suk, H.-I., Shen, D.: Sparse discriminative feature selection for multiclass Alzheimer's disease classification. In: Wu, G., Zhang, D., Zhou, L. (eds.) MLMI 2014. LNCS, vol. 8679, pp. 157–164. Springer, Cham (2014). doi:10.1007/978-3-319-10581-9_20
6. Yuan, M., Lin, Y.: Model selection and estimation in regression with grouped variables. J. Roy. Stat. Soc. Ser. B (Statistical Methodology) **68**(1), 49–67 (2006)
7. Xiang, S., Yuan, L., Fan, W., Wang, Y., Thompson, P.M., Ye, J., Alzheimer's Disease Neuroimaging Initiative, et al.: Bi-level multi-source learning for heterogeneous block-wise missing data. NeuroImage **102**, 192–206 (2014)

8. Argyriou, A., Evgeniou, T., Pontil, M.: Convex multi-task feature learning. Mach. Learn. **73**, 243–272 (2008)

9. Liu, J., Ji, S., Ye, J.: Multi-task feature learning via efficient $\ell_{2,1}$-norm minimization. In: Proceedings of the Twenty-Fifth Conference on Uncertainty in Artificial Intelligence, pp. 339–348. AUAI Press (2009)

10. Guerrero, R., Ledig, C., Schmidt-Richberg, A., Rueckert, D., Alzheimer's Disease Neuroimaging Initiative, et al.: Group-constrained manifold learning: application to AD risk assessment. Pattern Recogn. **63**, 570–582 (2017)

11. Zhu, X., Suk, H.I., Lee, S.W., Shen, D.: Subspace regularized sparse multitask learning for multiclass neurodegenerative disease identification. IEEE Trans. Biomed. Eng. **63**(3), 607–618 (2016)

12. Boyd, S., Parikh, N., Chu, E., Peleato, B., Eckstein, J.: Distributed optimization and statistical learning via the alternating direction method of multipliers. Found. Trends Mach. Learn. **3**, 1–122 (2011)

13. Yuan, L., Liu, J., Ye, J.: Efficient methods for overlapping group lasso. IEEE Trans. Pattern Anal. Mach. Intell. **35**(9), 2104–2116 (2013)

14. Zhou, J.: Multi-task learning in crisis event classification. Technical report. http://www.public.asu.edu/jzhou29

A Novel Deep Learning Based Multi-class Classification Method for Alzheimer's Disease Detection Using Brain MRI Data

Jyoti Islam[✉] and Yanqing Zhang

Department of Computer Science, Georgia State University, Atlanta, GA, USA
jislam2@student.gsu.edu, yzhang@gsu.edu

Abstract. Alzheimer's Disease is a severe neurological brain disorder. It destroys brain cells causing people to lose their memory, mental functions and ability to continue daily activities. Alzheimer's Disease is not curable, but earlier detection can help improve symptoms in a great deal. Machine learning techniques can vastly improve the process for accurate diagnosis of Alzheimer's Disease. In recent days deep learning techniques have achieved major success in medical image analysis. But relatively little investigation has been done to applying deep learning techniques for Alzheimer's Disease detection and classification. This paper presents a novel deep learning model for multi-Class Alzheimer's Disease detection and classification using Brain MRI Data. We design a very deep convolutional network and demonstrate the performance on the Open Access Series of Imaging Studies (OASIS) database.

Keywords: Alzheimer's disease · Deep learning · Convolutional Neural Network · MRI · Brain imaging

1 Introduction

Alzheimer's Disease affects people in a numerous way. Patients suffer from memory loss, confusion, difficulty in speaking, reading or writing. Eventually, they may forget about their life and could not recognize even their family members. They can forget how to perform daily activities such as brushing teeth or combing hair. As a result, it makes people anxious or aggressive or to wander away from home. Alzheimer's Disease can even cause death in elder people. There are three major stages in Alzheimer's Disease - very mild, mild and moderate. Detection of Alzheimer's Disease (AD) is still not accurate until the patient reaches a moderate AD. But early detection and classification of AD are critical for proper treatment and preventing brain tissue damage. Several things are needed for proper medical assessment of AD. Physical and neurobiological exams, Mini-Mental State Examination (MMSE), and patient's detailed history are required for accurate AD detection and classification. In recent years, doctors are using brain Magnetic Resonance Imaging (MRI) data for earlier detection of Alzheimer's Disease.

© Springer International Publishing AG 2017
Y. Zeng et al. (Eds.): BI 2017, LNAI 10654, pp. 213–222, 2017.
https://doi.org/10.1007/978-3-319-70772-3_20

Researchers have developed several computer-aided diagnostic systems for accurate disease detection. They have developed rule-based expert systems from the 1970s to 1990s and supervised models from 1990s [11]. The supervised systems are trained with feature vectors extracted from medical image data. Extracting the features needs human experts that often require a lot of time, money and effort. With the advancement of deep learning models, now we can extract features directly from the images without the engagement of human expert. So researchers are focusing on developing deep learning models for accurate disease detection and classification.

Deep learning models have been successfully applied for different medical image analysis such as MRI, Microscopy, CT, Ultrasound, X-ray, Mammography, etc. Deep models have shown a prominent result for organ and substructure segmentation, several disease detection and classification in areas of pathology, brain, lung, abdomen, cardiac, breast, bone, retina, etc. But there is little existing work for AD detection using deep learning models. From previous research in the medical domain, it has been proved that MRI data can perform a significant role for early detection of Alzheimer's Disease. For our research work, we plan to analyze brain MRI data using deep learning model for Alzheimer's Disease detection and classification.

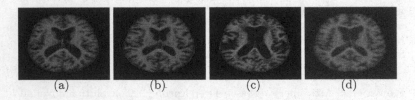

Fig. 1. Example of different brain MRI images presenting different AD stage. (a) Non-demented; (b) very mild dementia; (c) mild dementia; (d) moderate dementia.

Machine learning studies using neuroimaging data for developing diagnostic tools helped a lot for automated brain MRI segmentation and classification. Most of them use handcrafted feature generation and extraction from the MRI data. After that, the features are fed into machine learning models such as Support Vector Machine, Logistic regression model, etc. These multi-step architectures are complex and highly dependent on human experts. Besides, the size of datasets for neuroimaging studies is small. While image classification datasets used for object detection and classification has millions of image (for example, ImageNet database [18]), neuroimaging datasets typically have less than 1000 images. But to develop robust neural networks we need a lot of images. Because of the scarcity of large image database, it is important to develop models that can learn useful features from the small dataset. For our proposed system, we are using deep learning model which eliminates the need for hand-crafted feature generation. Deep learning models transform input to output and build a feature hierarchy from simple low-level features to complex high-level feature. The popular deep learning model used for image analysis is Convolutional Neural Network (CNN). We propose a very deep CNN model for analyzing the brain MRI images and classifying them into different AD stages.

Alzheimer's disease has a certain progressive pattern of brain tissue damage. It shrinks the hippocampus and cerebral cortex of the brain and enlarges the ventricles [19]. Hippocampus is the responsible part of the brain for episodic and spatial memory. It also works as a relay structure between our body and brain. While average reduction per year in the hippocampus is between 0.24 and 1.73%, Alzheimer's disease patients suffer shrinkage between 2.2 and 5.9% [1]. The reduction in hippocampus cause cell loss and damage specifically to synapses and neuron ends. So neurons can't communicate anymore via synapses. As a result, brain regions related to remembering (short term memory), thinking, planning, and judgment are affected [19]. The degenerated brain cells have low intensity in MRI images [5,29]. Figure 1 shows some brain MRI images presenting different AD stage.

Sometimes the signs that distinguish Alzheimer's disease MRI data can be found in normal healthy aged brain MRI data. Extensive knowledge and experience are required to distinguish the AD MRI data from the aged normal MRI data. A robust and effective automated machine learning model will help immensely the scientists and medical persons working for AD diagnosis and ultimately assist the timely treatment of the AD patients. A generic automated Alzheimer's Disease detection and classification framework is shown in Fig. 2. Our proposed deep CNN model can detect early stages of Alzheimer's disease and successfully classify the major three different stages. We have experimented the performance of the proposed model on the Open Access Series of Imaging Studies (OASIS) database [15] which provides T1-weighted MRI scans with demographics and clinical assessment data. Our main contributions are as follows:

- We propose a novel and faster framework for Alzheimer's disease detection analyzing brain MRI data.
- Our framework can classify three major stages of Alzheimer's disease.
- We demonstrate that utilizing hyper-parameters from a very deep image classifier CNN can help feature learning from small medical image dataset.

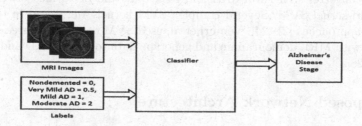

Fig. 2. Diagram of a generic Alzheimer's disease detection and classification framework.

The rest of the paper is organized as follows: Sect. 2 presents briefly about the related work. Proposed Alzheimer's Disease detection and classification framework is presented in Sect. 3. Experimental details and results are described in Sect. 4. Finally, we present future work and conclude the paper in Sect. 5.

2   Related Work

Developing an automated Alzheimer's Disease detection and classification model is a pretty challenging task. But there is some remarkable research work in this area. Dimensional reduction and variations methods were used by Aversen et al. [2] to analyze structural MRI data. They have used both SVM binary classifier and multi-class classifier to detect AD MRI images using Alzheimer's Disease Neuroimaging Initiative (ADNI) database [9]. Brosch et al. [3] developed a deep belief network model and used manifold learning for AD detection from MRI images. Katherine Gray developed a multi-modal classification model using random forest classifier to detect AD from MRI and PET data [6]. Gupta et al. have developed a sparse autoencoder model for AD, Mild Cognitive Impairment (MCI) and healthy control (HC) classification using ADNI dataset [7]. Hosseini-As et al. adapted a 3D CNN model for AD diagnostics [8]. Kloppel et al. used linear SVM to detect AD patients using T1 weighted MRI scan [10]. Liu et al. [12] developed a deep learning model using a subset of ADNI dataset and classified AD and MCI patients. Liu et al. have developed a multimodal stacked auto-encoder network using zero-masking strategy. Their target was to prevent loss of any information of the image data. They have used SVM to classify the neuroimaging features obtained from MR/PET data [13].

Magnin et al. utilized an anatomically labeled brain template to identify regions of interest from whole brain images and concluded that it could be used for early AD detection [14]. Morra et al. compared several model's performances for AD detection including hierarchical AdaBoost, SVM with manual feature and SVM with automated feature [16]. Payan et al. [17] trained sparse autoencoders and 3D CNN model to classify AD, MCI and HC patients using ADNI dataset. Sarraf et al. used fMRI data and deep LeNet model on ADNI dataset for AD detection [20]. Suk et al. developed an autoencoder network based model for AD detection. They have extracted features from magnetic current imaging (MCI) and MCI-converter structural MRI and PET data and performed classification using multi-kernel SVM. Several complex SVM kernels were used in their AD detection approaches [21–24]. Vemuri et al. used SVM to develop three separate classifiers with MRI, demographic and genotype data to classify AD and healthy patients [28].

3   Proposed Network Architecture

In this section, the proposed Alzheimer's disease detection and classification framework would be presented. The proposed model is shown in Fig. 3. Our model is inspired by Inception-V4 network [25]. After the preprocessing is done, the input is passed through a stem layer. A stem layer includes several 3 * 3 convolution layers, 1 * 1 convolution layer, and max pooling layer. There is seven 3 * 3 convolution layer connected in different stages and two filter-expansion layers (1 * 1 convolution layer). Inception-A module has four filter-expansion layers, three 3 * 3 convolution layer, and one average pooling layer. Inception-B module has four filter-expansion layers, four 1 * 7 convolution layer, two 7 * 1 convolution layer and one average pooling layer. Inception-C module has four filter-expansion layers, three 1 * 3 convolution layer, three 3 * 1 convolution layer and one average pooling layer. Reduction-A module has one filter-expansion layer, three 3 * 3 convolution layer, and one 3 * 3 max-pooling layer. The Reduction-B module has two filter-expansion layers, two 3 * 3 convolution layer, one 1 * 7 convolution layer, one 7 * 1 convolution layer and one 3 * 3 max pooling layer. The input and output of all these modules pass through filter concatenation process. We have redesigned the final softmax layer for Alzheimer's disease detection and classification. The softmax layer has four different output class: nondemented, very mild, mild and moderate AD. The network takes an MRI image as input and extracts layer-wise feature representation from the first stem layer to the last drop-out layer. Based

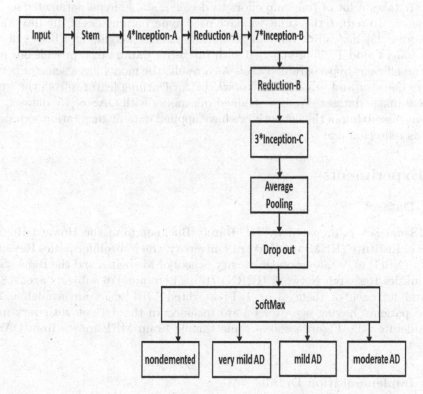

Fig. 3. Block diagram of proposed Alzheimer's disease detection and classification framework.

on this feature representation, the input MRI image is classified to any of the four output classes.

To measure the loss of the proposed network, we have used cross entropy. The Softmax layer takes the feature representation, f_i and interprets it to the output class. A probability score, p_i is also assigned for the output class. If we define the number of Alzheimer's disease stages as m, then we get

$$p_i = \frac{exp(f_i)}{\sum_i exp(f_i)}, i = 1, ..., m$$

and

$$L = -\sum_i t_i log(p_i),$$

where L is the loss of cross entropy of the network. Back propagation is used to calculate the gradients of the network. If the ground truth of an MRI image is denoted as t_i, then,

$$\frac{\partial L}{\partial f_i} = p_i - t_i$$

There is numerous possible combination for the hyper-parameters of a network. It takes a lot of time and effort to decide a stable hyperparameter set for a network. To reduce this time, we have used hyperparameters of the Inception-V4 model [25] instead of random initialization. The weights and biases of the inception-v4 model [25] pre-trained with ImageNet database [4] provide our network an efficient hyperparameter set. As a result, the model has a sense of better feature detector and can use that knowledge for learning features from the small medical image dataset. We have trained our model with OASIS [15] dataset. To prevent overfitting in the network, we have applied data augmentation technique such as reflection and scaling.

4   Experiments

4.1   Dataset

OASIS dataset is prepared by Dr. Randy Buckner from the Howard Hughes Medical Institute (HHMI) at Harvard University, the Neuroinformatics Research Group (NRG) at Washington University School of Medicine, and the Biomedical Informatics Research Network (BIRN) [15]. There are 416 subjects aged 18 to 96, and for each of them, 3 or 4 T1-weighted MRI scans are available. 100 of the patients having age over 60 are included in the dataset with very mild to moderate AD. Figure 4 shows some sample brain MRI images from OASIS dataset.

4.2   Implementation Details

We have implemented the proposed deep CNN model for Alzheimer's disease detection and classification using Tensorflow and Python on a Linux X86-64

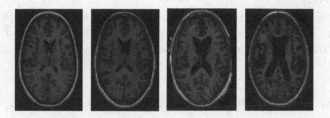

Fig. 4. Sample images from OASIS dataset.

machine with AMD A8 CPU, 16 GB RAM and NVIDIA GeForce GTX 770. We have applied data augmentations techniques - scaling and reflection on the images. Since the dataset is small, 5-fold cross validation is performed on the dataset. For each fold, We have used 70% as training data, 10% as validation data and 20% as test data. The input size of the Inception-V4 network [25] is 299 * 299 * 3. To fit the MRI data, we have designed the input size of our network as 299 * 299 * 1. We have modified the Inception B and C module so that they can accept the MRI data. The convolutional filter size of Inception-B is 1154 in the original network. We made it to 1152 to fit the MRI data. The convolutional filter size of Inception-C is 2048 in the original network. We made it to 2144 to fit the MRI data. The network is optimized with the RMSProp [27] algorithm and early-stopping is used for regularization. The decay of the network is 0.9 and batch size is 8. The base learning rate is set to 0.045.

4.3 Results

To our best knowledge, our approach is the first one for Alzheimer's disease detection and classification using deep learning method on OASIS dataset. So, we are not comparing it with previous traditional methods. The current accuracy of our method is 73.75%. The confusion matrix is presented in Table 1. The proposed model is much faster and takes less than 1 h to train and test the OASIS dataset for Alzheimer's disease detection and classification. This performance is superior than all previous traditional methods. It would take weeks for human experts to analyze and classify all the MRI data. We do not need any manual hand-crafting for feature generation in our model.

Table 1. Confusion matrix

AD stage	Nondemented	Very mild	Mild	Moderate
Nondemented	52	0	0	0
Very mild	2	4	0	0
Mild	7	0	8	0
Moderate	3	0	1	3

We have implemented another deep model with traditional inception module and 22 layers following GoogleNet [26] architecture and compared the performance with our proposed model. The performance comparison is presented in Table 2.

Table 2. Five-fold cross validation performance accuracy comparison on the oasis dataset

No. of epochs	Traditional inception network	Proposed model
5	60.00%	71.25%
10	64.25%	73.75%

5   Conclusion

An automated Alzheimer's disease detection and classification framework is crucial for the early detection and treatment of the AD patients. We have proposed a deep CNN model for automated Alzheimer's disease detection and classification. We have demonstrated the performance of the model on OASIS dataset. Our method is faster, and it does not need any handcrafted feature, and it can handle the small medical image dataset. We have provided an one step analysis for the brain MRI data for AD detection and classification. There are several improvements possible for the proposed approach. In future, we hope to work with other MRI AD dataset such as ADNI and achieve similar or better performance. We want to apply transfer learning and check if it produces better result than the proposed approach. Currently, we are working with different hidden layers and convolutional filters to do more optimization to find a more efficient model to get better results. Finally, we want to explore semi-supervised and unsupervised deep learning methods for multi-class Alzheimer's disease detection and classification.

References

1. Ali, E.M., Seddik, A.F., Haggag, M.H.: Automatic detection and classification of Alzheimer's disease from MRI using TANN. Int. J. Comput. Appl. **148**(9), 30–34 (2016)
2. Arvesen, E.: Automatic classification of Alzheimer's disease from structural MRI. Master's thesis (2015)
3. Brosch, T., Tam, R., for the Alzheimer's Disease Neuroimaging Initiative: Manifold learning of brain MRIs by deep learning. In: Mori, K., Sakuma, I., Sato, Y., Barillot, C., Navab, N. (eds.) MICCAI 2013. LNCS, vol. 8150, pp. 633–640. Springer, Heidelberg (2013). https://doi.org/10.1007/978-3-642-40763-5_78
4. Deng, J., Dong, W., Socher, R., Li, L.J., Li, K., Fei-Fei, L.: ImageNet: a large-scale hierarchical image database. In: IEEE Conference on Computer Vision and Pattern Recognition, CVPR 2009, pp. 248–255. IEEE (2009)

5. Grady, C.L., McIntosh, A.R., Beig, S., Keightley, M.L., Burian, H., Black, S.E.: Evidence from functional neuroimaging of a compensatory prefrontal network in Alzheimer's disease. J. Neurosci. **23**(3), 986–993 (2003)
6. Gray, K.R.: Machine learning for image-based classification of Alzheimer's disease. Ph.D. thesis, Imperial College London (2012)
7. Gupta, A., Ayhan, M., Maida, A.: Natural image bases to represent neuroimaging data. In: ICML, vol. 3, pp. 987–994 (2013)
8. Hosseini-Asl, E., Keynton, R., El-Baz, A.: Alzheimer's disease diagnostics by adaptation of 3D convolutional network. In: 2016 IEEE International Conference on Image Processing (ICIP), pp. 126–130. IEEE (2016)
9. Jack, C.R., Bernstein, M.A., Fox, N.C., Thompson, P., Alexander, G., Harvey, D., Borowski, B., Britson, P.J., Whitwell, J.L., Ward, C.: The Alzheimer's Disease Neuroimaging Initiative (ADNI): MRI methods. J. Magn. Reson. Imaging **27**(4), 685–691 (2008)
10. Klöppel, S., Stonnington, C.M., Chu, C., Draganski, B., Scahill, R.I., Rohrer, J.D., Fox, N.C., Jack, C.R., Ashburner, J., Frackowiak, R.S.: Automatic classification of MR scans in Alzheimer's disease. Brain **131**(3), 681–689 (2008)
11. Litjens, G., Kooi, T., Bejnordi, B.E., Setio, A.A.A., Ciompi, F., Ghafoorian, M., van der Laak, J.A., van Ginneken, B., Sánchez, C.I.: A survey on deep learning in medical image analysis. arXiv preprint arXiv:1702.05747 (2017)
12. Liu, F., Shen, C.: Learning deep convolutional features for MRI based Alzheimer's disease classification. arXiv preprint arXiv:1404.3366 (2014)
13. Liu, S., Liu, S., Cai, W., Che, H., Pujol, S., Kikinis, R., Feng, D., Fulham, M.J.: Multimodal neuroimaging feature learning for multiclass diagnosis of Alzheimer's disease. IEEE Trans. Biomed. Eng. **62**(4), 1132–1140 (2015)
14. Magnin, B., Mesrob, L., Kinkingnéhun, S., Pélégrini-Issac, M., Colliot, O., Sarazin, M., Dubois, B., Lehéricy, S., Benali, H.: Support vector machine-based classification of Alzheimer's disease from whole-brain anatomical MRI. Neuroradiology **51**(2), 73–83 (2009)
15. Marcus, D.S., Wang, T.H., Parker, J., Csernansky, J.G., Morris, J.C., Buckner, R.L.: Open access series of imaging studies (OASIS): cross-sectional MRI data in young, middle aged, nondemented, and demented older adults. J. Cogn. Neurosci. **19**(9), 1498–1507 (2007)
16. Morra, J.H., Tu, Z., Apostolova, L.G., Green, A.E., Toga, A.W., Thompson, P.M.: Comparison of adaboost and support vector machines for detecting Alzheimer's disease through automated hippocampal segmentation. IEEE Trans. Med. Imaging **29**(1), 30 (2010)
17. Payan, A., Montana, G.: Predicting Alzheimer's disease: a neuroimaging study with 3D convolutional neural networks. arXiv preprint arXiv:1502.02506 (2015)
18. Russakovsky, O., Deng, J., Su, H., Krause, J., Satheesh, S., Ma, S., Huang, Z., Karpathy, A., Khosla, A., Bernstein, M.: Imagenet large scale visual recognition challenge. Int. J. Comput. Vis. **115**(3), 211–252 (2015)
19. Sarraf, S., Anderson, J., Tofighi, G.: DeepAD: Alzheimer's disease classification via deep convolutional neural networks using MRI and fMRI. bioRxiv p. 070441 (2016)
20. Sarraf, S., Tofighi, G.: Classification of Alzheimer's disease using fMRI data and deep learning convolutional neural networks. arXiv preprint arXiv:1603.08631 (2016)
21. Suk, H.I., Lee, S.W., Shen, D., for the Alzheimer's Disease Neuroimaging Initiative: Hierarchical feature representation and multimodal fusion with deep learning for AD/MCI diagnosis. NeuroImage **101**, 569–582 (2014)

22. Suk, H.I., Lee, S.W., Shen, D.: Latent feature representation with stacked auto-encoder for AD/MCI diagnosis. Brain Struct. Funct. **220**(2), 841–859 (2015)

23. Suk, H.-I., Shen, D.: Deep learning-based feature representation for AD/MCI classification. In: Mori, K., Sakuma, I., Sato, Y., Barillot, C., Navab, N. (eds.) MICCAI 2013. LNCS, vol. 8150, pp. 583–590. Springer, Heidelberg (2013). https://doi.org/10.1007/978-3-642-40763-5_72

24. Suk, H.-I., Shen, D.: Deep learning in diagnosis of brain disorders. In: Lee, S.-W., Bülthoff, H.H., Müller, K.-R. (eds.) Recent Progress in Brain and Cognitive Engineering. TAHP, vol. 5, pp. 203–213. Springer, Dordrecht (2015). https://doi.org/10.1007/978-94-017-7239-6_14

25. Szegedy, C., Ioffe, S., Vanhoucke, V., Alemi, A.: Inception-v4, inception-resnet and the impact of residual connections on learning. arXiv preprint arXiv:1602.07261 (2016)

26. Szegedy, C., Liu, W., Jia, Y., Sermanet, P., Reed, S., Anguelov, D., Erhan, D., Vanhoucke, V., Rabinovich, A.: Going deeper with convolutions. In: Proceedings of the IEEE Conference on Computer Vision and Pattern Recognition, pp. 1–9 (2015)

27. Tieleman, T., Hinton, G.: RMSProp: divide the gradient by a running average of its recent magnitude. COURSERA: Neural networks for machine learning. Technical report, p. 31 (2012)

28. Vemuri, P., Gunter, J.L., Senjem, M.L., Whitwell, J.L., Kantarci, K., Knopman, D.S., Boeve, B.F., Petersen, R.C., Jack, C.R.: Alzheimer's disease diagnosis in individual subjects using structural mr images: validation studies. Neuroimage **39**(3), 1186–1197 (2008)

29. Warsi, M.A.: The fractal nature and functional connectivity of brain function as measured by BOLD MRI in Alzheimer's disease. Ph.D. thesis (2012)

A Quantitative Analysis Method for Objectively Assessing the Depression Mood Status Based on Portable EEG and Self-rating Scale

Zhijiang Wan[1,2,3], Qiang He[3], Haiyan Zhou[3], Jie Yang[5], Jianzhuo Yan[4], and Ning Zhong[1,2,3(✉)]

[1] Department of Life Science and Informatics, Maebashi Institute of Technology, Maebashi 371-0864, Japan
wandndn@gmail.com, zhong@maebashi-it.ac.jp
[2] Beijing Advanced Innovation Center for Future Internet Technology, Beijing University of Technology, Beijing 100124, China
[3] International WIC Institute, Beijing University of Technology, Beijing 100124, China
[4] College of Electronic Information and Control Engineering, Beijing University of Technology, Beijing 100124, China
[5] Beijing Anding Hospital of Capital Medical University, Beijing 100088, China

Abstract. In order to recognize the major depressive mood status of inpatients and achieve its daily change information, a POMS-BCN scale was used to rate the mood status. Meanwhile, a personalized quantified model based on portable EEG was built, which aimed at objectively assessing the major depressive mood status for each patient. 6 inpatients were recruited to join the experiment. The Principal Component Analysis method is used to extract first principal component curve from the POMS-BCN data. The feature extraction method is used to extract linear and nonlinear features from portable EEG data. The regression analysis based on Random Forest is adopted to build the personalized quantified model. The principal component analysis result shows that the first principal component curve is able to recognize the major emotional factor and depict its daily change information. Additionally, the expected quantitative value outputted from the personalized quantified model is highly correlated (the absolute value of correlation coefficient 0.7, P-value 0.05) with the actual first principal component data, which implies that the personalized quantified model can give an accurate objective assessment for the major depressive mood status.

Keywords: Depression quantitative analysis · Objective assessment · Depression mood status · Portable EEG · Self-rating scale

1 Introduction

Depression is a state of low mood and aversion to activity that can affect a person's thought, behavior, feelings and sense of well-being [1]. 121 million people worldwide affected by depression which makes it become one of the most common mental disorders. According to the investigation by the World Health Organization, depression will be the second major disability causing disease in the world by 2020 [2]. In order to ameliorate the status, it is significant to study the depression diagnosis and treatment methods.

In the traditional treatment methods for depression, rating scales and questionnaires, such as the Hamilton Depression rating scale with 17-items (HAM-D17), the Young Mania Rating Scale (YMRS), the Mini-International Neuropsychiatric Interview (MINI), the Beck Depression Inventory (BDI) and the Children's Depression Inventory (CDI) etc., are used to assess the depression severity. However, the scales mentioned above are subjective, which means the accuracy of the assessment result depends on the experience of clinician [3–5]. Moreover, due to the retest reliability of the scales [5,11,12], the most clinical scales are always used to assess the psychological status in a cross section manner. In this situation, those scales not only cannot assess the daily change of the psychological status in a low granularity, but also cannot reflect the overall tendency accurately. Meanwhile, the depression involves the numbing of emotions, especially grief, fear, anger and shame [13], the most clinical scales, which have pertinence for one specific depressive symptom, cannot assess the severity of various emotions and recognize the major emotional factor.

Corresponding to the subjective methods of depression assessment, the objective quantification analysis method based on physiological data, which can provide a unified and quantitative criterion, could be an effective adjunct tool for traditional depression assessment and diagnosis method. In the recent years, the studies focused on depression quantitative assessment and diagnosis have become a hot topic [3,6]. The researchers aim at finding out some effective physiological and behavioral markers to assess and diagnose depression [7–10]. Specifically, a various of measurements, such as fMRI, eye tracking, PET and EEG with multichannel etc., are used to collect physiological signal and extract trait-oriented features for diagnosing depression [1,4,7,9,14–18]. Among them, EEG device is widely used because of its low-cost and comparatively easily operate. The linear and nonlinear EEG features are adopted in classifying depressive and normal control [14,19–21]. Using the linear and nonlinear features extracted from EEG data, we can objectively assess and diagnose the depression [21]. Although they have shown no gold standards to accurately assess and diagnose depression, the objective quantitation analysis method based on physiological data is still promising. Rather than only use the subjective or objective assessment method, we prefer to use the combination of the subjective assessing method and the objective quantitation analysis method to improve the clinical effect of depression. Few studies tried to use the EEG features to assess the depression severity, and compare the assessment effectiveness with traditional rating scale. It is easy to understand that the objective assessment result is reliable if its effectiveness is as well as the assessment result made by traditional rating scale.

In this study, there are three questions we want to discuss and solved:

(1) How to recognize the major emotional factor based on a long term self-rating data?
(2) How to achieve the daily change information of the major emotional factor?
(3) How to build a reliable objective quantification model based on physiological data and self-rating scale data to assess the major emotional factor?

For question (1) and (2), an electronic self-rating scale named Profile of Mood Status and Brief Chinese Norm (POMS-BCN) is utilized to rate the depression mood status by the patients themselves. The Principle Component Analysis (PCA) is used to process the POMS-BCN data. For each patient, a curve called First Principle Component (FPC) is extracted to subjectively reflect the overall tendency and daily change information of the major depressive mood status. For question (3), the regression analysis method based on Random Forest (RF) is selected to build the objective quantification model, and assess the major emotional factor.

2   Material and Method

2.1   Experimental Design

6 inpatients were recruited from Beijing Anding hospital to join the experiment and every inpatient was asked to finish the data collection for 2 weeks. All depressive patients, which are right-handed and willingness to give written informed consent. The data we collected can be categorized into objective data and subjective data. The objective data is the portable EEG data collected from prefrontal lobe using B3 Band, a portable EEG device which is equipped with NeuroSky EEG biosensor (512 Hz sampling frequency and 12-bit ADC precision) [22]. The subjective data is the self-rating scale data evaluated by subjects themselves using an electronic POMS-BCN scale, which is adapted from the Profile of Mood Status (POMS) rating scale authored by Dr. Maurice Lorr et al. in 1971 [23]. In order to ensure the consistency of the data collection time, the portable EEG data and POMS-BCN data are collected in the same time (twice a day, morning and evening). Specifically, every subject is firstly asked to sit on a sofa and keep awake with eye closing for 5 min in a dimly illuminated, acoustically and electrically shielded room. And then, the subject is required to self-rate the current mood status via the electronic POMS-BCN. It is noteworthy that in order to illustrate the validity of the POMS-BCN, the HAM-D17 or YMRS is also adapted as reference to assess the depressive severity of patients. The patients are asked to self-rate the HAM-D17 or YMRS 3 times 2 weeks (the beginning of the first week, the end of the first week and the end of the second week).

2.2   Data Analysis

Portable EEG Data Analysis. Besides the abnormal situation, such as operating error or the patient wants to relax in the weekend, the ultimate sample size of the collected portable EEG data is 168. Every sample is a time series with the length of 153600 (512 * 5 * 60). In order to ensure the processed EEG data is collected in the resting state, the beginning 15 s and the ending 15 s of every EEG series are removed respectively. Thus, the length of every EEG series inputted into the feature extraction step is 138240 (512 * 4.5 * 60). Furthermore, every time series is divided into 135 segments (every segment contains data of 4 s) that overlapped by 50 A data de-noising method based on Discrete Wavelet Transform (DWT) is used to process the segment. DWT is used in two aspects: EEG data de-nosing and decomposition. For data de-nosing, a soft thresholding algorithm with db5 wavelet base is utilized, the formula is defined as follows.

$$f(i) = \begin{cases} c(i) - \tau * e^{1-(\frac{c(i)}{\tau})}, & c(i) \geqslant \tau \\ 0, & else \\ c(i) + \tau * e^{1-(\frac{c(i)}{\tau})}, & c(i) \leqslant \tau \end{cases} \tag{1}$$

where τ is threshold and $c(i)$ is the wavelet coefficients extracted from the raw EEG segment by DWT.

For data decomposition, an eight-layer DWT with $db5$ wavelet base is used to decompose the de-noised raw EEG segment into several subband components. Finally, 6 time series, which include the de-noised raw data and five subbands (delta (2–4 Hz), theta (4–8 Hz), alpha (8–16 Hz), beta (16–32 Hz) and gamma (32–64 Hz)), are extracted from the raw EEG segment. Furthermore, the linear, wavelet and nonlinear features are extracted from the 6 time series respectively. A feature vector with 176 dimensions (time domain: 54, frequency domain: 54, wavelet: 56 and nonlinear: 12) is extracted from the raw EEG segment.

POMS-BCN Scale Data. The sample size of the POMS-BCN scale data is the same with the portable EEG data. Because of the POMS-BCN contains 7 emotional factors and adopts multiple adjectives with similar meaning to limn the factor repeatedly, there are a number of scale items with approximate emotion ratings which are highly-correlated with each other. It would be better if we could attenuate the correlation, and recognize the major emotional factor from the POMS-BCN dataset of 2 weeks for each patient. In view of this, the PCA is utilized to process the POMS-BCN scale data. For each patient, based on the POMS-BCN dataset of 2 weeks, a t-by-40 matrix can be assembled. A FPC curve is extracted from the matrix, which can provide an intuitive observation for the overall tendency and daily change information of the major emotional factor in 2 weeks. For recognizing the major emotional factor, the corresponding component scores of each emotional factor, which are contained in the FPC curve, need to be aggregated into 7 vectors. We defined the emotional factor whose vector has the maximum sum as the major emotional factor.

Personalized Quantified Model Construction for Objectively Assessing the Major Emotional Factor. The regression analysis method based on RF is used to construct the objective quantization model. For each patient, a RF regression model is trained and built based on the EEG features (independent variable) and the FPC curve (dependent variable). In order to guarantee the objective quantification value can assess the major emotional factor accurately, the effectiveness of the model should be tested. The testing steps are listed as follows:

(1) Before running the regression analysis based on RF, the EEG features and FPC curve of each patient are split up into two parts respectively: training dataset and testing dataset.
(2) The RF regression model is trained by the training dataset, the personalized quantification model can be constructed.
(3) The testing dataset is adopted to test the effectiveness of the model. Specifically, the EEG features included in the testing dataset are inputted into the trained model, the expected data outputted by the model, which is the objective quantification value of major emotional factor.
(4) The correlation of actual data (the testing data of FPC curve) and the expected data can be calculated via correlation analysis. We are looking forward to getting such good result as the correlation coefficient r is close to 1 and the probability p-value is less than 0.05. In this situation, the regression model can be deemed to be able to objectively reflect the variation tendency of the major emotional factor accurately.

3   Results

3.1   Depressive Mood Status Assessment Based on POMS-BCN Data

Major Emotional Factor Recognition. The result of the major emotional factor recognition for each patient is illustrated in the Fig. 1. For each subfigure, the horizontal ordinate indicates the 7 emotional factors, the alphabet A-G means the angry, self-emotion, depress, nervous, flurry, fatigue and energy respectively. The vertical ordinate means the weight of each factor. Each weight value is calculated based on the FPC curve, which is the first column in the p-by-p ($p = 40$) principal component scores matrix. As seen from the Fig. 1, we can conclude that (1) depress is not the major emotional factor for unipolar depressive, other negative emotions (nervous, flurry, fatigue and angry) can also be the major emotional factor. For instance, the major emotional factor of patient 1 is nervous, the major emotional factor of patient 3 and 4 is depress, the major emotional factor of patient 6 is angry. (2) for the unipolar depressive patient with mania symptom (patient 2), the major emotional factor is energy, the subsidiary emotional factor is self-emotional relevant, the weights of other negative emotional factor are much less than them. (3) for the bipolar depressive patient with mania symptom (patient 5), the major emotional factor is self-emotional relevant, the subsidiary emotional factor is fatigue, and it is noteworthy that the depress and the fatigue have close emotional factor weight. It indicates that the negative emotional factors are always along with the energy or the self-emotional factor, which can be defined as positive emotional factor, for a bipolar depressive. The conclusions concluded above are consistent with common sense, and we also tried to verify the conclusions by discussing with clinicians, no objections were proposed.

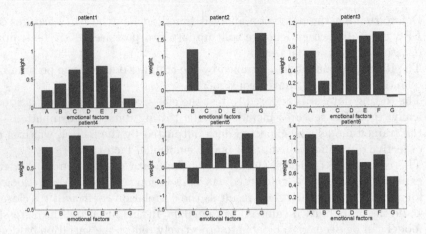

Fig. 1. Result of the major emotion factor recognition for each patient.

Long Term Dynamic Change of the Major Emotional Factor. The long term dynamic change of the major emotional factor can be exposited from the overall tendency and the local variation of the FPC curve respectively. The result can be seen from the Fig. 2. For each subfigure, the horizontal ordinate indicates the collecting time of POMS-BCN data, the ordinate means the FPC scores, which is the first column of the principal component scores matrix, every score implies the value of the major emotional factor rated every time. The red line is the FPC curve. We can conclude that for the patient 1, 2, 4 and 5, the overall tendency of the curve declined, which indicates the depressive symptom is relieved in the experiment stage. In contrast, for the patient 3 and 6, the overall tendency of the curve declined in the first week but rebounded in the second week, which means the depressive symptom is not remitted after receiving the drug therapy of 2 weeks. In order to further support the conclusions, the assessment result rated by other clinical scale (HAM-D17 or YMRS) is also illustrated in the Fig. 2. It is noteworthy that because of the second patient is a unipolar depressive with mania symptom, we use the YMRS scale to assess the mania symptom. The green line is utilized to show the assessment result of HAM-D17 or YMRS. As shown, the overall tendency of the assessment result of HAM-D17 or YMRS is congruent with FPC curve. For the local variation of the FPC curve, the bar graph is adopted to show and compare the relative change of every score in the FPC curve. The scores in the FPC curve are split into two kinds: the value of the major emotional factor rated in the morning, which is marked by the white bar, and value of the major emotional factor rated in the evening, which is marked by the blue bar. As shown, we can find out that in the most cases, the value rated in the morning is higher than the value rated in the evening, especially the value rated in the first week. To my knowledge, the feelings of depressives are often worse in the morning than in the afternoon or evening. The local variation of the FPC curve might be a sturdy evidence to exposit this perspective.

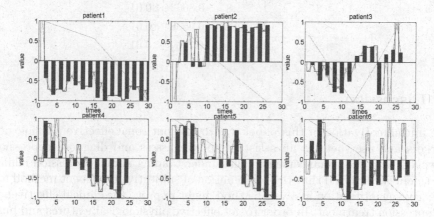

Fig. 2. Dynamic change of the major emotional factor in long term. (Color figure online)

3.2 Validity Test for the Personalized Objective Quantification Model

For each patient, we build a personalized objective quantification model to objectively assess the depressive mood status based on EEG features and FPC curve. According to the steps for testing the effectiveness of the personalized objective quantification model, the data of EEG feature and FPC curve of previous 8 days (16 times) are used to train the personalized quantification model. After inputting the rest of EEG features into the trained model, the expected quantitative value of the major depressive mood status can be outputted. The correlation analysis is adopted to calculate the correlation between the expected quantitative value and the actual quantitative value (the rest data of FPC curve). Table 1 shows the correlation analysis result based on the data of EEG feature and FPC curve of each patient. As shown in the table, TrS means the sample size of training data. TeS indicates the sample size of testing data. The correlation coefficient (CC) and P-value are acquired by calculating the correlation. The absolute CC values of 6 patients are above 0.7 and the P-values are below 0.05, which implies that the expected quantitative values of the major depressive mood status have a high correlation with the actual quantitative values. For each patient, the personalized quantified model based on the data of 8 days is able to quantify the major depressive mood status comparatively accurate.

Table 1. Correlation analysis result based on the data of EEG feature and FPC curve

Number	TrS	TeS	CC	P-value
1	16	14	0.7981	<0.01
2	16	10	0.7321	0.0161
3	16	10	−0.7480	0.0128
4	16	8	−0.8663	<0.01
5	16	12	−0.9598	<0.01
6	16	12	0.7564	<0.01

4   Discussion

The initial motivation of this paper was to find out some effective physiological markers and use them as an assistant tool to assess and diagnose depression. However, with the enhancement of the understanding about depression clinical treatment, we realized that the combination of subjective treatment method and objective quantification analysis method might improve the clinical effectiveness of depression treatment. In order to get effective physiological features and build an objective quantification model, the subjective treatment method is a good reference to guide the feature selection. We are looking forward that the objective quantification result for depression assessment based on the physiological feature can achieve the same effectiveness with the assessing result rated by traditional scale.

Based on the subjective rating data, we aims at using the POMS-BCN scale to recognize the major emotional factor and achieve the daily change information of the major emotional factor. Compare with the traditional scale usage, this work is original because of the traditional scales always assess the specific and single depressive symptom with a rough granularity. Based on the analysis result of POMS-BCN scale, we can conclude that (1) depress is not the major emotional factor for unipolar depressive, other negative emotions (nervous, flurry, fatigue and angry) can also be the major emotional factor; (2) for the unipolar depressive patient with mania symptom, the major emotional factor is energy, the subsidiary emotional factor is self-emotional relevant; (3) for the bipolar depressive patient with mania symptom, the negative emotional factors are always along with the energy or the self-emotional factor; (4) compare with the result of HAM-D17 or YMRS, the FPC curve is able to reflect the overall variation tendency; (5) the local variation of the FPC curve shows that the feelings of the depressives are often worse in the morning than in the afternoon or evening. The conclusions depicted above are accordance with the clinical experience of clinicians. The regression analysis based on RF is used to build personalized quantification model and objectively assess the depressive mood status. Our target is to make the effectiveness of assessing the depressive mood status based on physiological data as well as the assessment effectiveness of the traditional scale data. The expected quantitative values outputted from the RF regression model have a high correlation with the actual quantitative values. It indicates that the personalized quantization model based on the data of 8 days is able to quantify the major depressive mood status comparatively accurate.

However, there are some limitations in this study. For the POMS-BCN scale, the scale is adapted from the POMS rating scale authored by Dr. Maurice Lorr et al. in 1971. The rating method (from paper to electronic) and rating range of each item are changed, and we did not test its reliability and validity. For the portable EEG data, although the correlation result shows that the personalized quantification model based on the data of 8 days is able to quantify the major depressive mood status comparatively accurate, the small data sample size might limit our ability to get a persuasive correlation result. In the future, not only will the reliability and validity of the POMS-BCN scale be tested, but also the sample size of portable EEG data and scale data is worthy to be increased.

Acknowledgments. This work is partially supported by the National Basic Research Program of China (No. 2014CB744600), National Natural Science Foundation of China (No. 61420106005), Beijing Natural Science Foundation (No. 4164080), and Beijing Outstanding Talent Training Foundation (No. 2014000020124G039).

References

1. Chaudhury, D., Walsh, J.J., Friedman, A.K., Juarez, B., Ku, S.M., Koo, J.W., Ferguson, D., Tsai, H.-C., Pomeranz, L., Christoffel, D.J.: Rapid regulation of depression-related behaviours by control of midbrain dopamine neurons. Nature **493**(7433), 532–536 (2013)
2. Whiteford, H.A., Degenhardt, L., Rehm, J., Baxter, A.J., Ferrari, A.J., Erskine, H.E., Charlson, F.J., Norman, R.E., Flaxman, A.D., Johns, N.: Global burden of disease attributable to mental and substance use disorders: findings from the global burden of disease study 2010. Lancet **382**(9904), 1575–1586 (2013)
3. Guo, T., Xiang, Y.-T., Xiao, L., Hu, C.-Q., Chiu, H.F., Ungvari, G.S., Correll, C.U., Lai, K.Y., Feng, L., Geng, Y.: Measurement-based care versus standard care for major depression: a randomized controlled trial with blind raters. Am. J. Psychiatry **172**(10), 1004–1013 (2015)
4. Bilello, J.A.: Seeking an objective diagnosis of depression. Biomark. Med. **10**(8), 861–875 (2016)
5. Potvin, S., Charbonneau, G., Juster, R.-P., Purdon, S., Tourjman, S.V.: Self-evaluation and objective assessment of cognition in major depression and attention deficit disorder: implications for clinical practice. Compr. Psychiatry **70**, 53–64 (2016)
6. Ricken, R., Wiethoff, K., Reinhold, T., Schietsch, K., Stamm, T., Kiermeir, J., Neu, P., Heinz, A., Bauer, M., Adli, M.: Algorithm-guided treatment of depression reduces treatment costs, results from the randomized controlled German Algorithm Project (GAPII). J. Affect. Disord. **134**(1), 249–256 (2014)
7. Trivedi, M.H., Rush, A., Crismon, M., et al.: Clinical results for patients with major depressive disorder in thetexas medication algorithm project. Arch. Gen. Psychiatry **61**(7), 669–680 (2004)

8. Crowell, S.E., Baucom, B.R., Yaptangco, M., Bride, D., Hsiao, R., McCauley, E., Beauchaine, T.P.: Emotion dysregulation and dyadic conflict in depressed and typical adolescents: evaluating concordance across psychophysiological and observational measures. Biol. Psychol. **98**, 50–58 (2014)

9. Etkin, A., Bchel, C., Gross, J.J.: The neural bases of emotion regulation. Nat. Rev. Neurosci. **16**(11), 693–700 (2015)

10. Rosebrock, L.E., Hoxha, D., Norris, C., Cacioppo, J.T., Gollan, J.K.: Skin conductance and subjective arousal in anxiety, depression, and comorbidity: implications for affective reactivity, pp. 1–13 (2016)

11. Sano, A., Picard, R.W.: Stress recognition using wearable sensors and mobile phones. In: Affective Computing and Intelligent Interaction, pp. 671–676. IEEE Press, Geneva (2013)

12. Radloff, L.S.: The CES-D scale: a self-report depression scale for research in the general population. Appl. Psychol. Meas. **1**(3), 385–401 (1977)

13. Streiner, D.L., Norman, G., Cairney, J.: Health Measurement Scales: A Practical Guide to Their Development and Use, p. 156. Oxford University Press (2015)

14. Joormann, J., Quinn, M.E.: Cognitive processes and emotion regulation in depression. Depress. Anxiety **31**(4), 308–315 (2014)

15. Salle, D.L., Choueiry, J., Shah, D., Bowers, H., McIntosh, J., Ilivitsky, V., Knott, V.: Effects of ketamine on resting-state EEG activity and their relationship to perceptual/dissociative symptoms in healthy humans. Front. Pharmacol. **7** (2016)

16. Fekete, T., Beacher, F.D., Cha, J., Rubin, D., Mujica-Parodi, L.R.: Small-world network properties in prefrontal cortex correlate with predictors of psychopathology risk in young children: a NIRS study. NeuroImage **85**, 345–353 (2014)

17. Nunez, P.L., Srinivasan, R., Fields, R.D.: EEG functional connectivity, axon delays and white matter disease. Clin. Neurophysiol. **126**(1), 110–120 (2015)

18. Thakor, N.V., Tong, S.: Advances in quantitative electroencephalogram analysis methods. Annu. Rev. Biomed. Eng. **6**, 453–495 (2004)

19. Smart, O.L., Tiruvadi, V.R., Mayberg, H.S.: Multimodal approaches to define network oscillations in depression. Biol. Psychiatry **77**(12), 1061–1070 (2015)

20. Mohammadi, M., Al-Azab, F., Raahemi, B., Richards, G., Jaworska, N., Smith, D., de la Salle, S., Blier, P., Knott, V.: Data mining EEG signals in depression for their diagnostic value. BMC Med. Inform. Decis. Mak. **15**(1), 108 (2015)

21. Stewart, J.L., Coan, J.A., Towers, D.N., Allen, J.J.: Resting and task elicited prefrontal EEG alpha asymmetry in depression: support for the capability model. Psychophysiology **51**(5), 446–455 (2014)

22. Li, X., Hu, B., Sun, S., Cai, H.: EEG-based mild depressive detection using feature selection methods and classifiers. Comput. Methods Programs Biomed. **136**, 151–161 (2016)

23. Curran, S.L., Andrykowski, M.A., Studts, J.L.: Short form of the Profile of Mood States (POMS-SF): psychometric information. Psychol. Assess. **7**(1), 80 (1995)

Workshop on Affective, Psychological and Physiological Computing (APPC 2017)

Social Events Forecasting in Microblogging

Yang Zhou[1,2] , Chuxue Zhang[3] , Xiaoqian Liu[1] ,
Jingying Wang[1,2] , Yuanbo Gao[2] , Shuotian Bai[4] ,
and Tingshao Zhu[1(✉)]

[1] Institute of Psychology, Chinese Academy of Sciences, Beijing, China
tszhu@psych.ac.cn
[2] University of Chinese Academy of Sciences, Beijing, China
[3] Beihang University, Beijing, China
[4] School of Information Engineering, Hubei University of Economics,
Wuhan, Hubei, China

Abstract. Along with the popularization and rapid development of Internet, there is a growing interest in the research to identify the trend of social events on social media. Currently news could quickly spread on various social media (e.g. Sina Weibo) with a limited time, which may trigger the severity of the events that requires timely attention and responses from government. This paper proposes to predict the trend of social events on Sina Weibo, which is the most popular social media in China now. In this study, combining social psychology and communication sciences, we extracted comprehensive and effective features which may relate to the trend of social events on social media, and constructed the trend prediction models using three classical regression algorithms. The real social events data was used to verify the performance of our model, and the outstanding performance with precision of 0.56 and an f-measure of 0.71 demonstrate the efficiency of our features and models.

Keywords: Social events · Social media · Trend prediction

1 Introduction

The importance of predicting the trend of social events is unarguable. A recent study indicated that social media can capture vital information related to the events occurred in the real world [13]. We assume that if there is a sudden and evident increase in the amount of Weibo data related to an event, this event might be a significant event with great possibility. Based on this premise, in order to figure out the developmental mechanisms of social events, it is necessary to investigate the course of events on the social media.

Microblogs, such as Twitter and Weibo, are experiencing an explosive level of growth recently. Millions of worldwide microblog users broadcast their daily observations on an enormous variety of topics. The hot topics on Weibo could spread about in a short time, or finally turn into severe social events. If we can forecast the tendency of social events early and timely, it will be quite helpful for government to react promptly and efficiently.

© Springer International Publishing AG 2017
Y. Zeng et al. (Eds.): BI 2017, LNAI 10654, pp. 235–243, 2017.
https://doi.org/10.1007/978-3-319-70772-3_22

This paper proposes to predict the trends of social events on social media. In order to concurrently address all these technical challenges, we present a novel computational approach that combines the strengths of machine learning (e.g., LASSO regression) and theories of psychology and communication together. Based on social psychology and communication research, we extracted unprecedented amount of comprehensive and effective features which are relevant to the tendency of social events. Then we built predicting models utilizing regression algorithms in machine learning. The results of these experiments demonstrated the effectiveness of our proposed approaches.

2 Related Work

Currently, there are two main categories of research related to social events on microblogging: (1) theoretical research in social psychology, and (2) computer science.

In social psychology, several theories have been developed to explain the spreading of social events [3, 6, 12, 21]. Some of these theories provide clues on finding features relevant to predicting trends of social events. From a psychological perspective, there are two characteristics of collective action: the individual behaviors represents his group; the action aims at improving the current situation of the group [3, 22, 23]. Based on these two characteristics, events on social media can be treated as typical collective action, which involves a few theories of social psychology such as relative deprivation, group identity and group-based anger [18, 19]. Relative deprivation theory states that people will be motivated to undertake action if they recognize an unfavorable discrepancy between the expected group status and the real status [7]. In addition, social identity theory states that decisions to participate in social events are partly determined by perceptions of the in-group [18]. It has been proved that group-based anger has effects on collective action tendencies [20].

There are some research on prediction trends based on the data from the Internet. Khan [14] predicted a stock trend by mining news articles, and the accuracy was 70% by using KNN classifier. Achrekar et al. [1] introduced a data collection framework which predicted flu trends on social networks. Tung and Lu [17] proposed an event-driven warning model for predicting the depression tendency in web posts. Some research have been conducted on predicting the trends of social events on social media. Agarwal et al. [2] used the directed links of "*following*" on twitter to determine the flow of information and indicate each user's influence on others. They proposed certain features that could be used to classify "*trends*" and "*non-trends*" in earlier stages. Zhou et al. [26] analyzed the process of topic discussion according to three main factors (individual interest, group behavior, and time lapse). They assumed that the larger number of users presently involved in the discussion, meant the increased possibility of users participating in the discussion the next time. The longer the interval between present and peak time is, the less possibility that users participate in the discussion. Kwak et al. [16] tracked some hot topics on Twitter for nearly four months, and they found the temporal variation tendency of involved user's behaviors, active period and the number of relevant tweets posted. Zhao et al. [25] designed a new query expansion method to expand both keywords and key tweets by considering both semantic and social network relationships, and used the burstiness of key tweets to predict civil unrest events. Zhou et al. [28] predicting the trends of social events of 2015 on Sina Weibo.

Although there are some research about the tendency of events, most of them focus on theoretical issues. Some computer scientists paid more attention to improve performance of algorithms, and since they ignore the influential factors, the algorithms do not work very well [21].

This paper proposes more comprehensive features for predicting the trends of social events on social media. According to the current theories, following features were extracted from microblogs: the scale of events, emotional words, dynamic words, and opinion leaders. Additionally, our method was extensively evaluated on Sina Microblog data covering 69 social events with precision of 0.56 and an f-measure of 0.71. The prediction results of our method based on 5-fold cross-validation demonstrated our model's effectiveness.

3   Method

3.1   Feature Extraction

According to the theories and existing studies we have mentioned above, several types of features could be extracted from microblogs. Based on the social psychology theories (such as collective unconscious, contagion and conformity), the scale of event is an important factor in identifying whether an individual is willing to participate or not [8, 11]. As defined above, the emotional words should be considered as important features as well. We utilized Chinese affective lexicon proposed by Professor Lin of Dalian University of Technology to identify emotional words, which was based on the six sentiment categories proposed by Ekman [5, 24].

The number of microblogs containing predefined keywords was considered as an effective feature for our models. However, the language used in microblogs is highly dynamic and unstructured, which makes these fixed keywords to have only limited effect since they may not capture the fast-evolving expressions in Microblog. To handle the heterogeneity and dynamic challenge of microblogging data, DQE (Dynamic Query Expansion) dynamically generates a set of event-related key terms via a heterogeneous information network [25]. The key terms are exhaustively extracted and then weighted appropriately based on DQE's iterative process. This method has previously been applied in Sina Microblog for preliminary research, and it worked very well [27]. Specifically, the results of DQE include expansion microblogs (the number of microblogs which contain dynamic keywords) and emotional words (the number of emotional words contained in expansion microblogs).

According to opinion leader theory, opinion leaders' participation may make an event more serious. Influential users on Sina Microblog are usually verified, so we took the number of verified users that participated in events as one kind of features.

3.2 Feature Selection

To select the most effective feature set in our method, the above operations are utilized via an iterative feature selection algorithm, as shown in Algorithm 1. The details of the algorithm implementation are described as follows.

Given a feature set F which contains k features, denote $Y(F)$ as the F-measure of the model using F as the predictor variables. The best F-measure Y_{best} is initialized as $Y(F)$. The original feature set can be used to initialize effective subset S. At each iteration, one feature in S is eliminated, so the number of features in S_{temp} is $k - r$ in the r iteration. If $Y(S_{temp}) > Y_{best}$, it means S_{temp} is more effective than S, and S should be replaced by S_{temp}. If $Y(S_{temp}) < Y_{best}$, it means S is the most effective feature set, then the iterations will be terminated.

```
Algorithm 1 Feature Selection.
Input: Original feature F = {f₁,f₂,⋯,fₖ};
Output: Effective subset S;
Initialization: Set S = F; Y_best = Y(F);
1: while b = 1 do
2:      b = 0;
3:      for s ∈ S do
4:          S_temp = S-s;
5:          if Y(S_temp) > Y_best then
6:              Y_best = Y(S_temp);
7:              S = S_temp;
8:              b = 1;
9:          end if
10:     end for
11: end while
```

In order to improve prediction accuracy, we utilized backward elimination algorithm to remove redundant features [15].

First, for the original feature set, we eliminated one feature at a time in each iteration to produce a subset of features from all the features. Then we used the evaluation function to evaluate the feature subset and compared the evaluation result with stopping criterion. If the performance meets the stopping criterion, the iteration stops and the corresponding feature subset would be taken as the result of feature selection. Otherwise, another feature would be eliminated, and the new iteration continued. The general process of feature selection is depicted in Fig. 1.

3.3 Regression Methods

We constructed the prediction model based on three classical regression algorithms: linear regression, lasso regression, and ridge regression. Linear regression is an approach for modeling the relationship between a scalar dependent variable y and one or more explanatory variables (or independent variables) denoted X. LASSO (Least

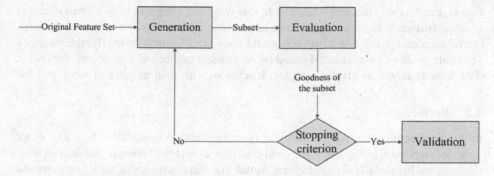

Fig. 1. The process of feature selection.

Absolute Shrinkage and Selection Operator) is a regression method based on minimizing the sum of the squares of error [9]. By means of least square method, unknown data can be easily obtained and minimize the sum of the squares of the error between real data and calculated data. LASSO is formulated as Eq. (1):

$$\operatorname{argmin} \left\| q^1 D^t + q^2 D^{t,2} + \cdots + q^{k+1} D^{t,k} - Y^t \right\|_2^2 + \rho \left\| q^l \right\|_1 \tag{1}$$

In which, ρ is the regularization parameter to prevent overfitting, and $q^l (l \in [1,k])$ is coefficient of regression.

Ridge regression uses least squares but shrinks the estimated coefficients towards zero [10]. This introduces some bias, but can greatly reduce the variance, resulting in a better mean-squared error. It performs particularly well when there is a subset of true coefficients that are small or even zero.

Elastic Net is the combination of ridge regression and lasso regularization [4]. It is a regularized regression method that linearly combines the L1 and L2 penalties of the lasso and ridge methods. Like lasso, elastic net can generate reduced models by generating zero-valued coefficients.

4 Experiment

4.1 Dataset and Labels

69 social events (46 in 2014, 23 in 2015) were selected from events list provided by People's Daily Online (http://yuqing.people.com.cn/GB/210114/) and Xinhua (http://www.xinhuanet.com/yuqing/index.htm). After thorough examination of these events, we summarized some keywords or hashtags for each one. The number of microblogs containing corresponding keywords were treated as outcome variables [13]. For each event, we collected one million users' microblogs posted from the day of the event.

In the case that the microblogs associated with certain sensitive events may be deleted by official organizations or users, the number of microblogs of some events were just few. In order to reduce invalid data, the events with the number of microblogs less than 100 were removed, and the final number of valid events was 51. For all the events, the data was taken as a sample. In this way, 255 samples (51 × 5) was obtained in total. But since some events did not last as long as five days, the number of related microblogs on the third or fourth day could have been lower. Obviously, the tendency of events couldn't be predicated based on insufficient number of data points, thus some data were removed as invalid samples, leaving us with final number of samples 186.

4.2   Performance

In practical application, people always pay more attention to whether the social event will develop into a big one. We analyzed the events in "*annual public opinion*" released by People's Daily online and found that there were more than one thousand microblogs (within 1 million active users' microblogs collected) related to each event. So we defined an event as a big one when the number of correlated microblogs is more than one thousand.

The events in 2014 were considered as training set. After collecting valid data, the features we mentioned above could be extracted from the data. In order to improve the effect of the model, irrelevant and redundant features should be removed by using backward elimination algorithm introduced before. By comparing the prediction results of three kinds of regression methods based on 5-fold cross-validation, lasso regression performed better with the precision 0.78 and the recall 0.88 (Table 1). We measure the mean absolute error and correlation coefficient between the prediction and true volume. A correlation of 0.68 and a mean absolute error of 0.42, achieved in the regression model, shows that our linear model serves as a good approximate to the actual model.

Table 1. Performance comparison among three regression methods.

Method	Precision	Recall	F-measure
Lasso	0.78	0.88	0.82
Ridge regression	0.73	0.85	0.79
Elastic net	0.66	0.77	0.71

This predictive model was used to apply to the events in 2015, and achieved a precision of 0.33, a recall of 0.43. Considering the negative influence could be caused by the interval between training set (events in 2014) and testing set (events in 2015), samples in 2015 is sorted in chronological order as 135–186 based on the 134 samples of 2014, sequenced as 1–134. By taking 10 as the sliding window length, the first ten samples of 2015 (135–144) was predicted by the samples of 2014 (1–134), and the prediction of eleventh to twentieth samples of 2015 (145–154) was by using the common training model of the samples in 2014 together with the first ten samples (1–144) in 2015. By this analogy, the results are shown as Table 2.

Figure 2 showed that the model achieved better effect with the increasing data of training set despite a slight decrease in the fifth test, and performed well prediction validity based on 0.71 f-measure, which illustrates the necessity to update the prediction model in time. In practice application, it's best to add the date of recent social events into the training set from time to time, to ensure a more accurate prediction in the future.

Table 2. The application results of the predictive model.

Train	Pred	Precision	Recall	F-measure
1–134	135–186	0.33	0.43	0.38
1–134	135–144	0.4	0.4	0.4
1–144	145–154	0.5	0.5	0.5
1–154	155–164	0.5	1	0.67
1–164	165–174	0.4	1	0.57
1–174	175–186	0.56	1	0.71

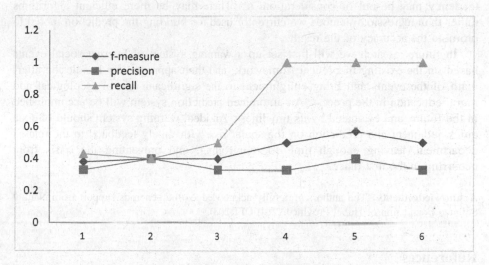

Fig. 2. The application results of the predictive model.

5 Conclusion

This paper is with distinctive feature by adopting the interdisciplinary research to combine psychology, informatics and sociology together, making a thorough study into the theory and method of the development tendency of social events. Under the background of big data, we innovatively put forward predicting the tendency of social events based on social-network data. Through batch downloading the user data, we achieve a real-time monitoring on the movement of public opinion and are able to make a prediction.

We first gain theoretical support through investigation on the research relevant to network events. Based on the solid theoretical foundation, this paper takes a series of more complete factors in order to predict the tendency of social events in Microblogging. Then we construct a platform to collect, store and calculate the big data, and downloaded active users' digital records utilizing APIs of Sina Weibo and collected actual events from People's Daily Online and Xinhua. Finally, using machine learning to acquire the predicting models,we built a model for predicting tendency of events based on selected features, which achieved a precision of 0.56 and an f-measure of 0.71. The results demonstrated that the features we obtained not only backed by theoretical basis, but also proved to be effective, which are of practical application value in predicting the tendency of events on Microblog.

Events tendency forecasting is of great significance in the early stage of the happening of Internet social events, while also a demanding and challenging work. Though certain achievement has been made, there are still limitations to this project. (a) Only the data of Sina Weibo has been used for research, despite the popularity of which in China, given a large deviations in the types of social media could bring more effectiveness to the result. (b) Factors will cause critical influence on the social events tendency may be out of consideration. (c) There may be more efficient algorithms rather than regression methods we currently used for building the prediction model to promote the accuracy of the result.

In future research we will first set up a warning system of Internet social events based on the existing theoretical framework, and then appropriately handle the aftermath of the events and bring enlightenment to the significant role of ideological and moral education in the process. An automated prediction system will be accomplished in the future, and evaluated by its timeliness. An ideal warning system should offers a quick judgment and prediction on the social risk with timely feedback to the related department, leaving enough time for controlling, and preventing the crisis from occurring in the first place.

Acknowledgements. The authors gratefully acknowledges the generous support from Natural Science Foundation of Hubei Province (2016CFB208).

References

1. Achrekar, H., Gandhe, A., Lazarus, R., Yu, S.-H., Liu, B.: Online social networks flu trend tracker: a novel sensory approach to predict flu trends. In: Gabriel, J., Schier, J., Van Huffel, S., Conchon, E., Correia, C., Fred, A., Gamboa, H. (eds.) BIOSTEC 2012. CCIS, vol. 357, pp. 353–368. Springer, Heidelberg (2013). doi:10.1007/978-3-642-38256-7_24
2. Agarwal, P.: Prediction of trends in online social netwok. Ph.D. thesis, Indian Institute of Technology New Delhi (2013)
3. Brunsting, S., Postmes, T.: Social movement participation in the digital age predicting offline and online collective action. Small Group Res. **33**(5), 525–554 (2002)
4. De Mol, C., De Vito, E., Rosasco, L.: Elastic-net regularization in learning theory. J. Complex. **25**(2), 201–230 (2009)
5. Ekman, P.: Facial expression and emotion. Am. Psychol. **48**(4), 384 (1993)
6. Glasbergen, P.: Global action networks: agents for collective action. Glob. Environ. Change **20**(1), 130–141 (2010)
7. Goio, F., Gurr, T.R.: Why Men Rebel. Princeton University Press, Princeton (1974)
8. Granovetter, M.: Threshold models of collective behavior. Am. J. Sociol. **83**, 1420–1443 (1978)
9. Hans, C.: Bayesian lasso regression. Biometrika **96**(4), 835–845 (2009)
10. Hoerl, A.E., Kennard, R.W.: Ridge regression: biased estimation for nonorthogonal prolems. Technometrics **12**(1), 55–67 (1970)

11. Hornsey, M.J., Blackwood, L., Louis, W., Fielding, K., Mavor, K., Morton, T., O'Brien, A., Paasonen, K.E., Smith, J., White, K.M.: Why do people engage in collective action? Revisiting the role of perceived effectiveness. J. Appl. Soc. Psychol. 36(7), 1701–1722 (2006)

12. Ivancevich, J.M., Matteson, M.T., Konopaske, R.: Organizational Behavior and Management. Bpi/Irwin (1990)

13. Kaleel, S.B., Abhari, A.: Cluster-discovery of Twitter messages for event detection and trending. J. Comput. Sci. 6, 47–57 (2015)

14. Khan, S.: Mining news articles to predict a stock trend (2014)

15. Kumar, A., Naughton, J., Patel, J.M., Zhu, X.: To join or not to join? Thinking twice about joins before feature selection. In: Proceedings of the 2016 ACM SIGMOD International Conference on Management of Data, SIGMOD, vol. 16 (2016)

16. Kwak, H., Lee, C., Park, H., Moon, S.: What is Twitter, a social network or a news media? In: Proceedings of the 19th International Conference on World Wide Web. pp. 591–600. ACM (2010)

17. Tung, C., Lu, W.: Analyzing depression tendency of web posts using an event-driven depression tendency warning model. Artif. Intell. Med. 66, 53–62 (2016)

18. Van Zomeren, M., Postmes, T., Spears, R.: Toward an integrative social identity model of collective action: a quantitative research synthesis of three socio-psychological perspectives. Psychol. Bull. 134(4), 504 (2008)

19. Van Zomeren, M., Spears, R.: Metaphors of protest: a classification of motivations for collective action. J. Soc. Issues 65(4), 661–679 (2009)

20. Van Zomeren, M., Spears, R., Fischer, A.H., Leach, C.W.: Put your money where your mouth is! Explaining collective action tendencies through group-based anger and group efficacy. J. Pers. Soc. Psychol. 87(5), 649 (2004)

21. Wan, M., Liu, L., Qiu, J., Yang, X.: Collective action: definition, psychological mechanism and behavior measurement. Adv. Psychol. Sci. 19(5), 723–730 (2011)

22. Wright, S.C.: The next generation of collective action research. J. Soc. Issues 65(4), 859–879 (2009)

23. Wright, S.C., Taylor, D.M., Moghaddam, F.M.: Responding to membership in a disadvantaged group: from acceptance to collective protest. J. Personal. Soc. Psychol. 58(6), 994 (1990)

24. Yu, X.L.L.H.P., Jianmei, R.H.C.: Constructing the affective lexicon ontology. J. China Soc. Sci. Tech. Inf. 2, 006 (2008)

25. Zhao, L., Chen, F., Dai, J., Hua, T., Lu, C.T., Ramakrishnan, N.: Unsupervised spatial event detection in targeted domains with applications to civil unrest modeling. PLoS ONE 9(10), e110206 (2014)

26. Zhou, Y., Guan, X., Zhang, Z., Zhang, B.: Predicting the tendency of topic discussion on the online social networks using a dynamic probability model. In: Proceedings of the Hypertext 2008 Workshop on Collaboration and Collective Intelligence, pp. 7–11. ACM (2008)

27. Zhou, Y., Lu, T., Zhu, T., Chen, Z.: Environmental incidents detection from chinese microblog based on sentiment analysis. In: Zu, Q., Hu, B. (eds.) HCC 2016. LNCS, vol. 9567, pp. 849–854. Springer, Cham (2016). doi:10.1007/978-3-319-31854-7_88

28. Zhou, Y., Zhang, L., Liu, X., Zhang, Z., Bai, S., Zhu, T.: Predicting the trends of social events on Chinese social media. In: Cyberpsychology, Behavior and Social Networking (accepted)

Study on Depression Classification Based on Electroencephalography Data Collected by Wearable Devices

Hanshu Cai[1], Yanhao Zhang[1,4], Xiaocong Sha[1], and Bin Hu[1,2,3(✉)]

[1] Gansu Provincial Key Laboratory of Wearable Computing,
School of Information Science and Engineering, Lanzhou University,
Lanzhou 730000, China
{caihsh13,zhangyanhao15,shaxc14,bh}@lzu.edu.cn
[2] Center of Excellence in Brain Science and Intelligence Technology Chinese
Academy of Sciences (CEBSIT), Shanghai 200031, China
[3] Beijing Institute for Brain Disorders, Beijing 100069, China
[4] Army Staff Department, Western Theater Command, PLA,
Lanzhou 730000, China

Abstract. Depression has become a disease, which may threaten millions of families' well-being. The current method of screening depression is subjective, labor-consuming and costly. Study on Electroencephalogram (EEG) has become a new direction to explore an objective, low-cost and accurate method to detect depression. In this paper, three-electrode EEG data of 158 subjects (90 depressed and 68 normal control) in resting state, and under audio stimulation (positive and negative) were collected and processed. After feature selection using Sequential Floating Forward Selection (SFFS), four popular classification methods were applied and classification accuracies were verified using 10-fold cross validation. Results have shown the accuracy of classification will be improved when male and female are classified separately. The highest accuracy of male and female classification are 91.98%, 79.76%, respectively, compare to 77.43% when the classification is processed as gender-free. The effective depressive features of male and female are also different, which may be caused by the differences of brain structure. This research suggests a possible pervasive method of depression classification for future clinical application.

Keywords: Depression · EEG · Pervasive · Health care

1 Introduction

Depression is a common mental disorder. At present, more than 300 million people have been suffering from depression worldwide [1]. Depression has brought great losses not only to individuals but also the society as a whole. However, depression could not be diagnosed using physiological data, such as blood pressure or body temperature. Current methods are scales-based interviews, which are carried out by psychiatrists. The whole process is labor-consuming and costly, and results are highly depended on psychiatrists' subjective experience. Therefore, recent studies are trying to

Y. Zeng et al. (Eds.): BI 2017, LNAI 10654, pp. 244–253, 2017.
https://doi.org/10.1007/978-3-319-70772-3_23

use Electroencephalogram (EEG) as a tool to explore a more objective, low-cost and accurate method to detect depression. After all, EEGs are ubiquitous bioelectrical signals, which don't subject to human subjective controls [2]. Recent studies have shown that men and women have significant differences in the brain structure, mental state, and cognitive methods [3]. And depressed patients have shown significant differences in relative convergences of EEGs of intra-left temporal and front-left temporal lobes at delta band between male and female depression patients [4]. Thus in this paper, we are trying to analyze and compare the results of depression classification of EEG data in different gender groups in resting state, and under audio stimulation (both positive and negative) using 64 different linear and nonlinear features, in order to explore the possibility to distinguish depressed patients with normal controls by valid EEG characteristics.

2 Related Work

The EEG signal can be divided into five bands according to the frequency, which are Delta wave (<4 Hz), Theta wave (4–8 Hz), Alpha wave (8–14 Hz), Beta wave (14–30 Hz), Gamma wave (30–50 Hz) [5]. In the EEG signal processing and analysis aspects, the main research methods include nonlinear dynamic analysis and frequency domain analysis, which includes frequency estimation and classical power spectrum estimation [6]. At present, the related researches on EEG mainly focus on three aspects: the absolute power and relative power of each band, the asymmetry of power spectrum between the hemispheres of the brain, and the correlation between each EEG signals. Knott analyzed power spectrum of EEG data of male depressive patients and normal controls in 2001. It was found that the relative power values of all brain regions of the depressive patients were significantly larger than those of normal subjects, and the Beta absolute power values in the front of the bilateral brain regions are significantly greater than those in normal subjects [7]. Pollock found that depressed patients have greater Alpha and Beta band power spectrum than normal in 1990 [8]. Lznak found that the slow wave of EEG signal will be increased and the fast wave will be weakened with the relief of depressive symptoms [9]. In addition to frequency domain analysis, many researchers have begun attempting to analyze EEG signals of depressive patients with nonlinear dynamics. Bachmann calculated fractal dimension directly in the time domain with nonlinear Higuchi's Fractal Dimension (HFD) method and detected a small (3%) increase in depression patients, which was significantly different in each lead. It indicated that the nonlinear features of EEG signal based on the chaotic theory had a rising trend in depression patient [10].

Some studies had shown that there were some difference in brain structure between male and female. Miller found that women with childhood depression had higher right mid-frontal alpha suppression, and men with childhood depression had higher left mid-frontal alpha suppression, relative to comparison subjects. At all scalp sites, women showed greater alpha power than men [11]. Davidson's results revealed significantly greater relative right-hemisphere activation during emotion versus non-emotion trials only in females; males showed no significant task-dependent shifts in asymmetry between conditions [12]. Trotman's experiment suggested a sex-related difference in the degree of lateralization of hemispheric function, with males having a more strict segregation of function [13].

3   Experiment Analysis

3.1   Experiment Design

In this paper, EEG data was collected by three-electrode EEG collector (Fig. 1), which was developed by Ubiquitous Awareness and Intelligent Solutions Lab (UAIS). Based on the international 10–20 standard [14], Fp1 and Fp2 electrode were selected as these electrodes were closely related with brain emotions and covered with no hair [15]. In the process of data collection, the semi-wet electrode was used, due to its low cost and easy to apply, unlike the traditional wet electrodes, which normally require a professional technician to setup up (Fig. 1).

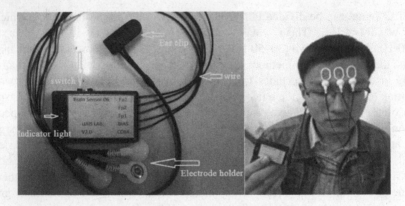

Fig. 1. Three-electrode EEG collector

3.2   EEG Data Collection

In this paper, three-electrode EEG data of 158 subjects in resting state, and under audio stimulation (both positive and negative) were collected (34 cases of male healthy individuals, 40 cases of male depressed patients, 34 cases of female healthy individuals and 50 cases of female patients with depression). Screened by using the Mini International Neuropsychiatric Interview (MINI) [16], the MINI scores of each depressed patient who was selected to experiment was in line with the American Diagnostic Classification of Diseases (DSM-IV) [17] Depression Diagnostic Criteria, and their PHQ-9 [18] score which was the depression module of the Patient Health Questionnaire (PHQ) was greater than or equal to 5 points.

The experiment was conducted in an indoor environment which was quiet, well ventilated, with no light exposure, and with no strong electromagnetic interference. The subjects were asked to be quietly eyes closed and try to remain relaxed, taking effect to make the interference, such as Electrooculogram (EOG) and Electromyography (EMG) to a minimum [19].

The resting state EEG was taken when subjects were asked to be relaxed with eyes closed and 60 s were recorded including 15,000 sample points. Two audio stimuli (positive and negative) were selected from the International Affective Digital Sounds (IADS) [20, 21]. Each piece of audio stimulation played for 6 s, which includes 1500 sample points, and followed by 6 s rest. Then, a total of 24 s of EEG data under audio stimulation were collected.

4 Methodology

The EEG data were preprocessed and features extracted to get a feature matrix. Then, Sequential Floating Forward Selection (SFFS) algorithm [22] was used to select most effective feature subsets and four popular classification methods including Support Vector Machine (SVM) [23], K-Nearest Neighbor (KNN) [24], Artificial Neural Networks (ANN) [25, 26], Decision Tree (DT) [27] were applied and classification accuracies were verified using 10-fold cross validation [28].

4.1 Data Preprocessing and Feature Extraction

The band-pass filter was firstly used to remove interference noises, such as EMG, ECG and from close by hydro power lines. Then the method based on wavelet transform which was proposed by Peng and others in 2011 [29] was used to remove the EOG. Thus more pure EEG data was prepared.

Table 1. Features used in FP1 and FP2

Features list			
1	Theta absolute power	17	Gamma relative power
2	Theta relative power	18	Gamma absolute center frequency
3	Theta absolute center frequency	19	Gamma relative center frequency
4	Theta relative center frequency	20	Gamma power spectrum entropy
5	Theta power spectrum entropy	21	Full band absolute power
6	Alpha absolute power	22	Full band relative power
7	Alpha relative power	23	Peak-to-peak
8	Alpha absolute center frequency	24	Variance
9	Alpha relative center frequency	25	Hjorth activity
10	Alpha power spectrum entropy	26	Inclination
11	Beta absolute power	27	Kurtosis
12	Beta relative power	28	C0 complexity
13	Beta absolute center frequency	29	Correlation integrals
14	Beta relative center frequency	30	Power spectrum entropy
15	Beta power spectrum entropy	31	Shannon entropy
16	Gamma absolute power	32	Kolmgolov entropy

A variety of linear and nonlinear features of EEG data were extracted using modern power spectrum estimation, and a total of 64-dimensional data feature matrix was established. The features used in this paper are shown in Table 1. Thus 32 features for each electrode-site were extracted and a 64-dimensional feature matrix was constructed.

4.2   Feature Selection and Classification

SFFS algorithm was firstly used as the core algorithm of feature selection. Then four classifiers (SVM, KNN, ANN, and DT) were applied. Afterward, 10-fold cross validation was used to verify the accuracy. Finally, highest accuracy model and most effective feature subset were obtained by comparison.

Figure 2 shows the flow of feature selection and classification. In this paper, the SFFS algorithm was used to search the feature matrix from the empty set, construct the feature subset, and then the feature subset was used to classify the data by SVM, KNN, ANN, and DT. The classifier returns the obtained classification result to the SFFS algorithm until the classification accuracy is no longer rise.

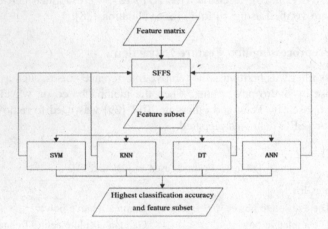

Fig. 2. Flow chart of feature selection and classification

Thus, the selected feature subset is a most effective depression feature selected by the corresponding classifier, and the classification accuracy is the best classification result obtained by this classifier. Finally, the classification results of four classifiers are compared, and the results of the classifier with the highest classification accuracy are selected as the final experimental results.

5   Results and Discussion

5.1   Results of Depression Classification Accuracy

KNN perform the best on the EEG Data when processed gender-free. As shown in Fig. 3, the classification accuracy of KNN was 76.83%, 71.07% and 77.43% in the resting state, under positive audio stimulation and under negative audio stimulation, respectively.

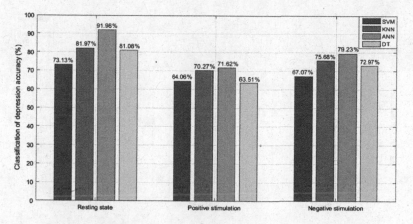

Fig. 3. The depression classification accuracy of male subjects

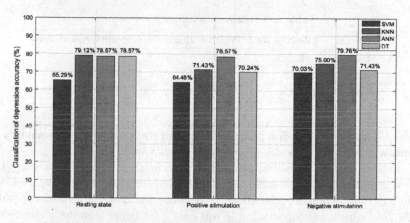

Fig. 4. The depression classification accuracy of female subjects

ANN perform the best on the male subject EEG Data. As shown in Fig. 4, the classification accuracy of ANN was 91.98%, 71.62% and 79.23% in the resting state, under positive audio stimulation and under negative audio stimulation, respectively.

KNN perform the best on the female subject EEG Data. As shown in Fig. 5, in the resting state, KNN got the highest accuracy, 79.12%. Under the audio stimulation, ANN got the highest accuracy (positive 78.57%, negative 79.76%).

As shown in Table 2, for male, female and gender-free, the accuracy of the highest classification of depression in resting state, under positive audio stimulation, and negative audio stimulation was compared.

The results showed the accuracy of depression classification increases when male and female are processed separately, and the accuracy of classification would be increased under negative audio stimulation and decreased under positive audio stimulation. Matter of fact, the accuracy of classification was the lowest when under positive audio stimulation, regardless whether gender is considered. This was consistent

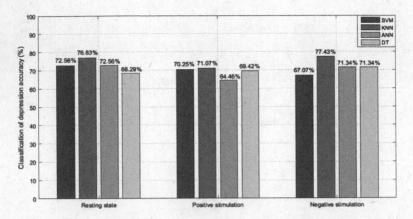

Fig. 5. The gender-free classification accuracy of depression

Table 2. Classification accuracy comparison

Subjects	Resting state	Positive stimulation	Negative stimulation
Male	91.98%	71.62%	79.23%
Female	79.12%	78.57%	79.76%
Gender-free	76.83%	71.07%	77.43%

with the theory that depression patients look at themselves and the surrounding environment was pessimistic, which proposed by Beck's study [30].

In male case, the depression classification accuracy in resting state was 91.98% which was higher than under audio stimulation (positive 71.62%, negative 79.23%). In female case, the depression classification accuracy under negative audio stimulation was 79.76% which was highest, but it was only a bit better than the accuracy of the other two cases (resting state 79.12%, positive stimulation 78.57%). Compared to 77.43% when classification was processed as gender-free, the accuracy rate of male classification would be greatly improved and the accuracy of female classification would be slightly improved when male and female were classified as depression separately. As the specific differences between the current male and female brain work had not yet been clarified, the specific causes of this difference in the results need further study.

5.2 Results of Feature Selection

SFFS algorithm founded the effective feature subset for each experiment. The effective feature subset was different from male to female, and it was also different from in resting state to under audio stimulation. As shown in Table 3, for male in resting state, the most effective feature subset was Gamma relative center frequency at Fp2, Theta relative center frequency and Beta absolute power at Fp1, but for female under negative audio stimulation, the most effective feature subset was Theta absolute power and Beta absolute center frequency at Fp1. This may be due to the different brain structure of male and female, and man and women were also reflected different in same external stimulation.

To explore the difference of EEG features between depressive patients and normal controls, the value of the most effective feature subset was calculated for male and female separately. As shown in Table 3. For male in resting state, the relative frequency of Gamma wave at Fp2 electrode of depressive patients was 9.52% higher than that in normal controls, and the absolute power of the Beta wave at Fp1 electrode was 12.67% higher, but the relative center frequency of Theta at Fp1 electrode was 12.57% lower. For female under negative audio stimulation, the absolute power of Theta wave at Fp1 electrode of depressive patients was 18.75% higher than that in normal controls, but the absolute center frequency of Beta at Fp1 electrode was 1.96% lower. Thus, the depressive features of both male and female were significantly varied in different states, so it was better for people of different gender to do depression diagnosis with different features-based models.

Table 3. Feature selection results and the differences of EEG features between depressive patients and normal controls

Feature selection results	Normal controls	Depressive patients	Difference rate
Male in resting state:			
Fp1: Theta relative center frequency	10.58	9.25	−12.57%
Fp1: Beta absolute power	1.38	1.58	12.67%
Fp2: Gamma relative center frequency	21.53	23.58	9.52%
Female under negative audio stimulation:			
Fp1: Theta absolute power	35.46	42.11	18.75%
Fp1: Beta absolute center frequency	19.86	19.47	−1.96%

6 Conclusion

Based on the study of the classification of depression in different sex groups, the necessity of classification of depression based on EEG data in different sex groups was expounded, which provides a new idea for the study of depression based on EEG data. For male: the best condition is in resting state. The most effective feature subset was Gamma relative center frequency at Fp2, Theta relative center frequency and Beta absolute power at Fp1. The highest classification accuracy was 91.98% compared to 77.43% when classification was processed as gender-free, and the accuracy increased by 14.55%. For female: the best condition is under negative audio stimulation. The most effective feature subset was Theta absolute power and Beta absolute center frequency at Fp1. The highest classification accuracy was 79.76% compared to 77.43% when classification was processed as gender-free, and the accuracy increased by 2.33%. The effective depressive features of male and female are also different, which may be caused by the differences of brain structure. This research suggests a possible pervasive method of depression classification for future clinical application.

Acknowledgment. This work was supported by the National Basic Research Program of China (973 Program) (No. 2014CB744600), the National Natural Science Foundation of China (Grant No. 61632014, No. 61210010), Program of Beijing Municipal Science & Technology Commission (No. Z171100000117005), the Program of International S&T Cooperation of MOST (No. 2013DFA11140).

References

1. Depression. [EB/OL]. http://www.who.int/mediacentre/factsheets/fs369/en/
2. Moruzzi, G., Magoun, H.W.: Brain stem reticular formation and activation of the EEG. Electroencephalogr. Clin. Neurophysiol. **1**(1–4), 455–473 (1949)
3. Gloor, P.: Discoverer of the brain wave. Science **168**, 562–563 (1970). (Book reviews: Hans Berger on the electroencephalogram of man. The fourteen original reports on the human electroencephalogram)
4. Ahmadlou, M., Adeli, H.: Spatiotemporal analysis of relative convergence of EEGs reveals differences between brain dynamics of depressive women and men. Clin. EEG Neurosci. **44**(3), 175 (2013)
5. Mantini, D., Perrucci, M.J., Del, G.C., Romani, G.L., Corbetta, M.: Electrophysiological signatures of resting state networks in the human brain. Proc. Natl. Acad. Sci. USA **104**(32), 13170 (2007)
6. Marple, S.L.J.: A tutorial overview of modern spectral estimation, vol. 4, pp. 2152–2157 (1989)
7. Knott, V., Mahoney, C., Kennedy, S., et al.: EEG power, frequency, asymmetry and coherence in male depression. Psychiatry Res. Neuroimaging **106**(2), 123–140 (2001)
8. Pollock, V.E., Schneider, L.S.: Quantitative, waking EEG research on depression. Biol. Psychiatry **27**(7), 757–780 (1990)
9. Iznak, A.F., Iznak, E.V., Sorokin, S.A.: Changes in EEG and reaction times during the treatment of apathetic depression. Neurosci. Behav. Physiol. **43**(1), 79–83 (2013)
10. Bachmann, M., Lass, J., Suhhova, A., et al.: Spectral asymmetry and Higuchi's fractal dimension measures of depression electroencephalogram. Comput. Math. Methods Med. **2013** (2013). doi:10.1155/2013/251638
11. Miller, A., Fox, N.A., Cohn, J.F., Forbes, E.E., Sherrill, J.T., Kovacs, M.: Regional patterns of brain activity in adults with a history of childhood-onset depression: gender differences and clinical variability. Am. J. Psychiatry **159**(6), 934–940 (2002)
12. Davidson, R.J., Schwartz, G.E., Pugash, E., Bromfield, E.: Sex differences in patterns of EEG asymmetry. Biol. Psychol. **4**(2), 119–138 (1976)
13. Trotman, S.C., Hammond, G.R.: Sex differences in task-dependent EEG asymmetries. Psychophysiology **16**(5), 429 (1979)
14. Jasper, H.H.: The ten-twenty electrode system of the international federation. Electroencephalogr. Clin. Neurophysiol. **10**, 367–380 (1985)
15. Petrantonakis, P.C., Hadjileontiadis, L.J.: Emotion recognition from EEG using higher order crossings. IEEE Trans. Inf. Technol. Biomed. **14**(2), 186 (2010). A Publication of the IEEE Engineering in Medicine & Biology Society
16. Rai, G.S.: The mini-mental state examination. J. Am. Geriatr. Soc. **41**(3), 346 (1993)
17. Sheehan, D.V., Janavs, J., Baker, R., et al.: MINI-Mini international neuropsychiatric interview -English version 5.0.0-DSM-IV. J. Clin. Psychiatry **59**, 34–57 (1998)
18. Kurt, M.D., Spitzer, R.L., Dsw, J.B.W.W.: The PHQ-9. J. Gen. Intern. Med. **16**(9), 606–613 (2001)

19. Fatourechi, M., Bashashati, A., Ward, R.K., Birch, G.E.: EMG and EOG artifacts in brain computer interface systems: a survey-clinical neurophysiology. Clin. Neurophysiol. **118**(3), 480–494 (2007)
20. International Affective Digital Sounds. [DB/OL]. https://www.iads.org/
21. Stevenson, R.A., James, T.W.: Affective auditory stimuli: characterization of the International Affective Digitized Sounds (IADS) by discrete emotional categories. Behav. Res. Methods **40**(1), 315–321 (2008)
22. Lopes, F.M., Martins, D.C., Barrera, J., Cesar, R.M.: SFFS-MR: a floating search strategy for GRNs inference. In: Dijkstra, T.M.H., Tsivtsivadze, E., Marchiori, E., Heskes, T. (eds.) PRIB 2010. LNCS, vol. 6282, pp. 407–418. Springer, Heidelberg (2010). doi:10.1007/978-3-642-16001-1_35
23. Huang, J., Shao, X., Wechsler, H.: Face pose discrimination using support vector machines (SVM). Int. Conf. Pattern Recognit. **1**(4), 154–156 (1998)
24. Guo, G., Wang, H., Bell, D., Bi, Y., Greer, K.: KNN model-based approach in classification. In: Meersman, R., Tari, Z., Schmidt, D.C. (eds.) OTM 2003. LNCS, vol. 2888, pp. 986–996. Springer, Heidelberg (2003). doi:10.1007/978-3-540-39964-3_62
25. Yao, X.: Evolving artificial neural networks. Proc. IEEE **87**(9), 1423–1447 (1999)
26. Meng, X.: Study on the fresh level analog of the meat with artificial neutral network. Chin. J. Spectrosc. Lab. **21**(5), 970–973 (2004)
27. Quinlan, J.R.: Induction on decision tree. Mach. Learn. **1**(1), 81–106 (1986)
28. Kohavi, R.: A study of cross-validation and bootstrap for accuracy estimation and model selection. Int. Joint Conf. Artif. Intell. **14**, 1137–1143 (1995)
29. Peng, H., Hu, B., Qi, Y., et al.: An improved EEG de-noising approach in electroencephalogram (EEG) for home care. In: International Conference on Pervasive Computing Technologies for Healthcare, pp. 469–474 (2011)
30. Beck, A.T., Steer, R.A., Garbin, M.G.: Psychometric properties of the Beck Depression inventory: twenty-five years of evaluation. Clin. Psychol. Rev. **8**(1), 77–100 (1988)

Corticospinal Tract Alteration is Associated with Motor Performance in Subacute Basal Ganglia Stroke

Jing Wang[1], Ziyu Meng[1], Zengai Chen[2], and Yao Li[1($\boxtimes$)]

[1] School of Biomedical Engineering, Shanghai Jiao Tong University,
Shanghai 200240, China
yaoli@sjtu.edu.cn
[2] Renji Hospital, School of Medicine,
Shanghai Jiao Tong University, Shanghai 200127, China

Abstract. Microstructural changes of corticospinal tract (CST) correlate with motor performance in ischemic stroke patients. However, the findings about CST structural alteration after stroke varied due to different lesion sites, recovery degree and different disrupted pathways. Basal ganglia (BG) plays an important role in motor control and execution. Despite the intimate anatomical relation between BG and CST, the impact of BG stroke lesion on CST integrity and its association with motor performance remains unclear. In this study, we recruited 10 stroke patients with lesion specifically in BG area and investigate the CST structural alteration 1–3 months post stroke using diffusion tensor imaging (DTI) methodology. The bilateral cerebral peduncle (CP), posterior limb of internal capsule (PLIC) and superior cornal radiation (sCR) areas were investigated and the regional DTI parameters were calculated. Our results showed a significant decline of ipsileional FA in CP, PLIC and sCR, which is in correlation with patient's concurrent Fugl-Meyer index (FMI) score. Moreover, the lateralization of FA in CP and PLIC negatively correlated with FMI. Our work showed that the CST structural alteration associated with motor function of BG stroke patients within subacute stage. The FA value and its lateralization served as informative markers for motor performance evaluation.

Keywords: Basal ganglia · Diffusion tensor imaging · Corticospinal tract · Motor performance · Stroke

1 Introduction

As a major neuronal pathway of the brain, corticospinal tract (CST) mediates voluntary fine movements. Fibers of CST originate from primary motor and premotor cortex, pass through corona radiata (CR), internal capsule (IC), and cerebral peduncle (CP), and reach to the brainstem area axially [1]. Microstructural changes of CST have been shown correlated with motor damage in ischemic stroke patients through a variety of diffusion imaging (DTI) studies [2–5].

Schaechter et al. [3] found positive correlation between fractional anisotropy (FA) value of bilateral CST and motor score in chronic stroke patients. Lindenberg et al. [4]

© Springer International Publishing AG 2017
Y. Zeng et al. (Eds.): BI 2017, LNAI 10654, pp. 254–260, 2017.
https://doi.org/10.1007/978-3-319-70772-3_24

found that the stroke patient motor performance was related to ipsilesional FA and radial diffusion (RD) value. Moreover, the patients with middle cerebral artery stroke showed lower FA in the ipsilesional CR and posterior limb of internal capsule (PLIC) 30 days after stroke onset [6]. However, the findings about CST structural alteration after stroke varied due to different lesion sites, recovery degree and different disrupted pathways [5, 7, 8].

Basal ganglia (BG) is composed of three major nuclei: caudate nucleus, lenticular nucleus including putamen and globus pallidus, and the amygdala. It plays an important role in motor control and execution [9]. Descending fibers of CST pass through the head and body of caudate nucleus, and the PLIC area is surrounded by thalamus and caudate nucleus, as a major part of CST with high density of fibers [10, 11]. Despite the intimate relation between BG and CST, the impact of BG stroke lesion on CST integrity and its association with motor performance remains unclear.

In this study, we recruited the BG stroke patients and investigated the CST structural alteration within subacute stage, i.e. 1–3 months, after stroke onset using DTI methodology. Specifically, we select superior corona radiata (sCR), PLIC and CP as three regions of interest (ROIs) to study their relation with patients' motor performance.

2 Method and Subjects

2.1 Patients

Ten patients with stroke in BG and surrounding area were recruited from Shanghai Renji Hospital, Shanghai, China. All of them were provided the written informed consent approval by the ethics committee of Renji Hospital. All patients suffered from unilateral lesion. All subjects were scanned 1–3 months after stroke onset and showed motor deficits during scanning. The demographical and clinical characteristics of stroke patients were summarized in Table 1.

Table 1. Demographic information, lesion information, and motor scores of all patients

Patient ID	Gender	Age(year)	Lesion hemisphere	Lesion site	FMI
1	M	55	R	BG	25
2	F	71	R	BG, CS	45
3	M	58	R	BG, CS	100
4	F	67	R	BG, CS	5
5	F	56	R	BG	43
6	F	66	R	BG	23
7	M	42	L	BG	98
8	M	64	L	BG, CS	95
9	M	59	L	BG, CS	77
10	F	47	L	BG	27

Abbreviation: M: male, F: female; R: right cortex, L: left cortex;
CS: Centrum semiovale.

2.2   Image Acquisition

DTI data were acquired on a Philips Achieva 3.0T MRI scanner (Philips Medical Systems, Best, The Netherlands) using a single shot echo planar imaging sequence. The scanning parameters are as follows: 16 gradient directions with b = 800 s/mm^2, 1 non-diffusion image (b = 0 s/mm^2), repetition time (TR) = 6.92 s, echo time (TE) = 77 ms, slice thickness = 3 mm, slice spacing = 3 mm, Flip angle = 90°, acquisition matrix = 256 × 256, field of view = 256 × 256 mm^2. Figure 1 shows the main CST pathway along the axial direction.

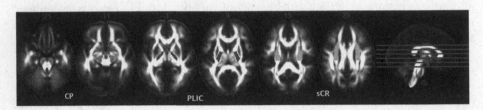

Fig. 1. Multislice presentation of CST pathway along the axial direction (red: CP, blue: PLIC, green: sCR) (Color figure online)

2.3   Motor Evaluation

Motor function was evaluated and measured quantitatively by Fugl-Meyer index (FMI). FMI has been applied widely in clinical examination for quantitative motor impairment in stroke. The score ranges from 0 to a maximal value of 100, in which the higher score reflects the better motor function [12]. For each patient, the FMI score was obtained before MR scanning.

2.4   Image Process

After converting raw DICOM data to NIFTI form using the software dcn2nii [13], the DTI data were processed using FMRIB Software Library (FSL 4.1, Oxford, UK) [14] with the following steps: (1) the diffusion-weighted images of each subject were registered to the corresponding b0 image for eddy current correction using FMRIB's Diffusion Toolbox (FDT); (2) a binary mask was extracted to remove non-brain matter using Brain Extraction Tool (BET) [14]; (3) FA and the diffusion tensor eigenvalues of each voxel ($\lambda1$, $\lambda2$ and $\lambda3$) were calculated by a linear least-square fitting algorithm [15]. We also calculated axial diffusion (AD), RD, and mean diffusion (MD) based on the following equations:

$$AD = \lambda1 \tag{1}$$

$$RD = (\lambda2 + \lambda3)/2 \tag{2}$$

$$MD = (\lambda1 + \lambda2 + \lambda3)/3; \tag{3}$$

(4) All the DTI maps were registered to FMRIB58_FA_1mm [14] so that they could be compared in the same space; (5) Six ROIs, e.g. CP, PLIC and sCR of bilateral hemispheres, were exacted using the atlas from JHU-ICBM-labels-1mm.nii.gz [14] and the averaged DTI parameters were calculated as an average within these ROIs. Moreover, the laterality index (LI) of the parameters were calculated as follows:

$$LI_{FA} = \left(\sum FA_{CH} - \sum FA_{IH}\right) / \left(\sum FA_{CH} + \sum FA_{IH}\right); \qquad (4)$$

where CH denotes the contralesional hemisphere and IH denotes the ipsilesional hemisphere.

2.5 Statistical Analysis

For group-level analysis, the data of subject with left hemispheric lesion were flipped along the mid sagittal plane. The statistical analysis was conducted using paired-samples t-test with IBM SPSS 20 (SPSS Inc, Chicago, USA). The correlation between FMI and DTI parameters was evaluated using Pearson correlation analysis for all ROIs. The significant level was set at $p < 0.05$ (two tailed test).

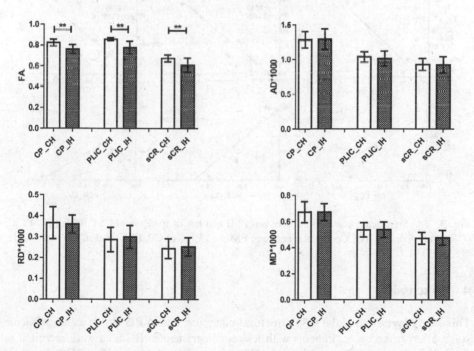

Fig. 2. Comparison of all DTI parameters between contralesional and ipsilesional hemisphere (**: $p < 0.01$.)

3   Results

In this study, six regions were segmented from the CST as the most affected areas after BG stroke. The FA, AD, RD, and MD values obtained from these areas were compared between contralesional and ispilesional hemispheres. The association of white matter integrity with FMI was examined to show the clinical relevance of DTI parameters.

Firstly, when comparing all DTI parameters in six ROIs, it is shown that FA of non-affected hemisphere was consistently higher than that of lesion side in all ROIs. However, other parameters (AD, RD and MD) did not show significant difference between hemispheres, as displayed in Fig. 2. Moreover, the FA of ipsilensional CP (r = 0.7214, p = 0.0179), PLIC (r = 0.8810, p = 0.0008) and sCR (r = 0.6789, p = 0.0309) was significantly correlated with FMI, respectively. The higher the ipsilesional FA value is, the better motor function the patients have (Fig. 3A). Finally, the correlation of FMI and LI of all DTI parameters were investigated. Interestingly, a negative correlation pattern was founded between FMI and LI of FA in CP (r = 0.7461, p = 0.0132) and PLIC (r = 0.8836, p = 0.0010) significantly (Fig. 3B). Yet for all other DTI parameters, no significant correlation with FMI was observed.

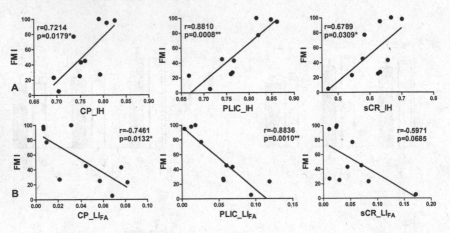

Fig. 3. (A) Significant correlations between FMI and FA of ipsiletional CST ROIs (CP, PLIC, sCR) were founded; (B) Correlation between FMI and LI_{FA} of CP, PLIC and sCR. (*: p < 0.05, **: p < 0.01.)

4   Discussion

This study investigated the CST structural alteration due to BG stroke at the subacute stage after stroke onset. Patients with lesion constrained in BG area were recruited in the study to investigate the changes of CST white matter integrity after BG lesion. Our results showed a significant decline of ipsileional FA in CP, PLIC and sCR at 1–3 months after BG stroke onset. Moreover, FAs of all three ROIs in affected hemisphere as well as the LI of FA in CP and PLIC were correlated with patient's' concurrent FMI score, indicating their roles in motor performance evaluation.

Previous studies showed that the lesion location is one of the critical factors contributing to motor performance deterioration [5, 16]. In patients suffering from middle cerebral artery stroke, lower FA in the ipsilesional CR and PLIC were found at 30 days after stroke onset. In addition, PLIC at acute stroke stage was related to poor motor outcome at chronic stage [5]. Koyama et al. [17] performed DTI study in subcortical hemorrhage patients within 1 month after stroke and found neural degeneration in CP, PLIC and CR. They also found that FA in CP correlated with motor performance. Yu et al. [18] investigated patients with BG lesion at 1–3 months after stroke and found both reconstructed CST fiber and CP had lower FA in ipsilesional side compared to the contralesional one. A DTI-fMRI study on chronic stroke patients quantifies structural integrity of the CST by the number of fibers passing through the reconstructed CST. The results showed that less fiber bundles passed through CST within the CP area compared with normal subjects [2].

FA lateralization index served as a predictive marker of motor recovery. The parameters applied included FA ratio (FA_{IH} /FA_{CH}), ΔFA (FA_{CH} –FA_{IH}) and LI_{FA} in longitudinal stroke research for motor recovery prediction. Liu et al. [8] found that both FA and FA ratio of ischemic stroke patients within 2 weeks post onset were positively correlated with the patients' motor performance, in consistency with our results. Groisser et al. [19] found in ischemic stroke patients suffering from middle cerebral artery infarction a reduced FA of CST at 1–2 months after stroke and that the ΔFA of CST at subacute stage was predictive for future motor functions of the upper limb.

In conclusion, our work showed that the CST structural alteration was closely related to motor function of BG stroke patients. The results indicate FA value obtained using DTI served as an informative marker for motor performance evaluation within subacute stage of stroke.

References

1. Ino, T., Nakai, R., Azuma, T., Yamamoto, T., Tsutsumi, S., Fukuyama, H.: Somatotopy of corticospinal tract in the internal capsule shown by functional MRI and diffusion tensor images. NeuroReport **18**, 665–668 (2007)
2. Schaechter, J.D., Perdue, K.L., Wang, R.: Structural damage to the corticospinal tract correlates with bilateral sensorimotor cortex reorganization in stroke patients. NeuroImage **39**, 1370–1382 (2008)
3. Schaechter, J.D., Fricker, Z.P., Perdue, K.L., Helmer, K.G., Vangel, M.G., Greve, D.N., Makris, N.: Microstructural status of ipsilesional and contralesional corticospinal tract correlates with motor skill in chronic stroke patients. Hum. Brain Mapp. **30**, 3461–3474 (2009)
4. Lindenberg, R., Zhu, L.L., Ruber, T., Schlaug, G.: Predicting functional motor potential in chronic stroke patients using diffusion tensor imaging. Hum. Brain Mapp. **33**, 1040–1051 (2012)
5. Puig, J., Blasco, G., Daunis, I.E.J., Thomalla, G., Castellanos, M., Figueras, J., Remollo, S., van Eendenburg, C., Sanchez-Gonzalez, J., Serena, J., Pedraza, S.: Decreased corticospinal tract fractional anisotropy predicts long-term motor outcome after stroke. Stroke **44**, 2016–2018 (2013)

6. Puig, J., Pedraza, S., Blasco, G., Daunis, I.E.J., Prats, A., Prados, F., Boada, I., Castellanos, M., Sanchez-Gonzalez, J., Remollo, S., Laguillo, G., Quiles, A.M., Gomez, E., Serena, J.: Wallerian degeneration in the corticospinal tract evaluated by diffusion tensor imaging correlates with motor deficit 30 days after middle cerebral artery ischemic stroke. AJNR Am. J. Neuroradiol. **31**, 1324–1330 (2010)

7. Stinear, C.M., Barber, P.A., Smale, P.R., Coxon, J.P., Fleming, M.K., Byblow, W.D.: Functional potential in chronic stroke patients depends on corticospinal tract integrity. Brain : J. Neurol. **130**, 170–180 (2007)

8. Liu, X., Tian, W., Qiu, X., Li, J., Thomson, S., Li, L., Wang, H.Z.: Correlation analysis of quantitative diffusion parameters in ipsilateral cerebral peduncle during Wallerian degeneration with motor function outcome after cerebral ischemic stroke. J. Neuroimaging **22**, 255–260 (2012)

9. Martin, R.F., Bowden, D.M.: A stereotaxic template atlas of the macaque brain for digital imaging and quantitative neuroanatomy. NeuroImage **4**, 119–150 (1996)

10. Mavridis, I., Boviatsis, E., Anagnostopoulou, S.: Anatomy of the human nucleus accumbens: a combined morphometric study. Surg. Radiol. Anat. **33**, 405–414 (2011)

11. Kim, Y.H., Kim, D.S., Hong, J.H., Park, C.H., Hua, N., Bickart, K.C., Byun, W.M., Jang, S.H.: Corticospinal tract location in internal capsule of human brain: diffusion tensor tractography and functional MRI study. NeuroReport **19**, 817–820 (2008)

12. Gladstone, D.J., Danells, C.J., Black, S.E.: The fugl-meyer assessment of motor recovery after stroke: a critical review of its measurement properties. Neurorehabilitation and Neural Repair **16**, 232–240 (2002)

13. NITRC Tools & Resources. https://www.nitrc.org/projects/dcm2nii/

14. FSL. https://fsl.fmrib.ox.ac.uk/fsl/fslwiki

15. Basser, P.J., Pierpaoli, C.: Microstructural and physiological features of tissues elucidated by quantitative-diffusion-tensor MRI. J. Magn. Reson. B **111**, 209–219 (1996)

16. Ward, N.S., Brown, M.M., Thompson, A.J., Frackowiak, R.S.: Neural correlates of motor recovery after stroke: a longitudinal fMRI study. Brain : J. Neurol. **126**, 2476–2496 (2003)

17. Koyama, T., Tsuji, M., Nishimura, H., Miyake, H., Ohmura, T., Domen, K.: Diffusion tensor imaging for intracerebral hemorrhage outcome prediction: comparison using data from the corona radiata/internal capsule and the cerebral peduncle. J. Stroke Cerebrovasc Dis. **22**, 72–79 (2013)

18. Yu, C., Zhu, C., Zhang, Y., Chen, H., Qin, W., Wang, M., Li, K.: A longitudinal diffusion tensor imaging study on Wallerian degeneration of corticospinal tract after motor pathway stroke. NeuroImage **47**, 451–458 (2009)

19. Groisser, B.N., Copen, W.A., Singhal, A.B., Hirai, K.K., Schaechter, J.D.: Corticospinal tract diffusion abnormalities early after stroke predict motor outcome. Neurorehabilitation and Neural Repair **28**, 751–760 (2014)

Detecting Depression in Speech Under Different Speaking Styles and Emotional Valences

Zhenyu Liu[1], Bin Hu[1,2(✉)], Xiaoyu Li[1], Fei Liu[1], Gang Wang[3],
and Jing Yang[4]

[1] Gansu Provincial Key Laboratory of Wearable Computing,
School of Information Science and Engineering, Lanzhou University,
Lanzhou, China
{liuzhyl2,bh,xylil5,fliul4}@lzu.edu.cn
[2] CAS Center for Excellence in Brain Science and Intelligence Technology,
Shanghai Institutes for Biological Sciences, Chinese Academy of Sciences,
Shanghai, China
[3] Beijing Anding Hospital of Capital Medical University, Beijing, China
gangwangdoc@gmail.com
[4] Lanzhou University Second Hospital, Lanzhou, China
yangdoctor2007@126.com

Abstract. Detecting depression in speech is a hot topic in recent years. Some inconsistent results in previous researches imply a few important influence factors are ignored. In this paper, we investigated a sample of 184 subjects (108 females, 76 males) to examine the influence of speaking style and emotional valence on depression detection. First, classification accuracy was used to measure the influence of these two factors. Then, two-way analysis of variance was employed to determine interactive acoustical features. Finally, normalized features by subtracting got higher classification accuracies. Results show that both speaking style and emotional valence are important factors. Spontaneous speech is better than automatic speech and neutral is the best choice among three emotional valences in depression detection. Normalized features improve the detection performance.

Keywords: Speech · Depression · Speaking style · Emotional valence

1 Introduction

Depression is a common mental disease, which is characterized as prominent and persistent low mood, decreasing interest, even hallucinations [1] and suicide tendency [2]. Depressive disorder affects individual physical and mental health, social communication, vocational ability and body activity [3]. Globally, an estimated 350 million people of all ages suffer from depression [4]. Current diagnosis depends heavily on patients' self-report (e.g., Self-Rating Depression Scale (SDS) [5]) and clinicians' experience. Subjective biases during diagnosis lead to a high misdiagnosis rate. Therefore, an objective accurate convenient method for depression detection is necessary. Various

© Springer International Publishing AG 2017
Y. Zeng et al. (Eds.): BI 2017, LNAI 10654, pp. 261–271, 2017.
https://doi.org/10.1007/978-3-319-70772-3_25

behavioral markers have been discussed for this goal, such as facial actions [6], vocal prosody [7, 8], audiovisual behavior [9] and eye movement [10]. Among them, speech is an attractive candidate for its advantages: non-invasive, fast, convenient and economic.

It has been verified that a speaker's affect and cognition can alter speech production [11], via amount of variations to the somatic and autonomic nervous systems [12]. Additionally, researchers support the feasibility and validity of acoustical measures on depression [13, 14]. Originally, researchers aimed at determining the correlation between depression and some particular speech features, for instance, prosodic features, cepstral features and Linear Prediction Cepstrum Coefficient (LPCC) [7, 15, 16]. Later, some new features like glottal waveform and Teager Energy Operation (TEO) also have shown correlation with depression. In recent years, researchers try to find out suitable combination of features for better detection performance [15, 17]. Ooi presented a multi-feature approach using Gaussian Mixture Model (GMM) and reported a binary classification accuracy of 73%. Cummins [18] and Kinnunen [19] built a 5-class Support Vector Machine (SVM) using spectral features and Mel Frequency Cepstral Coefficients (MFCC) combination with shifted Delta Coefficient displayed high accuracy.

However, there are some inconsistent results among these studies. For example, Breznitz [20] reported there were correlations between both F0 range and F0 average and depression severity, while Alpert [21] got an opposite view on Breznitz. Similarly, Mundt indicated first formant variability (F1) was not significantly correlated with depression whilst second formant (F2) was mildly correlated with it [8]. However, in the follow-up study, both F1 and F2 were not significantly correlated with depression [13]. There are some possibilities leading to this inconsistent: small sample size, different speaking styles, different experiment procedure and so on. Due to depressed individuals' cognitive dysfunction [22] and conversational negativity [23], speaking style and emotional valence will be discussed in this paper as key factors.

Reading [16], counting [21], interview [16, 21], picture description, these speaking styles are often used for speech signals collection in depression detection. Alpert [21] figured out reading speech might not strongly reflect the acoustic effects of depression. Alghowinem [16] reported spontaneous speech (e.g., interview, picture description) gave better results than automatic speech (e.g., reading, counting). The emotional valence of stimulus may affect depressed patients' speech production because of disturbances in emotion regulation. Shankayi [24] figured out scientific texts are better than emotional texts on detecting depression. Alghowinem showed the "Sadness Characteristic" questions performed better than others. However, there still few studies to discuss it although speaking style or emotional valence may potentially lead to inconsistent experiment results.

In this paper, we examine whether speaking style and emotional valence influence the detection of depression. Emotion can be regarded as three dimensions: valence, arousal and dominance. Only valence is the major concern in previous researches due to negative bias of depressed patients' emotion. Therefore, we employ three speaking styles (interview, reading, picture description) under three emotional valences (positive, neutral, negative) separately in the experiment. First, we observe the classification accuracy to determine if the speaking style or emotional valence influences depression detection. Second, we examine interactive feature sets and try to explore how speaking style and emotional valence bring influence. Third, we normalize features under different speaking styles and emotional valences for a higher classification accuracy.

2   Method

2.1   Participants

Experimental recordings were collected from an ongoing study in Beijing and Lanzhou, China. Participants were 184 adults (108 females, 76 males) with the age range of 18–55, and were matched by gender, age and education level. These participants were examined by psychiatrists following Diagnostic and Statistical Manual of Mental Disorders (DSM-IV). The Patient Health Questionnaire-9 (PHQ-9) [25] is also used as inclusion and exclusion criteria (health control: <5, patient: ≥ 5). In addition, the patients who were psychotic disorder in the past or current, severe somatic disease, alcohol and drug abusers, pregnant woman were excluded. Each participant was asked to sign informed consent and fill out basic information. The details are presented in Table 1.

Table 1. Basic information of participants

Gender	Group	Number of participants	Age (years)	PHQ-9 score	Recordings	Tasks
Male	Healthy	38	34.5 ± 8.7	2.2 ± 1.6	27	Interview
	Depressed	38	35.4 ± 9.9	16.0 ± 5.8		Reading
Female	Healthy	54	35.7 + 10.8	1.1 ± 1.6		Picture
	Depressed	54	36.3 ± 10.9	17.2 ± 6.6		Description

2.2   Experiment

Two factors were examined in the experiment: speaking style and emotional valence. Three speaking styles were involved in the study: interview, reading and picture description. Each one had three kinds of emotional valences: positive, neutral and negative. The order of speech with different emotional valences were assigned randomly to counteract the sequence effect. The experiment language was Chinese and the experiment lasts about 25 min. Each participant responded 18 questions (6 positive, 6 neutral and 6 negative) in interview section. These questions were designed based on DSM-IV and other depression scales such as Hamilton Depression Rating Scale (HDRS) [26]. For instance: "what kind of TV programs do you like?", "what makes you feel desperate", etc. 6 groups Chinese words composed reading section. Positive words (e.g. glorious, victory) and negative words (e.g. heart-broken, pain) were selected from affective ontology corpus created by Hongfei Lin [27], and neutral words (e.g. village, center) were selected from Chinese affective words extremum table [28]. Participants were asked to read these words in their common ways. Three facial expression pictures were used in picture description section, which were from Chinese

Facial Affective Picture System (CFAPS) [29]. Speakers are asked to describe these pictures freely according to prompts.

2.3 Data Collection

Data collection was conducted in an isolated, quiet and soundproof room without electromagnetic interference. Participants were asked not to touch any equipment and keep a distance between mouth and microphone about 20 cm. The microphone was Neumann TLM102. A RME FIREFACE UCX audio card with 44.1 kHz sampling rate and 24-bit sampling depth was used for collecting speech signals. Speech recordings were saved as uncompressed WAV format. Ambient noise should be lower than 60 dB. For each participant, 27 recordings were used to analysis in this study.

2.4 Data Preprocessing and Feature Extraction

Speech recordings were segmented and labeled manually. The preprocessing was performed on frame which is 25 ms length with 50% overlap. A band-pass filter (60–4500 Hz) was employed for remove irrelevant signals. Endpoint detection was used for some particular features' extraction. Voice characteristics contain two categories: acoustical and linguistic features [30]. The later one was not considered in this paper since we are discussing objective biomarker for depression detection. Open-source software openSMILE [31], VOICEBOX [32] and Praat [33] were used to extract features. The all-feature set was 1753 dimensions containing common acoustical features, such as MFCC, LPCC, fundamental frequency (F0), TEO and so on.

2.5 Data Analysis

2.5.1 Classification on All-Feature Set

Classification was conducted on different speaking styles and emotional valences using all-feature set (1753-dimension) in this part to examine their influence. Five widely used classifiers were employed in this paper: Naïve Bayes (NB), Support Vector Machine (SVM) with a radial basis kernel, meta-Bagging (Bagging), k-Nearest-Neighbors (kNN, k = 5) and Random Forest (RF) and average accuracies of five classifiers are the present results. Leave-One-Out Cross Validation (LOOCV) was used in classification in this paper.

2.5.2 Classification on Interactive Feature Set

To figure out whether some specific features vary due to influence factors, the interaction between groups and speaking styles, between groups and emotional valence separately are considered. Two-way analysis of variance [34] with fisher test was employed to determined interactive features with significant level (<0.05) from 1753 features. For simplicity, three speech recordings with positive, neutral and negative valence were selected from interview, reading and picture description respectively. For each speaking style, 3 interactive feature sets were determined between each pair of emotion valence: positive-neutral (Pos-Neu), positive-negative (Pos-Neg), neutral-negative (Neu-Neg). Similarly, for each emotional valence, 3 interactive feature sets between each pair of speaking styles: interview-reading (Int-Rea), interview-picture (Int-Pic), reading-picture (Rea-Pic). An interactive feature means the feature value vary due to both participant groups and another influence factor (speaking styles or emotional valences).

2.5.3 Classification on Normalized Feature Set

To get higher classification accuracy, we normalize the interactive features by subtract another speech sample. For instance, to interview, an interactive feature subset is based on two corresponding speech data with different valences (e.g., positive and neutral). Then, the difference value between positive and neutral speech signals is regarded as a normalized feature. This kind of normalization can reduce the fluctuation of acoustical features and get higher and stable recognition rate.

3 Results

3.1 Performance on All-Feature Set

For observing the influence of speaking style and emotional valence on depression detection, classification accuracy using all-feature set was employed. Table 2 showed the average classification accuracies of five classifiers on speech recordings. For speaking styles, interview and picture description performed better than reading, and picture description was slightly higher than interview on both male and female. In detail, the same trend was showed on both genders: picture description > interview > reading. For emotional valence, the order of accuracies on female were: neutral > negative > positive, and for male were: neutral > positive > negative. These showed that neutral valence is a better choice for depression detection. To sum up, reading is worse than interview and picture description, neutral emotional valence performs better than positive and negative.

Table 2. Average classification accuracy under different speaking styles and emotional valences

Gender	Emotional valence	Speaking style			AVG
		Interview	Reading	Picture description	
Female	Positive	0.554	0.531	0.546	0.544
	Neutral	0.612	0.531	0.637	0.593
	Negative	0.579	0.521	0.641	0.580
	AVG	0.582	0.528	0.608	0.572
Male	Positive	0.653	0.641	0.671	0.655
	Neutral	0.680	0.616	0.687	0.661
	Negative	0.654	0.568	0.663	0.629
	AVG	0.662	0.608	0.674	0.648

3.2   Performance on Interactive Feature Set

Tables 3 and 4 showed the amounts and classification results on interactive features under different situations. Table 3 demonstrated the performance of interactive features between speaking styles and participant groups and Table 4 presented the results of interactive features between emotional valences and participant groups. From Table 3, it is clear that the interactive features amounts are much smaller than 1753. Especially, on positive valence, they are all less than 82 for both genders. This implies that positive valence is a key factor under this condition. In other words, positive emotional valence is bad for distinguishing depression on any speaking styles. On neutral and negative valences, Int-Rea and Int-Des are with more interactive features relatively compared to Rea-Des. That means on neutral and negative valences, interview is very different with others. In Table 4, for reading and picture description, the range of interactive feature amounts is from 13 to 45, while the accuracies are higher than the corresponding all-feature set. Overall, although the dimensions of interactive feature sets are much smaller than all-feature set, they often get higher accuracy. These interactive features play an important role on depression detection.

Table 3. Classification accuracies and amounts of interactive features between speaking styles and participant groups

Emotion valence	Speaking style	Interactive features for female			AVG	Interactive features for male			AVG
		Int-Rea	Int-Des	Rea-Des		Int-Rea	Int-Des	Rea-Des	
Positive	Amount	51	46	33	–	40	81	81	–
	Interview	0.669	0.646	–	0.657	0.739	0.689	–	0.714
	Reading	0.578	–	0.585	0.582	0.666	–	0.608	0.637
	Description	–	0.570	0.643	0.607	–	0.716	0.705	0.711
Neutral	Amount	485	712	163	–	366	385	84	–
	Interview	0.707	0.707	–	0.707	0.761	0.742	–	0.751
	Reading	0.569	–	0.611	0.590	0.629	–	0.608	0.618
	Description	–	0.606	0.678	0.642	–	0.597	0.684	0.641
Negative	Amount	375	554	355	–	263	319	31	–
	Interview	0.670	0.631	–	0.651	0.716	0.711	–	0.713
	Reading	0.572	–	0.607	0.590	0.637	–	0.711	0.674
	Description	–	0.639	0.628	0.633	–	0.684	0.676	0.680

Table 4. Classification accuracies and amounts of interactive feature between emotional valences and participant groups

Speaking style	Emotion valence	Interactive features for female			AVG	Interactive features for male			AVG
		Pos-Neu	Pos-Neg	Neu-Neg		Pos-Neu	Pos-Neg	Neu-Neg	
Interview	Amount	607	436	20	–	238	20	172	–
	Positive	0.580	0.606	–	0.593	0.666	0.658	–	0.662
	Neutral	0.709	–	0.678	0.694	0.742	–	0.732	0.737
	Negative	–	0.680	0.663	0.671	–	0.737	0.721	0.729

(continued)

Table 4. (*continued*)

Speaking style	Emotion valence	Interactive features for female			AVG	Interactive features for male			AVG
		Pos-Neu	Pos-Neg	Neu-Neg		Pos-Neu	Pos-Neg	Neu-Neg	
Reading	Amount	13	32	45	–	19	20	19	–
	Positive	0.589	0.646	–	0.618	0.661	0.661	–	0.661
	Neutral	0.702	–	0.707	0.705	0.761	–	0.671	0.716
	Negative	–	0.604	0.602	0.603	–	0.687	0.661	0.674
Description	Amount	15	32	17	–	14	14	14	–
	Positive	0.661	0.596	–	0.629	0.697	0.747	–	0.722
	Neutral	0.631	–	0.657	0.644	0.605	–	0.634	0.620
	Negative	–	0.611	0.685	0.648	–	0.634	0.611	0.622

3.3 Performance on Normalized Feature Set

Classification accuracies of normalized feature sets are presented in Tables 5 and 6 respectively. Compared with all-feature set, classification performance improved apparently for both female (0.081) and male (0.052) on average. In Table 5, Nor-Rea-Des has higher accuracies for female in all three emotional valences and for male only in negative stimuli. Nor-Int-Rea has better performance for male in positive and neutral stimuli. These imply reading speech is a proper baseline for feature normalization. Similarly, normalized features from different emotional valences get better performance than all-feature set, the increase for female and male are 0.097 and 0.080 respectively. The highest accuracies in each speaking style appear in Nor-Pos-Neu and Nor-Neu-Neg, which always are correlation with neutral valence. Normalized feature sets improved the classification performance, and normalization on emotional valences is better than on speaking styles for both genders.

Table 5. Classification accuracies of normalized features obtained from the difference between each pair of speaking styles

Gender	Emotion valence	Normalized features			AVG
		Nor-Int-Rea	Nor-Int-Des	Nor-Rea-Des	
Female	Positive	0.693	0.696	**0.711**	0.700
	Neutral	0.694	0.704	**0.719**	0.706
	Negative	0.641	0.641	**0.650**	0.644
	AVG	0.676	0.680	0.693	0.683
Male	Positive	**0.739**	0.708	0.689	0.712
	Neutral	**0.787**	0.750	0.732	0.756
	Negative	0.671	0.658	**0.732**	0.687
	AVG	0.732	0.705	0.718	0.718

Table 6. Classification accuracies of normalized features obtained from the difference between each pair of emotional valences

Gender	Speaking style	Normalized features			AVG
		Nor-Pos-Neu	Nor-Pos-Neg	Nor-Neu-Neg	
Female	Interview	0.730	0.681	**0.759**	0.723
	Reading	0.657	0.674	**0.744**	0.692
	Description	0.676	0.680	**0.693**	0.683
	AVG	0.688	0.678	0.732	0.699
Male	Interview	**0.763**	0.734	0.737	0.745
	Reading	**0.803**	0.779	0.782	0.788
	Description	0.703	0.700	**0.716**	0.706
	AVG	0.756	0.738	0.745	0.746

4 Discussion

Speaking style and emotional valence are important factors to depression detection in speech. For speaking styles, the results showed spontaneous speech (interview and picture description) performs better than automatic speech (reading). This is consistent with the conclusions of Alpert [21], Ellggring [35] and Alghowinem [16]. This may be associated with the clinical depression symptoms. Depressive patients usually endure psychomotor retardation [30] and memory impairment [31], which lead to reduced initiative language and communication difficulties. And spontaneous speech requires more complex organization of language than automatic speech, which makes larger variation in vocal features between healthy controls and patients in free speech. Emotional valences influence the recognition accuracy of depression also. Neutral valence performs best for both male and female. These results are accordant with research of Shankayi [24]. He figured out that scientific texts presented better performance than emotional texts. A research conducted by C. Naranjo indicated that depressed patients prefer to interpret neutral voices and facial as a negative one. That is to say misjudgment to emotional valence may enlarge the distinction between depressive patients and healthy controls.

Interactive features are determined through analysis of variance. These features' values vary due to two factors: participant groups and speaking styles (or emotional valences). These results may partly explain the inconsistent results among previous researches. In other words, if a feature is an interactive feature, the trends change according to speaking style or emotional valence.

We utilize two speech samples for each participant to create normalized features to reduce individuals' differences. The results showed normalized features get higher detection accuracies compared all-feature set. One probable cause is features subtraction may enlarge the gap between two groups. And reading speech is a good baseline because the words are exactly the same. Feature normalization lead to smaller individual variability and stable recognition rate.

Gender difference is clearly displayed in above results. Recognition accuracies for male are higher than for female, and Hönig [32] and Smolak [33] have drawn the similar conclusion. In the latter study, the regression analysis examining voice and depression indicated that low voice is more strongly related to depression in men than in women. Women are more affected by surroundings, thus their voices can't represent the essential states totally. Moreover, in our study, the amount of male is fewer than female, this may lead to a higher but unstable classification accuracy.

5 Conclusion

This study aims at exploring the influence of speaking style and emotional valence on depression detection in speech. By examining the classification accuracies of three kinds of feature sets (1753-dimension feature set, interactive feature set and normalized feature set), three points were concluded: First, speaking style and emotional valence are important factors to the depression detection. Interview and picture description show better results than reading speech, and neutral emotion performs better than positive and negative emotion stimuli. Second, interactive feature sets have a stronger performance with fewer features than 1753-dimension feature set. Third, feature normalization can improve classification accuracies on both genders.

Acknowledgments. This work was supported by the National Basic Research Program of China (973 Program) (No. 2014CB744600), the National Natural Science Foundation of China (Grant No. 61632014, No. 61210010), Program of Beijing Municipal Science & Technology Commission (No. Z171100000117005), the Program of International S&T Cooperation of MOST (No. 2013DFA11140). Grateful acknowledgement is made to· Xiang Gao, Tianyang Wang, Lihua Yan, Huanyu Kang, for experimental implementation.

References

1. Toh, W.L., Thomas, N., Rossell, S.L.: Auditory verbal hallucinations in bipolar disorder (BD) and major depressive disorder (MDD): a systematic review. J. Affect. Disord. **184**, 18–28 (2015)
2. Zhang, Y., Zhang, C., Yuan, G., Yao, J., Cheng, Z., Liu, C., et al.: Effect of tryptophan hydroxylase-2 rs7305115 SNP on suicide attempts risk in major depression. Behav. Brain Funct. **6**, 1 (2010)
3. Angeleri, F., Angeleri, V.A., Foschi, N., Giaquinto, S., Nolfe, G.: The influence of depression, social activity, and family stress on functional outcome after stroke. Stroke **24**, 1478–1483 (1993)
4. http://www.who.int/mental_health/management/depression/en/
5. Zumg, W., Richards, C., Short, M.: Self-rating depression scale in an outpatient clinic: further validation of the SDS. Arch. Gen. Psychiatry **13**, 508–515 (1965)
6. Cohn, J.F., Kruez, T.S., Matthews, I., Yang, Y., Nguyen, M.H., Padilla, M.T., et al.: Detecting depression from facial actions and vocal prosody. In: 3rd International Conference on Affective Computing and Intelligent Interaction and Workshops. ACII 2009, pp. 1–7 (2009)
7. Cummins, N., Epps, J., Breakspear, M., Goecke, R.: An investigation of depressed speech detection: features and normalization. In: Interspeech, pp. 2997–3000 (2011)

8. Mundt, J.C., Snyder, P.J., Cannizzaro, M.S., Chappie, K., Geralts, D.S.: Voice acoustic measures of depression severity and treatment response collected via interactive voice response (IVR) technology. J. Neurolinguist. **20**, 50–64 (2007)

9. Scherer, S., Stratou, G., Morency, L.-P.: Audiovisual behavior descriptors for depression assessment. In: Proceedings of the 15th ACM on International Conference on Multimodal Interaction, pp. 135–140 (2013)

10. Kupfer, D., Foster, F.G.: Interval between onset of sleep and rapid-eye-movement sleep as an indicator of depression. Lancet **300**, 684–686 (1972)

11. Davidson, R.J., Pizzagalli, D., Nitschke, J.B., Putnam, K.: Depression: perspectives from affective neuroscience. Annu. Rev. Psychol. **53**, 545–574 (2002)

12. Ozdas, A., Shiavi, R.G., Silverman, S.E., Silverman, M.K., Wilkes, D.M.: Investigation of vocal jitter and glottal flow spectrum as possible cues for depression and near-term suicidal risk. IEEE Trans. Biomed. Eng. **51**, 1530–1540 (2004)

13. Mundt, J.C., Vogel, A.P., Feltner, D.E., Lenderking, W.R.: Vocal acoustic biomarkers of depression severity and treatment response. Biol. Psychiatry **72**, 580–587 (2012)

14. Nilsonne, A., Sundberg, J., Ternstrom, S., Askenfelt, A.: Measuring the rate of change of voice fundamental frequency in fluent speech during mental depression. J. Acoust. Soc. Am. **83**, 716–728 (1988)

15. Moore, E., Clements, M.A., Peifer, J.W., Weisser, L.: Critical analysis of the impact of glottal features in the classification of clinical depression in speech. IEEE Trans. Biomed. Eng. **55**, 96–107 (2008)

16. Alghowinem, S., Goecke, R., Wagner, M., Epps, J., Gedeon, T., Breakspear, M., et al.: A comparative study of different classifiers for detecting depression from spontaneous speech. In: 2013 IEEE International Conference on Acoustics, Speech and Signal Processing (ICASSP), pp. 8022–8026 (2013)

17. Ooi, K.E.B., Lech, M., Allen, N.B.: Multichannel weighted speech classification system for prediction of major depression in adolescents. IEEE Trans. Biomed. Eng. **60**, 497–506 (2013)

18. Cummins, N., Epps, J., Ambikairajah, E.: Spectro-temporal analysis of speech affected by depression and psychomotor retardation. In: 2013 IEEE International Conference on Acoustics, Speech and Signal Processing (ICASSP), pp. 7542–7546 (2013)

19. Kinnunen, T., Lee, K.-A., Li, H.: Dimension reduction of the modulation spectrogram for speaker verification. In: Odyssey, p. 30 (2008)

20. Breznitz, Z., Share, D.L.: Effects of accelerated reading rate on memory for text. J. Educ. Psychol. **84**, 193 (1992)

21. Alpert, M., Pouget, E.R., Silva, R.R.: Reflections of depression in acoustic measures of the patient's speech. J. Affect. Disord. **66**, 59–69 (2001)

22. Calev, A., Nigal, D., Chazan, S.: Retrieval from semantic memory using meaningful and meaningless constructs by depressed, stable bipolar and manic patients. Br. J. Clin. Psychol. **28**, 67–73 (1989)

23. Vanger, P., Summerfield, A.B., Rosen, B., Watson, J.: Effects of communication content on speech behavior of depressives. Compr. Psychiatry **33**, 39–41 (1992)

24. Shankayi, R., Vali, M., Salimi, M., Malekshahi, M.: Identifying depressed from healthy cases using speech processing. In: 19th Iranian Conference of Biomedical Engineering (ICBME), pp. 191–194 (2012)

25. Kroenke, K., Spitzer, R.L., Williams, J.B.: The PHQ-9: validity of a brief depression severity measure. J. Gen. Intern. Med. **16**, 606–613 (2001)

26. Hamilton, M.: A rating scale for depression. J. Neurol. Neurosurg. Psychiatry **23**, 56–62 (1960)

27. http://ir.dlut.edu.cn/Group.aspx?ID=4

28. http://www.datatang.com/data/43216
29. Gong, X., Huang, Y., Wang, Y., Luo, Y.: Revision of the Chinese facial affective picture system. Chin. Ment. Health J. **25**, 40–46 (2011)
30. Martinot, M.-L.P., Bragulat, V., Artiges, E., Dollé, F., Hinnen, F., Jouvent, R., et al.: Decreased presynaptic dopamine function in the left caudate of depressed patients with affective flattening and psychomotor retardation. Am. J. Psychiatry **158**, 314–316 (2001)
31. Clark, L., Chamberlain, S.R., Sahakian, B.J.: Neurocognitive mechanisms in depression: implications for treatment. Annu. Rev. Neurosci. **32**, 57–74 (2009)
32. Hönig, F., Batliner, A., Nöth, E., Schnieder, S., Krajewski, J.: Automatic modelling of depressed speech: relevant features and relevance of gender. In: Fifteenth Annual Conference of the International Speech Communication Association, pp. 1248–1252 (2014)
33. Smolak, L., Munstertieger, B.F.: The relationship of gender and voice to depression and eating disorders. Psychol. Women Q. **26**, 234–241 (2002)
34. Low, L.S.A., Maddage, N.C., Lech, M., Sheeber, L.B., Allen, N.B.: Detection of clinical depression in adolescents' speech during family interactions. IEEE Trans. Biomed. Eng. **58**, 574–586 (2011)
35. Ellgring, H., Scherer, K.R.: Vocal indicators of mood change in depression. J. Nonverbal Behav. **20**, 83–110 (1996)

Scientific Advances on Consciousness

Yinsheng Zhang[1,2(✉)]

[1] Capital Normal University, Beijing 100048, China
2573384707@qq.com
[2] Institute of Scientific and Technical Information of China,
Beijing 100038, China

Abstract. The article summarizes scientific advances on consciousness up to the present (the year 2017). The remarkable milestones of experimental research on consciousness, in particular those in response to some philosophic meanings, are selected. These chosen achievements are within more than half centuries and narrowed on five fields: (1) modeling consciousness, (2) analysis on consciousness quantum indeterminacy, (3) finding core-consciousness-function cells, (4) brain-machine interface, (5) brain research plans on brain information access and analysis with large-scale. The main conclusions cover that (1) Piaget consciousness model (PCM), which asserts that consciousness is the homomorphism between functional cells and their mapped objects in respective laws of motion, is a universal frame defining the consciousness in philosophic, scientific ways; (2) Receptive Field, Place Cell and Grid Cell, and some functional brain cells which specially make decisions are PCM instances; (3) consciousness has not been confirmed to be related to quantum states, but some tentative plans to be confirmed have been suggested; (4) brain-machine interface shows PCM too by physical or artificial ways. Meanwhile, some important data, analog or relevant technologies about the above achievements are described, and philosophic explanations are tried to be given.

Keywords: Brain · Consciousness · Homomorphism · Piaget · Quantum indeterminacy

1 Scientific Questions and Relevant Conclusions on Consciousness

At the turn of 21 century, consciousness was still kept open to large extent. A typical question concerning consciousness is expressed as follows:

"What is the biological basis of consciousness?" (hereafter "Question 1") [1]?

The following 5 fields are extracted for they are full of solutions to Question 1 especially in philosophical aspects: (1) modeling consciousness; (2) analysis on consciousness quantum indeterminacy; (3) finding core-consciousness-functional cells; (4) brain-machine interface; (5) widely implementing new brain projects in neurons information access and analysis with large-scale.

© Springer International Publishing AG 2017
Y. Zeng et al. (Eds.): BI 2017, LNAI 10654, pp. 272–281, 2017.
https://doi.org/10.1007/978-3-319-70772-3_26

2 Modeling Consciousness

Are there mathematical methods to describe consciousness? There is still skepticism on it now. After all, many things against consciousness are unknown. But if consciousness is physical as many scientists believe, why could it not be described mathematically just like physical formulas? A model depicting consciousness should satisfy that it can

Determine essential features of consciousness;
Conform to common understanding between philosophy, daily life and the science;
Cover various activities of consciousness; or, all the activities can be determined or described by the model;
Be physical, even executive by artificial simulation.

In facts, many scholars extracted the physical features of consciousness, and mathematically characterize some local or single sections of consciousness.

Rene Descartes took consciousness as non-material things without space and as free to physical laws [2]. The model cannot be tested and be accepted by the science.

A.M. Turing ascribed thinking (as the main activity of consciousness) to computation, of which the mechanism is running step by step an input-output function to approach to a logic truth or grammar generation [3, 4]. In this profile, consciousness is algorithms, or simply a Turing machine. Turing's model, being called "computationalism" or "reductionism", i.e., wholly physical, expressed mathematically and executive technically, has been controversial to some extent, e.g., refuted by Chinese Room Refutation.

Husserl wholly summarized core characteristics of consciousness [5], which were understood technically even for reproduction or simulation by a robot producer, as follows [6]:

(1) In first person; (2) Intention; (3) Dualism between activity and results; (4) Expectation; (5) Determination and Believing; (6) Embodiment; (7) Awareness; (8) Thinking with emotion; (9) With random; (10) Emotion.

Farber and Churchland [7], Baars and Edelman came up with their extractions of theories about consciousness [8].

The recent advances on consciousness modeling included Piaget psychological model (PCM) interpreted by Zhang Yinsheng, and characterized by homomorphism between the material units, symbols and its objective world. The interpreter verified that PCM exactly responded to the subjective ("psychology" in Piage) which referred philosophically, scientifically and in everyday to as the term "consciousness", and the model enabled some important puzzles such as Turing Test, Chinese Room about how to confirm whether a thing owns consciousness to be accounted reasonably. A mathematical expression of PCM had been given (see (1)) as a quantified description frame, which can be factorized into its operator, arguments and levels of the algebraic structure, and computation units such as neurons or machines functional parts. Up to now, all the conscious activities determined by discrete mathematics, can be modeled by PCM [9–11].

Given two algebraic systems $(X, \bullet)$, $(Y, *)$, X is a set of elements in an subjective system, $Y \bullet$ is a set external to X; $\bullet$ and $*$ are operators of X and Y; $f : X \to Y$ is a map. For any elements $a, b \in X$, if

$$f(a \bullet b) = f(a) * f(b) \tag{1}$$

holds, we say that system X has the consciousness, or X is aware of Y. (1) is the homomorphic model, i.e. PCM.

3   Discovery of Functional Cells for Consciousness

It is the scientific reasoning for monism that if consciousness is generated from or acted as cells activities, then some cells might play different roles in constituting consciousness. In other words, cells should functionally be different in term of the difference of activity types of consciousness. Many functional cells relevant to the various function kinds of consciousness have been discovered.

A number of functional cells—space-recognition relevant cells were unveiled. Time and space are not only objective parameters in physical laws, but also subjective functions in consciousness by Kant, who systematically brought up that reason is reorganizing sensory information by time and space [12]. Therefore, if we founded some cells playing a role relating to space recognition, it would mean that these cells were constituting reason like a organizer of experience in the consciousness.

In 1959, David Hubel and Torsten Wiesel discovered that in various cat's striate cortex areas ("Receptive Field", RF) discharged accordingly with the angles of stimuli, so the certain cells were space-feature related [13].

Extracted from its general meaning, RF is expanded to many disciplines. Especially, it is simulated by Artificial Intelligence (AI). As an example, the convolutional neural network algorithm draws lessons from RF in Artificial Neural Network (ANN). For one thing, ANN method, introduced by Warren McCulloch and Walter Pitts (MP Model) in 1943, showed consciousness' ubiquity, which could be simulated by technology systems [14]. An important innovation on ANN, deep learning, set more than 3 levels of the nodes to adjust weights between levels, which proposed by Rumelhart and McClenland [15], and recently by Hinton [16–18]. Deep learning had made great advance, e.g., the application named AlphaGo (Go, a Chinese ancient game regarded as one requesting high complex competitive decisions) [19]. Hence, ANN typically stimulated neurons in recursion, a mathematical computation model of Turing Machine. Thus, consciousness had been expressed not only as cells' structure and connections but also machine's algorithms. For another, machine learning not only functioned in processing data, but also in constituting methods, which had been treated as human unique capability, among data chaos. So machine learning were believed to be a mechanism linked to computer science, philosophy of mathematics and philosophy of science [20]. Therefore, from advance of ANN, as a direct lesson from experimental science of consciousness, AI demonstrated the analog to living consciousness not only in structure, but also in functions even high functions like constituting methods.

Another notably discovered space cognition cells was the cells especially for positioning, including Place Cell and Grid Cell, discovered by John O'Keefe in 1971, and by May-Britt Moser and Edvard Moser in 2005. Place Cell, were validated by different positions to perceive direction, position and velocity. Similarly, Grid Cell, played a coordinate role for searching roads [21–24].

Space cognition cells implies that some neurons functions in recognizing space frames, which should function in organizing experience as Kant considered to do, serve as the typical subjective activities. So a bridge, between physics and meta-physics, and between materials with their functions and philosophical reference to "consciousness" are established. Moreover, discover of space cognition cells illustrated PCM by checking a direct map between *Objects* (space elements) and definite *Units* (neurons).

The more abstractive function cells are "computation cells", which refer to some cells responsive to the quantity changes of some other activated cells. "computation cells" seem to calculate the quantity changes, rather than a hard matter, like stable cells.

The speculated cells doing computation were in left posterior (dorsolateral prefrontal cortex, DLPFC) by H.R. Heekeren, S. Marrett et al. who conducted an experiment making the subjects to decide if the stimuli were a face or a house given an image with noise as shown in Fig. 1 [25]. The left Fig. 1 shows the strength of brain cells responding to face (yellow) or house (green); the right shows four groups decision correspondence strengths' comparison between the two objects (face or house).

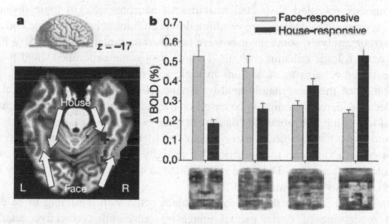

Fig. 1. Functional magnetic resonance imaging (fMRI) response Functional magnetic resonance imaging (fMRI) response to the stimuli which are characterized by both a face and a house. Where, the areas colored by yellow is cognized to be responding to face; green, to face. (Color figure online)

holds, we say that system X has the consciousness, or X is aware of Y. (1) is the homomorphic model, i.e. PCM.

Figure 2 shows a perceptual decision-making in poster DLPFC with high response to the gaps between the two strengths responded to the two characteristics of objects

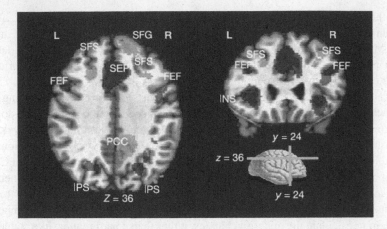

Fig. 2. Computation cells, which are speculated by a fMRI signals proportional to the gaps of cognition cells of face or house (yellow and green in Fig. 1), are considered to be generated by a reduction (a primary computation) of the noise between the cognition.

(face or house, see yellow and green in the Fig. 1). So the decision-making cells are making a reduction similar to the recursion function which performs by primary computations.

The another revealed abstracted activities of neurons were of logic deductions. Several experiments of fMRI were conducted to scan human brain in logic deductions [26]. Based on analysis, some theories were launched to explain what happens in brain neurons when a logic calculus making. Table 1 lists some experiments and their theories guessing mechanisms of neurons in logic activities.

Mental proof theories, mainly held by Braine, Henle, Rips, Osherson, et al. since 1970's decades, deny reasoning to be based on mental imagery, but on some reasoning rules, just like computer processing data by languages. By contrast, denying relying on rules-based way, mental model theories (spatial mental theories) maintain that reasoning is construction and manipulation by spatially situational organization such as shape, colors, textures etc. The main representatives include Laid [27], Byrne, Knauff et al. [28], latter to the mental proof theories.

Visual mental imagery theory proposes brain activity in reasoning to be definite kinds of processing of visual mental image by representing objective details like locations, colors, shapes, distance and so on. The processing leads to new information finding in implicit premises. The points were raised by Kosslyn [29] and Finke.

A theory conciliating mental proof theories and mental model theories is dual mechanism theory, which was established in the end of the 20 centuries, basically by Evans, Goel [30, 31] and Sloman. The theory confirms that the stimuli of the forms of the instances of syllogisms basically activate common areas and different areas by fMRI experiments, which should demonstrate dual mechanisms on rules and space-feature elements processing.

Table 1. Brodmann areas fMRI response on logic

Theories	Logic types		
	Judgment	Syllogistic reasoning	Relational reasoning
(Spatial) mental model			Prefrontal cortex (BA 6, 9); cingulate gyrus (BA 32); the superior and inferior parietal cortex (BA 7, 40), the precuneus (BA 7), the visual association cortex (BA 19)
Dual mechanism		The left anterior frontal lobe (BA 44), left fusiform gyrus (BA 18), right fusiform gyrus (BA 37), bilateral basal ganglia	
Visual mental imagery theories	Visual reflex zones (occipital lobe), lateral prefrontal part of the ridge (BA 8/9)		

Displayed equations are centered and set on a separate line.

4 Quantum Uncertainty of Consciousness

Quantum uncertainty, since it was discovered by Heisenberg in 1927, has been guessing to be an instinct of consciousness. In 1953, Eccles predicted that synapses in neurotransmitters pass mechanism might be quantum-uncertain, and compatible to Cartesian dualism [32, 33]. Quantum uncertainty in synapses was proposed to be like clouds covered consciousness transiting to cells-physical worlds.

This prediction faces up a challenge that synaptic vesicles—the most possible to be quantum—go at a lower velocity with t = 10 ms to complete the energy flux change E for carrying out calcium or else. To satisfy Heisenberg principle, which asserts.

$$\Delta E \times \Delta t > h/2\pi \tag{2}$$

$h = 6.63 \times 10^{-34}$ Js is Planck's constant, the energy change of a vesicle ΔE should be 5.2×10^{-30} J, which more than Van der Waals force (10^{-24} J, the weakest chemical bond) by about 200 000 times, namely, far more than the energy scale of Heisenberg [34].

Also, there exists a presupposition about consciousness uncertainty put forward by Penrose and Hameroff [35], who suggested the tubulin in cells might generate super-position in effect of Van der Waals force. However, the expectation of superposition is still more than the scale determined by Heisenberg principle about 10^{16} times (according to Max Tegmark in 2000 [33]).

Recent studies, say Matthew Fisher, pointed out that brains are probable to contain some special molecules in superposition for a long time, say, a spin could be structured by two phosphorus nuclei, and the spin might be maintained for about 1 day if 6 phosphates construct Posner a molecule, which has not be confirmed yet [36]. Though the uncertainty conjecture of consciousness appeared troubled, a new experiment plan, however, was set aiming at conforming it. The plan was based on a new computation of time superposition limit by Shan Gao in 2008 [37]:

$$\tau_c = \frac{\hbar E_p}{(\Delta E)^2} = (\frac{2.8\,\text{Mev}}{100\,\text{Mev}})^2 \approx 1\,\text{ms} \tag{3}$$

Where:

τ_c is average time of neurons for collapse of quantum superposition;
$\hbar$ is reduced Plank constant, $\hbar = h/2\pi$;
E_p is Plank energy, $E_p \approx 10^{19}$ GeV;
ΔE is Plank energy level of quantum superposition.

The study shows that a single cell, with membrane potential at 10^{-2} V, transmitting 10^6 Na ion through membrane, charges 10^4 eV energy difference in an expected superposition, which produces $\Delta E = 0.01$ meV. If a consciousness event sparks 10^4 neurons, which makes contrast neurons in equal quantity, it would create energy difference by 10^8 V, i.e., 100 meV. It would set apart 1 ms, a time human can perceive, to meet Heisenberg principle. So Shan Gao claimed that a confirmation test of superposition in vesicles could be made. He further proposed to make use of superposition photons as stimuli to radio eyes, testing whether the response time different from the response to the stimuli of non-superposition photons. If the responses were different, it would demonstrate a resonance response to the superposition, that is, a correlation between quantum superposition and consciousness. Although the plan seems to be practicable, there has not been an implement to be report yet.

Therefore, consciousness quantum uncertainty still stays conjectured so far.

5   Interface to Brain in Many Ways

In recent years, Man-Machine Fusion expanded to direct brain-computer interaction. The ways included brain-computer interface (BCI), computer-brain interface (BCI), brain-brain interface (BBI)—the composing order of the two words ("brain", "computer") basically indicates input and output order.

Interface between brains and computers makes consciousness more integrated between human and machines, as well as between materialism and idealism. Nowadays, a visual image can be rewritten by brain-responding signal recognition and interpretation, namely, which proves homomorphism between *Units* and consciousness (*Symbols*) in the way of physical signals decoded by the computer with the sensor.

Kay et al.'s experiment and engineering can be introduced as a typical BCI advance [38], which elaborates a serious process expounding brain signals to be a roughly primary stimuli image by translating patterns of the signals to features of the image. That is, an image stimulated a brain to generate a serial of fMRI which input into a computer to be recognized and roughly reproduced the primary stimuli by the computer. The effects illustrated that consciousness is no more than the attributes (indicated by patterns of fMRI) in term of physical laws by verifying that the physical signals visually mapped the mind images like seeing. The effects also confirm consciousness visually meets PCM, here, the stimuli-images belong to *Objects*; the recognized patterns, *Symbols*; the brain cells, *Units*.

Now that brain signals as the consciousness themself are physical, they could not only be received, recorded, but also be artificially transplant, produced, controlled and applied for a technical machine system. Consciousness, as material's feature, or some ways of existence of material, is universal in natural and artificial systems as well as their connections. Take a typical instance of the advance of Schwartz's team [39, 40], which elaborates this idea. Using grids of fine electrodes implanted in the primary motor cortex of monkeys, the engineers trained a monkey to generate patterns of brain activity to control an anthropomorphic robot arm that had a shoulder join, and a claw-like gripper "hand". After a few days the monkey was capable to make the robot-arm reaching out to a tasty treat such as a piece of fruit [41].

Brain-Brain Interface (BBI) intends to carry out direct signal communication from one brain to another. Carles Grau et al.'s experiment realized this goal. The experiment applied conscious perception of phosphenes (light flash) through neuroavigated, robotized transcanial magnetic stimulation (TMS) especially on block sensory (tactile, visual, or auditory) cues, receiving imagery-electroencephalographic (EEG) in combining BCI and CBI. The results performed one person's word emerging in mind to be transmitted in remote to another one; who understood it through the remote communication by encoding and decoding.

References

1. Kennedy, D.: What is the biological basis of consciousness? Science **309**(5731), 1–204 (2005)
2. Descartes, R.: Minds and bodies as distinct substances. In: Heil, J. (ed.) Philosophy of Mind, pp. 298–320. Oxford University Press, Oxford (2004)
3. Turing, A.M.: Intelligent Machinery. Mechanical Intelligence. Collected Works of A.M. Turing, pp. 107–128. North-Holland Press, Amsterdam, London, New York, Tokyo, Holland (1992)
4. Turing, A.M.: Intelligent Machinery. Mechanical Intelligence. Collected Works of A.M. Turing, pp. 133–160. North-Holland Press, Amsterdam, London, New York, Tokyo, Holland (1992)
5. Husserl, E.: Logische Untersuchungen. Edmund Husserl Gesammelte Werke (Husserliana). Band XIX/1, 2. Martinus Nijhoff Publishers (1984)
6. Takeno, J.: Creation of a Conscious Robot. Mirror Image Cognition and Self-awareness, p. 64. Pan Stanford Publishing Pte. Ltd., Singapore (2013)

7. Farber, I., Churchland, P.S.: Consciousness and the neurosciences: philosophical and theoretical issues. In: The Cognitive Neurosciences. The MIT Press, Cambridge (1995)
8. Baars, B.J., Edelman, D.B.: Consciousness, biology and quantum hypotheses. Phys. Life Rev. **9**(3), 285–294 (2012)
9. Piaget, J.: Le Structuralisme. Presss Universites de France, Paris (1979)
10. Piaget, J.: Biology and Knowledge. The University of Chicago Press, Chicago (1971)
11. Zhang, Y.: The mathematic model of consciousness. In: Proceedings of Second Asia International Conference on Modeling and Simulation, pp. 574–578. IEEE Press, Piscataway (2008)
12. Kant, I.: Critique of Pure Reason, Translated by Particia Litcher. Hackett Publishing Company, Inc., Cambridge (1996)
13. Hubel, D., Wiesel, T.: Receptive fields of single neurones in the cat's striate cortex. J. Physiol. **148**, 574–591 (1959)
14. McCulloch, W., Pitt, W.: A logical calculus of ideas immanent in nervous activity. Bull. Math. Biophys. **5**(1–2), 99–115 (1943)
15. Rumelhart, D.E., Mcclelland, J.L.: Parallel Distributed Processing: Exploration in the Microstructure of Cognition, vol. 1, 2. MIT Press, Cambridge (1986)
16. Hinton, G.E.: Learning multiple layers of representation. Trends Cogn. Sci. **11**(10), 428–434 (2007)
17. Hinton, G.E., Salakhutdinov, R.: Reducing the dimensionality of data with neural networks. Science **313**(5786), 428–434 (2006)
18. Goldberg, D.E., Holland, J.H.: Genetic algorithms and machine learning. Mach. Learn. **3**(2), 95–99 (1988)
19. Silver, D., et al.: Mastering the game of go with deep neural networks and tree search. Nature **529**(7587), 484–489 (2016)
20. Williamson, J.: A dynamic interaction between machine learning and the philosophy of science. Mind Mach. **14**(4), 539–549 (2004)
21. Okeefe, J., Nadel, L.: The Hippocampus as a Cognitive Map. Oxford University Press, Oxford (1978)
22. Okeefe, J.: A review of the hippocampal place cells. Prog. Neurobiol. **13**(4), 419–439 (1979)
23. Best, P.J., White, A.M., Mina, A.A.: Spatial processing in the brain: the activity of hippocampal place cells. Annu. Rev. Neurosci. **24**(1), 459–486 (2001)
24. Moser, I., Kropff, E., Moser, M.: Place cells, grid cells, and the brain's spatial representation system. Annu. Rev. Neurosci. **31**(1), 69–89 (2008)
25. Heekeren, H.R., et al.: A general mechanism for perceptual decision-making in the human brain. Nature **431**, 859–862 (2004)
26. Zhang, Y.: The Summation of the research on physical features of brain mapping logic in fMRI and the excavation of the relevant philosophical significance. J. Med. Philos. **28**(11), 27–29, 55 (2008)
27. Johnsonlaird, P.N.: Mental models and deduction. Trends Cogn. Sci. **5**(10), 434–442 (2001)
28. Knauff, M., et al.: Spatial imagery in deductive reasoning: a functional MRI study. Cogn. Brain Res. **13**(2), 203–212 (2012)
29. Kosslyn, S.M.: Image and Brain: The Resolution of the Imagery Debate. MIT Press, Cambridge (1994)
30. Goel, V., et al.: Dissociation of mechanisms underlying syllogistic reasoning. NeuroImage **12**(5), 504–514 (2000)
31. Goel, V., Dolan, R.J.: Functional neuroanatomy of three-term relational reasoning. Neuropsychologia **39**(9), 901–909 (2001)
32. Eccles, J.C.: The Neurophysiological Basis of Mind. The Principles of Neurophysiology. Oxford University Press, Oxford (1953)

33. Clarke, P.G.H.: Neuroscience, quantum indeterminism and the Cartesian soul. Brain Cogn. **84**(1), 109–117 (2014)

34. Sudhof, T.C.: The synaptic vesicle cycle. Annu. Rev. Neurosci. **27**(1), 509–547 (2004)

35. Penrose, R., Hameroff, S.: Consciousness in the universe: a review of the 'Orch OR' theory. Phys. Life Rev. **11**(1), 39–78 (2014)

36. Fisher, M.P.A.: Quantum cognition: the possibility of processing with nuclear spins in the brain. Ann. Phys. **362**, 593–602 (2015)

37. Gao, S.: A quantum theory of consciousness. Mind. Mach. **18**(1), 39–52 (2008)

38. Kay, K.N., Naselaris, T., Prenger, R.J., Gallant, J.L.: Identifying natural images from human brain activity. Nature **452**(7185), 352–355 (2008)

39. Velliste, M., Perel, S., Spalding, M.C., Whitford, A.S., Schwartz, A.B.: Cortical control of a prosthetic arm for self-feeding. Nature **453**(7198), 1098–1101 (2008)

40. Kalaska, J.F.: Neuroscience: brain control of a helping hand. Nature **453**(7198), 994–995 (2008)

41. Grau, C., et al.: Conscious brain-to-brain communication in humans using non-invasive technologies. PLOS ONE **9**(8), e105225 (2014). PMID: 25137064

Workshop on Big Data and Visualization for Brainsmatics (BDVB 2017)

BECA: A Software Tool for Integrated Visualization of Human Brain Data

Huang Li[1,2], Shiaofen Fang[1(✉)], Bob Zigon[1], Olaf Sporns[3],
Andrew J. Saykin[2], Joaquín Goñi[4,5,6], and Li Shen[1,2]

[1] Computer and Information Science, Purdue University Indianapolis,
Indianapolis, IN, USA
shfang@iupui.edu
[2] Radiology and Imaging Sciences, Indiana University School of Medicine,
Indianapolis, IN, USA
[3] Psychological and Brain Sciences, Indiana University Bloomington,
Bloomington, IN, USA
[4] School of Industrial Engineering, Purdue University, West-Lafayette, IN, USA
[5] Weldon School of Biomedical Engineering, Purdue University,
West-Lafayette, IN, USA
[6] Purdue Institute for Integrative Neuroscience, Purdue University,
West-Lafayette, IN, USA

Abstract. Visualization plays an important role in helping neuroscientist understanding human brain data. Most publicly available software focuses on visualizing a specific brain imaging modality. Here we present an extensible visualization platform, BECA, which employ a plugin architecture to facilitate rapid development and deployment of visualization for human brain data. This paper will introduce the architecture and discuss some important design decisions in implementing the BECA platform and its visualization plugins.

Keywords: Visualization · Human brain data · Plugin platform · Software architecture

1 Introduction

The advancement of medical imaging technologies produces a large amount of neuroimaging data which is capable of showing different brain properties. In addition, a number of measurements are calculated along with the raw data. To give neuroscientist an intuitive understanding of the raw data and their measurements, visualization plays a vital role. There arises the need to design and build a visualization software tool to visualize all kinds of raw data and measurements in the context of their spatial structures.

In collaboration with neuroscientist, four types of brain imaging and genomics data are integratively employed and visualized in the software: diffusion tensor imaging (DTI) fiber tracts, structural magnetic resonance imaging (sMRI) data, functional magnetic resonance imaging (fMRI) data, and genomic data.

© Springer International Publishing AG 2017
Y. Zeng et al. (Eds.): BI 2017, LNAI 10654, pp. 285–291, 2017.
https://doi.org/10.1007/978-3-319-70772-3_27

A few visualization software tools have been released for visualizing neuroimaging data. But they often focus on visualizing few types of brain data. For example, MRIcron [1] is developed to show 2D slice of sMRI or fMRI data. TrackVis [2] can be used to visualize DTI fiber tracts and sMRI data. FSLview [3] is capable of visualizing sMRI and fMRI data. To facilitate an interactive exploration of multimodal brain image data, it is of high value to build a framework that can visualize more imaging modalities.

To address this issue, we present BECA, an extensible visualization platform that accepts plugins for visualizing different types of human brain data. Those types are managed by different plugins which share the same platform.

2   Design and Implementation

The challenge of designing BECA comes from two aspects: firstly, different dataset requires different visualization methods. Even for just one dataset, domain expert may want to use different visualization techniques in different occasions. For example, for the a sMRI dataset, we can use either volume rendering to visualize MRI dataset for voxel-based analysis or isosurface rendering for surface analysis. On the other hand, from the view of software engineering, in the process of iterative software development, we need to deliver the software periodically in order to retrieve feedbacks from domain experts. Therefore, we introduce a plugin architecture which greatly facilitates the development and deployment of BECA.

2.1   Architecture

An overview of the software architecture is shown in Fig. 1. The Qt library [4] is used to build the user interface, and VTK [5] is used for 3D visualization. The plugin architecture is based on the classic Model-View-Controller (MVC) [6] design pattern. The Model component holds the data from the application domain, such as sMRI data as 3D image or fMRI data as 4D image, which are rarely changed in the development of the BECA. In contrast, the View and Controller components are much more unstable

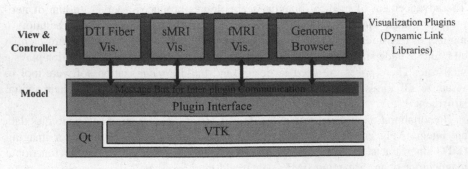

Fig. 1. Software architecture of BECA.

because they are responsible for representing the visualizations and responding to user interactions which vary from one visualization to another. Furthermore, in the iterative software development model, the visualization and interaction are very likely to be changed based on the feedbacks from neuroscientists. Therefore, we designed the plugin architecture to decouple the unstable View and Controller from the stable Model component. Different visualization plugins have different View and Controller component, but they share the same Model component. With the plugin architecture, each visualization plugin can be added or modified easily without changing the underlying infrastructure or other plugins. We also implemented a message bus for message passing between different visualization plugins, so that plugins can communicate with each other for implementing functions such as synchronizing the views in different visualizations. We have released the message bus as an open-source project hosted on Github (https://github.com/lheric/libgitlevtbus).

2.2 DTI Tractography Visualization

The pipeline for visualizing DTI tracts is shown in Fig. 2. Tracts are accessed via the plugin interface. We write a fiberColorFilter to color each fiber tract by its own property such as direction, length, or distance between endpoints.

Suppose a tract consist of $N+1$ points P_0 (x_0, y_0, z_0), P_0 (x_0, y_0, z_0), ..., P_n (x_n, y_n, z_n). The direction is encoded as a RGB color by following formula:

$$R = |x_0 - x_n|/d(P_0, P_n)$$
$$G = |y_0 - y_n|/d(P_0, P_n)$$
$$B = |z_0 - z_n|/d(P_0, P_n)$$

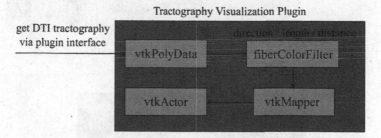

Fig. 2. VTK pipeline for visualizing DTI tractography

where function $d(p, q)$ computes the Euclidean distance between p and q. The length or distance of a tract is a scalar value which can be mapped into color via a lookup table. An example result is shown in Fig. 3.

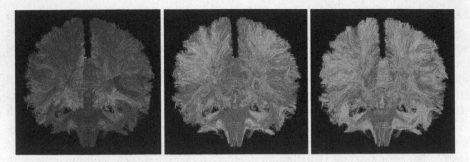

Fig. 3. Map direction (left), length (middle), and distance (right) as tract color.

2.3   sMRI Visualization

For sMRI visualization, the pipeline is shown in Fig. 4. The visualization plugin takes sMRI data as a 3D image. Then each region of interest (ROI) is extracted with vtkDiscreteMarchingCubes class in the VTK library. Each ROI get its own vtkActor so that they can have different colors or textures, as shown in Fig. 5.

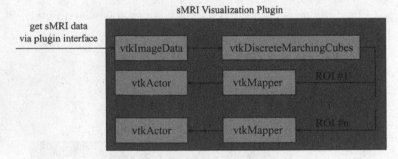

Fig. 4. VTK pipeline for visualizing sMRI

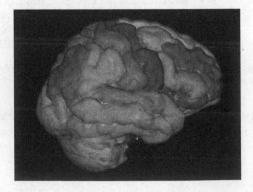

Fig. 5. Result of sMRI visualization.

2.4 fMRI Visualization

We use the texture mapping approach to visualize fMRI data on their corresponding ROI surfaces. We need to first encode this time-series data on a 2D texture image. We propose an offset contour method to generate patterns of contours based on the boundary of each projected ROI. The offset contours are generated by offsetting the boundary curve toward the interior of the region, creating multiple offset boundary curves. The technical detail is depicted in [7].

The implementation of the fMRI visualization pipeline is very similar to that in sMRI visualization, except that vtkTexture class to map the generated texture onto the surface of each ROI. Figure 6 shows an example output of fMRI visualization.

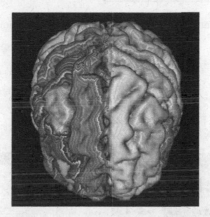

Fig. 6. An example of fMRI data visualization

2.5 GPU Accelerated Genome Browser

In BECA, we also implemented a browser for neuroimaging genomic data that implemented analysis of variance (ANOVA statistical test) and genome-wide association studies (GWAS). Different from previous visualization, the genome browser is computational intensive, which require GPU acceleration for real-time interaction. We used the server-client model in the genome browser, as shown in Fig. 7. The server is responsible for computation using CUDA [8] with NVIDIA GPUs, while the client is for display. The client and server communicate through a TCP/IP socket. There are mainly two benefits by employing the server-client model to decouple visualization from computation: firstly, the client can run on a terminal without a NVIDIA graphics card, which lower the hardware requirement to run the genome browser; secondary, different clients can share the same server, which further reduce the cost on deployment.

Figure 8 shows the primary components of the user interface. In the lower left hand corner, a 3-dimensional model of a reference brain is displayed, color-mapped with the p-value of the association between each voxel and the current SNP or gene. At the top of the user interface is the SNP or gene explorer. This region displays the $-\log 10$ (p-value) of the association between each SNP or gene and the current voxel.

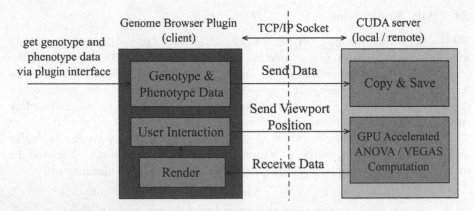

Fig. 7. Client-server model for GPU accelerated genome browser

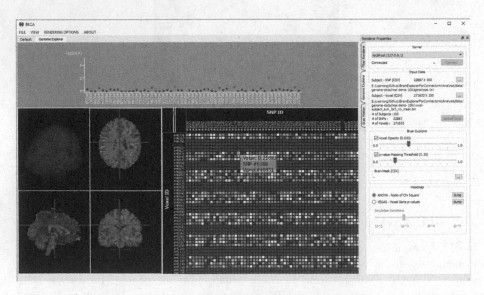

Fig. 8. User interface for the neuroimaging genome browser, which consist of SNP explorer (top), brain explorer (bottom left), and heat map (bottom right).

3   Conclusion

In this paper, we presented the design of our visualization platform, BECA, for integrated visualization of human brain data. The software is freely available at http://www.iu.edu/~beca/. It provided a plugin platform for rapid implementation of different visualizations for all kinds of brain data. In corporation with domain experts of neuroscience, work is still in progress to support more data formats and visualization techniques.

Acknowledgements. This work was supported by NIH R01 EB022574, R01 LM011360, U01 AG024904, RC2 AG036535, R01 AG19771, P30 AG10133, UL1 TR001108, R01 AG 042437, and R01 AG046171; DOD W81XWH-14-2-0151, W81XWH-13-1-0259, and W81XWH-12-2-0012; NCAA 14132004; and IUPUI ITDP Program.

References

1. Wang, R., Van, J.: Diffusion toolkit: a software package for diffusion imaging data processing and tractography. Proc. Intl. Soc. Mag. Reson. Med. 15 (2007)
2. Neuropsychology Lab: MRIcron. http://people.cas.sc.edu/rorden/mricron/index.html. Accessed 30 July 2017
3. Jenkinson, M., et al.: FSL. NeuroImage **62**, 90–782 (2012)
4. The Qt Company: Qt Framework. https://www.qt.io. Accessed 30 July 2017
5. Kitware: The Visualization Toolkit. http://www.vtk.org. Accessed 30 July 2017
6. Fowler, M.: Patterns of Enterprise Application Architecture. Addison-Wesley Professional, Boston (2003)
7. Li, H., Fang, S., Cortes, G.J., Contreras, J., Liang, L., Cai, C., West, J., Risacher, S., Wang, Y., Sporns, O., Saykin, A., Shen, L.: The ADNI integrated visualization of human brain connectome data. In: Guo, Y., Friston, K., Aldo, F., Hill, S., Peng, H. (eds.) BIH 2015. LNCS, vol. 9250, pp. 295–305. Springer, Cham (2015). doi:10.1007/978-3-319-23344-4_29
8. Nickolls, J.: Scalable Parallel Programming with CUDA. Queue **6**, 2 (2008)

Workshop on Semantic Technology for eHealth (STeH 2017)

Knowledge Graphs in the Quality Use of Antidepressants: A Perspective from Clinical Research Applications

Weijing Tang[1], Yu Yang[1], Zhisheng Huang[2(✉)], and Xiaoli Hua[1(✉)]

[1] Department of Pharmacy, Union Hospital, Tongji Medical College,
Huazhong University of Science and Technology, Wuhan 430022, China
stefenyhxl@aliyun.com
[2] Department of Computer Science, VU University Amsterdam,
Amsterdam 1081 HV, The Netherlands
zhg300@vu.nl

Abstract. The incidence of depression has increased dramatically in recent years and the quality use of antidepressants has drawn extensive concerns worldwide. In this paper, firstly, we analyze current barriers regarding the use of antidepressants. Secondly, a new informative system which is developed using knowledge techniques is proposed, and its role in the management of antidepressants is discussed. Three main functions of the current version of Knowledge Graphs for Depression (DepressionKG, version 0.6) are presented. Besides, we conduct a semi-structured focus group interview to evaluate this system. The results indicate that DepressionKG has the potential to be a feasible, evidence-based tool for healthcare professionals.

Keywords: Knowledge graphs · Quality use of medicines · Antidepressants · Intelligent monitoring

1 Introduction

Depression is a common mental disorder and is also the third leading cause of global disease burden worldwide [1]. The incidence of depression has been increasing dramatically over the past ten years. According to WHO, there is an increase of 18% of depression patients from 2005 to 2015 and currently more than 300 million people suffer from depression all over the world [2]. In China, more than 54 million people are living with depression [3]. Because of stigma, the actual incidence of depression in china might be much higher than the official data. Meanwhile, Depression is one of the leading cause of disability globally as it can lead to suicide [3]. Therefore, it is crucial that depressive patients could receive effective treatment to prevent them from getting worse and committing suicide.

At present, especially in moderate and severe depression, patients are normally treated with combination of two main approaches, pharmacological interventions (antidepressants) and psychological interventions (e.g. cognitive behavior therapy) [4]. As antidepressants are widely prescribed, their efficacy and safety has attracted

© Springer International Publishing AG 2017
Y. Zeng et al. (Eds.): BI 2017, LNAI 10654, pp. 295–303, 2017.
https://doi.org/10.1007/978-3-319-70772-3_28

extensive concern worldwide. However, healthcare professionals seem to have controversial opinions regarding prescribing antidepressants [5, 6]. Meanwhile, currently, there is no well-established informative systems to evaluate the appropriateness of prescribing antidepressant to target patient. In other words, an evidence-based, validated and feasible informative medical system specifically designed for antidepressants is lacking.

Knowledge graphs is a systematic structured domain knowledge which represented in a formal language of the Semantic Web technology. It has become one of the most important formats for web-oriented knowledge representation [7]. At present, knowledge graphs have been applied in various areas, including Internet search engines (i.e. Baidu, Google), geographic information, life sciences and medicine, multimedia, publication industry, engineering, social science etc. Therefore, knowledge graphs may have the potential to provide rational solutions for quality use of antidepressants.

In this paper, we will analyze issues relating to the development of an informative system for quality use of antidepressants from the perspective of clinical research. We will also introduce a new approach which is a knowledge-graph-based system for the rational use of antidepressants. The results of a preliminary study on this system will also be presented and discussed.

The contributions of this paper are:

- We analyses current conditions especially the barriers of quality use of antidepressants as well as the advantages of information management approaches regarding rational use of antidepressants.
- We introduce a new informative approach for the rational use of antidepressant - the knowledge graphs, which has been successfully applied in other areas.
- We present three main functions of our newly developed knowledge graph system specifically designed for antidepressants management.
- We conduct a preliminary qualitative study to evaluate the feasibility of current knowledge graphs model using thematic analysis.

2   Quality Use of Antidepressants and Information Management Approaches

The quality use of medicines (QUM) or the rational use of medications consists of four aspects which require a prescribed medication to be used judiciously, appropriately, safely and efficaciously [8]. In terms of the quality use of antidepressants, it could be interpreted as: antidepressants are prescribed appropriate to patient's clinical needs, in doses that meet patient's individual treatment requirement, be prescribed for an adequate period of time, and at the lowest cost to patients and the community. To satisfy the above requirements, it is vital that doctors can choose the most suitable antidepressants for patients at early stage of clinical interventions. However, it is reported that only one third of patients with depression benefit from the first antidepressants they try, while others may struggle from trying other antidepressants for years [9]. This condition may be the consequence of some unique features of antidepressants. The feathers can be summarized as unpredictable therapeutic effects, severe drug interactions and intolerable withdraw effects [4].

1. Unpredictable therapeutic effects: Antidepressants do not always improve patients' mood. Patients' response to the same medication may vary dramatically. For example, SSRIs are recommended as first-line medications worldwide for their favorable risk-benefit ratio, but they may also increase suicide thoughts in some patients.
2. Severe drug interactions: Antidepressants interact with many classes of medications including other antidepressants, opioids, stimulants etc. Combine use of two or more medicines from these categories can result in excessive serotonin production and cause serotonin syndrome. In severe situations, serotonin syndrome is life-threatening. Patients suffer from serotonin syndrome may experience side effects such as agitation, mental confusion, hyperthermia, tremor or even coma.
3. Intolerable withdraw effects: Withdraw effects occur when antidepressants are ceased abruptly. Common side effects ranging from flu-like symptoms, sleep disturbances, tremors, mood disturbances to cognitive disturbances. These unwanted effects are generally not additive but unbearable. So, if patients want to switch from one antidepressant to another, they must taper down current medication dose slowly and at the same time increase the new antidepressants' dose gradually.

At present, for a new patient, doctors often choose medications which are based on guideline recommendations, their professional experiences and the patient's profile (e.g. current disease conditions and medical conditions). This common strategy works well for most of chronic diseases, such as hypertension, diabetes, hyperlipidemia etc. However, as discussed above, for antidepressants, it remains confusing that how doctors could prescribe antidepressants more wisely.

Modern technology, especially information management approaches, could be one of promising solutions for this problem. At present, many electronic tools have been developed, including medication databases (e.g. MIMS, Uptodate), Internet search engines (e.g. PRNinfo) and electronic/CD form of guidelines (e.g. e-AMH, eTG). However, there is no tools specially designed for antidepressants.

3 Implementation of Knowledge Graphs

Based on the above discussions, in this study we developed a system of Knowledge Graphs of Depression (DepressionKG for short) for the quality use of antidepressants using knowledge techniques [11, 12]. Knowledge graphs of Depression is a set of integrated knowledge/data sources concerning depression. It provides the data infrastructure which can be used to explore the relationship among various knowledge/data sources of depression and support for clinical/medical decision support systems. DepressionKG is represented with the format RDF/NTriple, a semantic web standard. It integrates knowledge and data from different types of resources (Table 1) including medical guidelines of depression, clinical trials of depression, PubMed/Medline on Depression, Wikipedia/DBPedia Antidepressant, Drugbank, Medicine Instructions, SIDER (database for adverse drug reactions), SNOMED CT (database for medical terminology), depression diagnostic criteria (i.e. ICD10. DSM V) and databases for life sciences. Data from all these resources were integrated to generate the DepressionKG system using semantic approaches (Fig. 1). At present, this system has been developed with the system design for comprehensive analyses of antidepressants' adverse drug reactions. The current system consists of three main functions.

Table 1. Medical resources for DepressionKG

Knowledge resource	Number of data item	Number of triple
Clinical trial	10,190 trials	1,606,446
PubMed on depression	46,060 papers	1,059,398
Medical guidelines	1 guideline	1,830
Drugbank	4,770 drugs	766,920
Drugbook	264 antidepressants	13,046
Wikipedia antidepressant side effects	17 antidepressants	6,608
SIDER	1,169 drugs	193,249
SNOMED CT		5,045,225
Patient data	1,000 patients	200,000
Total		**8,892,722**

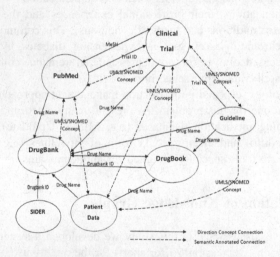

Fig. 1. Data integration of depressionKG

3.1 Analyze Adverse Drug Reactions for Single Antidepressants

This function is a primary session of the current system. Adverse effects of different antidepressants were analyzed and classified according to their frequency of occurrence (i.e. very common, common, uncommon, rare, very rare and unknown). Because we integrated and retrieved data from a large amount of different recourses, the frequency result of DepressionKG is more accurate than other databases.

Here is the example of the side effects of an antidepressant which are represented in the RDF Ntriple format in the Knowledge Graphs:

> wkipedia:Lorazepam rdf:type med:Drug.
> wkipedia:Lorazepam med:hasEnglishName "Lorazepam"@en.
> wkipedia:Lorazepam med:hasName "Lorazepam"@en.
>
> wkipedia:Lorazepam med:hasName "劳拉西泮"@cn.
>
> wkipedia:Lorazepam med:hasDrugBankID "DB00186".
> wkipedia:Lorazepam med:hasSideEffects wkipedia:Lorazepam-sfs.
> wkipedia:Lorazepam-sfs med:hasSideEffect wkipedia:Lorazepam-sfsvc1.
> wkipedia:Lorazepam-sfsvc1 med:hasSideEffectFrequency "verycommon".
> wkipedia:Lorazepam-sfsvc1 med:hasSymptom "sedation".
> wkipedia:Lorazepam-sfsvc1 med:hasSymptomIDwkipedia:sedation.
> wkipedia:Lorazepam-sfs med:hasSideEffect wkipedia:Lorazepam-sfsvc2.
> wkipedia:Lorazepam-sfsvc2 med:hasSideEffectFrequency "verycommon".
> wkipedia:Lorazepam-sfsvc2 med:hasSymptom "depression".
> wkipedia:Lorazepam-sfsvc2 med:hasSymptomIDwkipedia:depression.
> wkipedia:Lorazepam-sfs med:hasSideEffect wkipedia:Lorazepam-sfsvc3.
> wkipedia:Lorazepam-sfsvc3 med:hasSideEffectFrequency "verycommon".
> wkipedia:Lorazepam-sfsvc3 med:hasSymptom "emotional lability".
> wkipedia:Lorazepam-sfsvc3
> med:hasSymptomIDwkipedia:emotional_lability.
> wkipedia:Lorazepam-sfs med:hasSideEffect wkipedia:Lorazepam-sfsvc4.
> wkipedia:Lorazepam-sfsvc4 med:hasSideEffectFrequency "verycommon".
> wkipedia:Lorazepam-sfsvc4 med:hasSymptom "confusion".
> ......

From the statements above, we can also see that the antidepressant is annotated with its DrugBank ID for the data integration.

3.2   Analyze Single Adverse Drug Reactions

Based on the previous function analysis, we can further provide a reverse searching function which allows healthcare professionals to search antidepressants for a targeted adverse effect. For example, if a doctor wants to know which antidepressants have lower possibility to cause weight change. He only need to select "Influence of medicines on weight change" in the menu column and an analysis report will be generated for him immediately (Fig. 2). From this report, doctors can easily figure out which antidepressants are more likely to cause weight change. In function, any side effect related to antidepressants can be searched.

3.3   Analyze Adverse Drug Reactions of Polypharmacy

Polypharmacy is quite common in treatment for patients with depression. Aside with antidepressants, patients are often prescribed additional medications for symptomatic relief, for example, sedative mediations like benzodiazepines for treatment of insomnia. In this case, evaluation of single medication is no longer adequate to predict which

药物（Drug）	症状（Symptom）	频率（Frequency）
阿莫沙平	Weight gain	uncommon
阿托莫西汀	Weight loss	common
西酞普兰	Weight loss	common
西酞普兰	Weight gain	uncommon
氯米帕明	Weight gain	verycommon
氯米帕明	Weight loss	common
度硫平	Weight loss	uncommon
度硫平	Weight gain	uncommon
度洛西汀	Weight gain	uncommon
氟西汀	Weight loss	common
米氮平	Weight gain	verycommon
去甲替林	Weight gain	uncommon
去甲替林	Weight loss	uncommon
帕罗西汀	Weight gain	common
帕罗西汀	Weight loss	common
舍曲林	Weight loss	common
舍曲林	Weight gain	common
曲唑酮	Weight change	common
文拉法辛	Weight loss	common
文拉法辛	Weight gain	uncommon

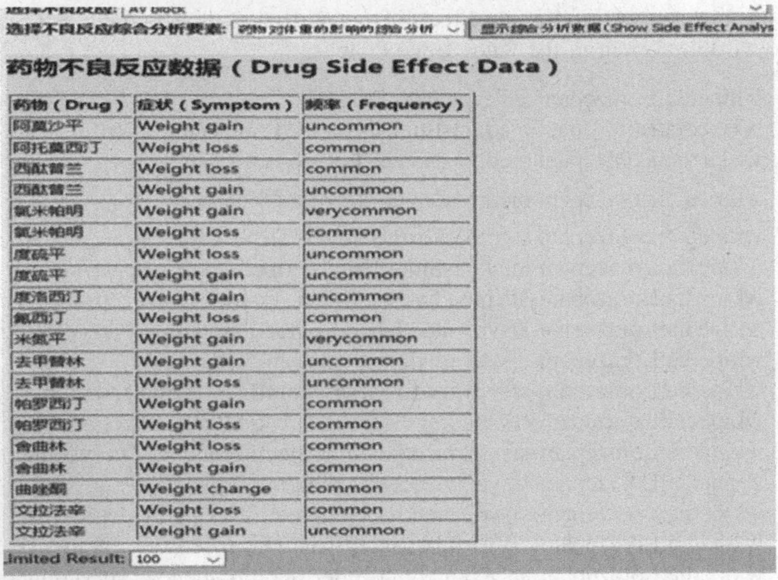

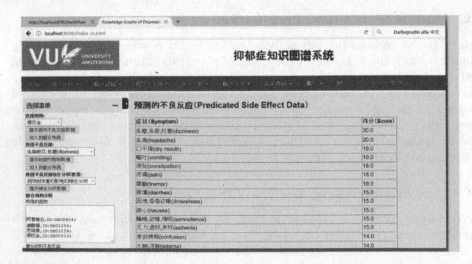

Fig. 2. DepressionKG antidepressants adverse effect report for weight change

adverse effects patients are more likely to suffer from. Therefore, we developed a new function which could analysis the potential adverse effect for multiple medications and give every single side effect a score according to their probability of occurrence. For example, if doctor would like to know the predicted side effects of valproate, lithium, quetiapine with alprazolam, he only need to input the medication names and the system will automatically analyze and present results with scores. As indicated in Fig. 3, if

症状(Symptom)	得分(Score)
头晕,头昏,眩晕(dizziness)	30.0
头痛(headache)	20.0
口干燥(dry mouth)	19.0
呕吐(vomiting)	19.0
便秘(constipation)	18.0
疼痛(pain)	18.0
震颤(tremor)	18.0
腹泻(diarrhea)	15.0
困倦,昏昏欲睡(drowsiness)	15.0
恶心(nausea)	15.0
嗜睡,欲睡,嗜眠(somnolence)	15.0
无力,虚弱,衰弱(asthenia)	15.0
意识模糊(confusion)	14.0
水肿,浮肿(edema)	14.0

Fig. 3. Predicated side effect data for the combination use of multiple drugs

these medications are prescribed together, the top three potential side effects are dizziness (30 points), headache (20 points), dry mouth (19 points) and vomiting (19 points).

4 System Evaluation

To further evaluate this system, we conducted a semi-structured focus group interview with a purposive sample of pharmacists (n = 5), physicians (n = 6), medical administration officers (n = 2) and medical engineers (n = 3) in May 2017. Participants were recruited from hospital practice settings. During the interview, All the participants were asked to comment on the design, feasibility and potential implementation of this sample report, especially in hospital settings. Discussions continued until no new themes emerged. The interview was recorded in a written form to be entered into QSR NVivo (Version 11.0). Each participant in the interview was assigned a number for anonymity. The data were analyzed using thematic analysis, themes and subthemes were derived using conventional content analysis. Emerging subthemes were developed by repeated study and coding of the transcripts. A summary of the themes and subthemes, with sample quotes, relating to the perspectives of healthcare professionals on DepressionKG (Version 0.6) is presented in Table 2.

Table 2. Thematic analysis of perspectives on the application of DepressionKG in antidepressants management: themes and subthemes with examples of quotes

Theme	Subtheme and example of quotes
Superiorities	*Future roles in clinical practice* "I like the reverse searching function because sometimes, I do have to consider the uncommon side effect when prescribing for certain patients. With this system, I don't have to look up medical instructions one drug by one another. This system will save my time" (Physician 04) "This system seems easy to use. Then we do not need to spend lots of time training doctors and pharmacist to use this system in the future" (Medical affairs officers 01)
	Use as a quick reference for patient education "As a clinical pharmacist, I have to spend lots of time on patient education. I think, this system, maybe in the future, could be more portable and give me a quick reference when consulting patients" (Pharmacist 02) "The presentation format is good, I mean, its straightforward. Actually, I can show the results to patients so they may have a better understanding of the antidepressant they are going to take" (Physician 01)
	User-friendly result presentation format "It is great that side-effects are presented according to occurrence of side effects. The scores are very straightforward for me." (Specialist 03)
Barriers	*Evaluation in clinical setting* "I will not use this system until the information proved to be accurate. At present, I prefer to use databases and searching engines." (Physician 06)

(*continued*)

Table 2. (*continued*)

Theme	Subtheme and example of quotes
	"I am not good at computer science, I have no idea if it is evidence-based or not. Maybe we need to test it for some period in the hospital in the future." (Pharmacist 04)
	Issues regarding hospital management "I am afraid that doctors might rely much on this system. As a medical administration officer, I don't want doctors to rely on electronic tools." (Medical administration officers 02) "I am afraid that doctors might rely much on this system. How can we monitor this? As a medical administration officer, I don't want doctors to rely on electronic tools." (Medical administration officers 02) "I am wondering if this system is open-licensed or not. It would be great if the system could be further modified based on our hospital's requirement." (Medical engineer 03)

5 Discussion

The quality use of antidepressant is a fundamental issue in the management of patients with depression. In this paper, the current issues related to quality use of antidepressants was discussed. A new information system model (DepressionKG) designed to assist healthcare professionals with rational use of antidepressants was introduced. In addition, the result of a preliminary semi-structured focus group interview was presented.

Current the DepressionKG system include three main functions. The first function is a primary function which allows healthcare professionals to search a single antidepressant's adverse reactions easily. Although this function is also available in other medication databases, our system's result is more accurate and straightforward because it retrieved from various medical resources and presented in a table format according to the frequency of occurrence. The second function provides a new option which enables healthcare professionals to search antidepressants for a targeted adverse effect. In traditional approaches, most of time doctors make decisions based on their own experience, or search drug database one antidepressant by another. Such approaches are neither strong evidenced-based nor efficient. With this new function, doctors could get a quick reference to rule out antidepressants which might cause unwanted side effects to targeted patients. Furthermore, we also developed a new function which could analysis the potential adverse effect for multiple medications and present analysis results with scores accordingly. With this function, doctors could have a holistic view of potential side effects when prescribing antidepressants with other medications.

In terms of the evaluation of DepressionKG, we received diverse views from healthcare professionals. Semi-structured focus group interview allows to collect data that is in rich detail and to examine the relationships to develop theories from observed events and observations [10]. The strength of using focus group interview is that we could obtain valuable views from health professionals who are going to be target users of this system. It is indicated that most participants provided supportive comments on the clinical implement of the system itself with only few considerations regarding the validity of the system. Other considerations were mainly focus on the hospital management issues. Overall, the current system was accredited by healthcare professionals as feasible with a promising future for its implementation in hospital settings. Based on the discussions above, future research will emphasize on the following aspects:

1. Improve current functions. More medical resources will be added to this system to provide more precise and evidence-based results for healthcare professionals. The interface will also be improved to be more user-friendly.
2. Expansion of joint functionality model. In future study, we hope to make this system more patient-focused and be able to generate individualized reports. We will integrate current medical resources and patients' profile together using semantic approaches.
3. Evaluate DepressionKG in hospital settings. We will cooperate with more hospitals and involve more experts and frontline clinical staff to test and evaluate our system comprehensively.

References

1. Whiteford, H.A., Degenhardt, L., Rehm, J., Baxter, A.J., Ferrari, A.J., Erskine, H.E., et al.: Global burden of disease attributable to mental and substance use disorders: findings from the global burden of disease study 2010. The Lancet **382**, 1575–1586 (2013)
2. WHO: fact sheet: depression. http://www.who.int/mediacentre/factsheets/fs369/en/. Accessed 22 July 2017
3. WHO: "Depression: let's talk" says WHO, as depression tops list of causes of ill health. http://www.who.int/mediacentre/news/releases/2017/world-health-day/en/. Accessed 22 July 2017
4. Australian Medicines Handbook: Psychotropic Drugs: Antidepressants, pp. 765–770. Australian Medicines Handbook Pty Ltd. Adelaide, SA (2015)
5. Spence, D.: Are antidepressants overprescribed? Yes. BMJ **346**, f191 (2013)
6. Reid, I.C.: Are antidepressants overprescribed? No. BMJ **346**, f190 (2013)
7. Chen, C.: Searching for intellectual turning points: progressive knowledge domain visualization. Proc. Nat. Acad. Sci. **101**, 5303–5310 (2004)
8. Australian government department of health (2017) quality use of medicines. http://www.health.gov.au/internet/main/publishing.nsf/content/nmp-quality.htm
9. Trivedi, M.H., Rush, A.J., Wisniewski, S.R., Nierenberg, A.A., Warden, D., Ritz, L., et al.: Evaluation of outcomes with citalopram for depression using measurement-based care in STAR* D: implications for clinical practice. Am. J. Psychiatry **163**, 28–40 (2006)
10. Huston, S.A., Hobson, E.H.: Using focus groups to inform pharmacy research. Res. Soc. Adm. Pharm. **4**, 186–205 (2008)
11. Huang, Z., Yang, J., van Harmelen, F., Hu, Q.: Constructing disease-centric knowledge graphs: a case study for depression (short version). In: Proceedings of the 2017 International Conference on Artificial Intelligence in Medicine (2017)
12. Yang, J., Huang, Z., Hu, Q., Wang, G.: Integration of side effect knowledge of antidepressants with knowledge graph of depression and its applications. J. China Digital Med. **12**(6), 2–4 (2017)

Using Knowledge Graph for Analysis of Neglected Influencing Factors of Statin-Induced Myopathy

Yu Yang[1(✉)], Zhisheng Huang[2], Yong Han[1], Xiaoli Hua[1],
and Weijing Tang[1]

[1] Department of Pharmacy, Union Hospital, Tongji Medical College,
Huazhong University of Science and Technology, Hubei, China
y505023647@126.com
[2] Department of Computer Science, VU University Amsterdam,
Amsterdam 1081 hv, The Netherlands
zhg300@vu.nl

Abstract. Statins have been widely used for the treatment of cardiovascular diseases. However, the most severe adverse effect of statins is myotoxicity, in the form of myopathy and other similar ones. Identifying whether it is a statins-induced muscle symptoms plays an important role in the use of statins. In this paper, we propose an approach to analyse the neglected influencing factors of statin-induced myopathy in a coronary heart disease case by using the technology of knowledge graphs. Through the n-of-1 trial, we can verify the accuracy of the knowledge graphs for this task. Furthermore, Knowledge graph of adverse reactions and symptoms is expected to assist physicians in determining adverse events in the future.

Keywords: Statin · Myopathy · Knowledge graph

1 Background

Statins are among the most widely prescribed drugs, which can reduce the risk of cardiovascular events, and adherence to statin therapy correlates with reduced cardiovascular mortality [1, 2]. Muscle pain, as the most commonly reported adverse events, approximately 60% of adults stop to take statins as the primary reason [3]. And contribute bad influence to cardiovascular events control.

A recent international survey regarding statin-associated symptoms indicated that 72% of overall adverse events were muscle-related [4]. Myalgia is the most common side effect of statins as a rate of 38.6%. In other muscle symptoms, rates for Myopathy, Myositis, Rhabdomyolysis, and Joints and Tendons were, respectively, 3.3%, 2.4%, 18.2%, 32%, and 5.5% as high [5].

As muscle symptoms are not only relevant to the use of statins [6], but also have a potential relationship with many factors, such as sleep disorders, neurotransmitter abnormalities, immune disorders, angina pectoris, et al. If muscle pain accompanied by elevated CK, it may induce myositis even rhabdomyolysis. This could happened in strenuous exercise, progressive muscular atrophy, skeletal muscle injury, meningitis, hypothyroidism, the use of penicillin and so on. As these situations sometimes happened together, it is obviously unreasonable simply due to statins-induced muscle damage, when it is impossible to accurately identify the role of factors.

Since many of the information about the statin-related myalgia is fragmentary, lacking a large randomized controlled study of thousands of patients to elaborated the relationship of statins and myalgia [7], this evaluation would be also untrustworthy. We try to show the problem by provide a case with arm pain in coronary heart disease patient using statins. This case was attributed to statin-related myalgia while using Lipitor. The Adverse Drug Reaction Probability (Naranjo) score was 3 [8], while the proposed statin myalgia clinical index score was 6 in this case [9], which means the arm pain was unlikely associated with statins. Considering not enough information was provided, other possible factors may be ignored.

Knowledge Graphs are a set of large scale semantic network consisting of entities and concepts as well as the semantic relationships among them, using representation the knowledge representation languages such as RDF and RDF Schema in the Semantic Web. Knowledge Graphs have been shown to be useful tools for integrating multiple medical knowledge sources, and to support such tasks as medical decision making. Linked Open Data (LOD)[1] and Linked Life Data (LLD)[2] are several well-known large scale Knowledge Graphs. In this paper we propose an approach to use the technology of Knowledge Graph to detect various relationships of muscle pain and influencing factors. With the support of the Knowledge Graphs, we make an analysis of the neglected influencing factors of statin-induced myopathy in coronary heart diseases.

2   Case Presentation

A 58-year-old yellow man presented to the emergency cardiology department with complaints of coronary heart disease and hypertension for a long time.At that time, His creatine kinase(CK) was 54U/L (normal range 38-174U/L). Other laboratory indicators were within normal limits. Physical examination was normal. He had a reported allergy to fish and Hellfish (rash). He was subsequently admitted for coronary heart disease with hypertension history.

Before admission, the patient suffered from a headache for more than one year, occasionally, with tinnitus and dizzy feeling. The head MRI showed multiple lacunae infarct, brain atrophy, with a normal head MRA, in 2015 June.

In February 2017, the coronary CTA shown no obvious plaque and luminal stenosis in the right coronary artery. No obvious stenosis shown in left main artery. Proximal-middle segments of left anterior descending could see thickening wall and non-calcified plaque, and the most obvious luminal stenosis was in the middle segments, where was narrow about 50–60%. No significant plaques and stenosis were found in the left circumflex coronary artery. The left ventricular myocardium was homogeneously enhanced.

---

[1]   http://linkeddata.org/.

[2]   http://linkedlifedata.com/.

After admission (March 15, 2017), he was treated with Aspirin Enteric-coated Tablets 100 mg once a day, Plaix 75 mg once a day, Panlisu 40 mg once a day, Lipitor 20 mg once a night, Valsartan and Hydrochlorothiazide Tablets one tablet a day.

On the third day of hospitalization, the patient received coronary arteriongraphy for coronary heart disease diagnosis in the afternoon. And there were no complaints before angiography.

Day 4 (March 18), the patient complainted of right arm myalgia, chest tightness, precordial discomfort with stress, accompanied by sweating without any physical activities. Nitroglycerin was taken without apparent relief and lasted approximately an hour after gradual remission, while the ECG was still normal. And the quick check shown a CK of 1091U/L, a CK-MB of 7.0 ng/ml, and Troponin T of 8.3 pg/ml within normal limits. On the next day, CK raised to 1348U/L, and Lipitor was stopped immediately.

Day 6 (March 20), CK decreased to 1154U/L, while the CK-MB was normal. After three days, CK turned to normal as 104U/L, without complaints of discomfort. And then, Fluvastatin capsule was taken to treat coronary heart disease without any myalgia complaint (Table 1).

Table 1. The change of myocardial enzymes

	March 18	March 19	March 20	March 21	March 22
TNI(pg/ml)	8.3	7.0	4.3	5.7	7.0
CK(U/L)	1091	1348	1154	666	104
CK-MB(ng/ml)	7.0	2.8	0.8	0.6	0.5
AST(U/L)	24	32	30	24	14
LDH(U/L)	163	188	178	174	185

His SLCO1B1 (T521C) genotype was TT, while the SLCO1B1 (A388G) genotype was AA. Overall, the patient belongs to *1a/*1a SLCO1B1 gene type.

3 Analysis

In this case, myalgia and the elevated CK was likely attributed to Lipitor. We need to analysis the probability of statin-associated muscle symptoms(SAMS) by a reliable method.

Statin-related muscle toxicity mainly are myalgia, myopathy, myositis, muscle necrosis, rhabdomyolysis [10]. According to the definitions and the patient's Creatine kinase(CK), it should be myositis. However, there are two large meta-analysis come to a conclusion that statin therapy did not found association with CK elevations compared with placebo [11, 12]. Considering the proposed statin myalgia clinical index score was 6 [9], and the potential predisposing factors of this patient are only Asian and chronic pain (headache over 1 years) [10], what is more, he did not have any muscle symptoms after changed to Fluvastatin [13], we have grounds to believe that the possibility of SAMS was small. So, we try to re-analyze the possibility from other neglected factors.

The patient had basic disease without a history of large amount of activities, and CK increased significantly, which was considered to be a pathological elevation. As known, Creatine kinase mainly exists in the skeletal muscle and myocardium, followed by brain tissue, and a very small amount of distribution in smooth muscle, red blood cells, liver and so on. Considering he had no history of encephalitis, brain disease or head injury, brain-derived CK elevation was excluded. The patient had a coronary angiography recently, and the MB isoenzyme of creatine kinase increased, accompanied by chest tightness and other clinical manifestations, so myocardial damage was not excluded. And there was not enough evidence to support that was a skeletal muscle injury.

When come to the drug-induced factors, many drugs may lead to CK increasing, for example, statins. As reported [14–16], patients with a *5 or *15 genotypes had higher rates of drug side effects than those with *1a or *1b genotypes when using statins. Compared to the result of gene detection, the patient had a relatively low risk of SAMS.

As for the pain, the patient had suffered an intravenous infusion of L-carnitine and Alprostadil in the morning which can lead to phlebitis. But the patient surface did not seen any symptoms of phlebitis, so this possibility could be rule out. He also had a coronary angiography in the last night, and the contrast agent could be the cause of arteritis, which can not be seen on the surface of the skin, manifested as arm pain. And this pain could be considered as myalgia subjectively. No evidence of neuropathic pain was found.

In order to determine the correctness of the preliminary analysis, we use the knowledge graph to coordinate all the relevant information. We use the following SPARQL query to search over the knowledge graph of Linked Life Data,

```
PREFIX skos: <http://www.w3.org/2004/02/skos/core#>
PREFIX lifeskim:
<http://linkedlifedata.com/resource/lifeskim/>
PREFIX pubmed:
<http://linkedlifedata.com/resource/pubmed/>
PREFIX umlsid:
<http://linkedlifedata.com/resource/umls/id/>
SELECT distinct ?doc ?title ?prefLabel
WHERE {
  ?doc lifeskim:mentions
<http://linkedlifedata.com/resource/umls/id/C0750863>.
  ?doc lifeskim:mentions ?concept .
  ?concept rdf:type
<http://linkedlifedata.com/resource/semanticnetwork/id/T0
47>.
  ?concept skos:prefLabel ?prefLabel .
  ?doc pubmed:articleTitle ?title
}
LIMIT 100
```

In that SPARQL query, we use the UMLS concept ID C0750863 to refer to the medical concept "Finding of creatine kinase level", and use the URI http://linkedlifedata.com/resource/semanticnetwork/id/T047 to refer to a symptom or a disease in Linked Life Data.

That semantic query returns the following answers (Table 2):

Table 2. Semantic query return results

Doc	Title	PrefLabel
pubmed-article: 859617	Muscle histology and creatine kinase levels in the foetus in Duchenne muscular dystrophy	Muscular Dystrophy, Duchenne
pubmed-article: 14248444	Creatine kinase levels in women who carry genes for three types of muscular dystrophy	Muscular Dystrophy
pubmed-article: 863458	Use of normal daughters' and sisters' creatine kinase levels in estimating heterozygosity in Duchenne muscular dystrophy	Muscular Dystrophy, Duchenne
pubmed-article: 3772449	CSF brain creatine kinase levels and lactic acidosis in severe head injury	Acidosis, Lactic
pubmed-article: 7053734	Diagnostic problem in acute myocardial infarction: CK-MB in the absence of abnormally elevated total creatine kinase levels	Acute myocardial infarction
.....		
pubmed-article: 3414544	Relation of peak creatine kinase levels during acute myocardial infarction to presence or absence of previous manifestations of myocardial ischemia (angina pectoris or healed myocardial infarction)	Acute myocardial infarction

In particular, we can find the following relationships between elevated Creatine Kinase (CK) and myocardial injury:

In 70 patients with confirmed transmural myocardial infarction the sensitivity of creatine kinase and creatine kinase MB at different times after admission was investigated by 364 measurements. Beside the number of correctly positive results the optimized standard method hitherto used was compared with the new optimized standard method with N-acetylcystein as activator. During the first 12 h after admission the percentage of correctly positive results was 64,4% for creatine kinase (GSH) and 86,7% for creatine kinase (NAC). The sensitivity of cretine kinase MB, however, was found to be 71,1% (GSH) and 86,7% (NAC). With respect to the poor specifity of creatine kinase the sensitivity of creatine kinase MB, especially when using the new optimized standard method is superior. Similar results were established 24 and 48 h after admission. [17]

The hypothesis that acute myocardial infarction (MI) is more extensive in patients without previous angina or healed MI was evaluated in 177 patients with documented recent acute MI. Ninety-nine patients (56%) had no previous angina or healed MI (negative history group), and the remaining 78 patients (44%) had a previous history of angina or healed MI (positive history group). The mean peak creatine kinase (CK) level in the negative history group was 784 compared with 419 IU in the positive history group (p less than 0.0001). The mean peak CK-MB level in the negative history group was 128 compared with 76 IU in the positive history group (p less than 0.001). The mean peak CK-MB level was higher in the negative history group after controlling for age, streptokinase administration, previous coronary artery bypass grafting or treatment with beta-blocking agents. Despite the high frequency of healed MI in the positive history group (73%), the rates of in-hospital complications were similar for the 2 groups. Patients with acute MI without previous angina or healed MI have substantially higher peak CK and CK-MB levels; this implies a larger MI than in patients with previous angina or healed MI. [18]

Therefore, we find that elevated CK may be associated with myocardial injury. Similarly, through the search over Linked Life Data, we can find that the arm pain may be related to the use of contrast agent with high likelihood.

4   Conclusion

In this case, the risk factors of arm pain and elevated CK are complicated. When monism fails to explain all the adverse symptoms, the interaction of influence factors may lead to miscarriage of justice.

Knowledge graph use the computer to read the latest authoritative medical literature and electronic medical records, build the knowledge base, give the base reasoning ability, and finally assist doctors to find out the correlation degree of adverse reactions and influencing factors.

The difficulty lies in data preprocessing and structuring. Because the knowledge graph is structured knowledge, it is made up of entity and entity relation. For example, for adverse reactions and symptoms, the relationship can be "inclusion relationship", "exclusive relationship", and even "gold standard relationship" (for example, all inflammation can lead to fever, which is a gold standard). According to the relationship between adverse reactions and symptoms, it is described in the literature as (X adverse reaction) have (Y symptoms), the clinical manifestations of (X adverse reaction) is (Y symptoms), (X adverse reaction) can cause (Y symptoms) and so on.

In this case, knowledge is shown by knowledge graph, and there are some contradictions in the description. So we add a large number of fault tolerance mechanisms, in the process of inference calculation. According to the statistical distribution of knowledge, assume the learning samples. Reduce the opportunities of errors knowledge unfolded through some sort of strategy, supplemented by manual checking, finally, to obtain a more reliable conclusion. For example, we use an approach to identifying the conflicting evidence [19], classified as A1, A2, B, C or D. And the result shown as below. PMID 24613429 is A1, PMID 26192349 is A1, PMID 17560286 is A1, PMID 9185636 is A2, PMID 25282031 is A2, PMID 27942805 is C, PMID 16286544 is B, PMID 24389208 is C. And we come to a conclusion that statins reduce all-cause mortality by 14% and reduce major adverse cardiac events by 20%, even they can reduce the CoQ10 which is the heart-needed mitochondrial activator.

We use the N-of-1 test [20] to verified the accuracy of the knowledge graph. Knowledge graph of adverse reactions and symptoms is expected to assist physicians in determining adverse events in the future.

Acknowledgement. The authors thank Dr. Tao for technical assistance. No conflict of interest has been declared.

References

1. Bruckert, E., Ferrières, J.: Evidence supporting primary prevention of cardiovascular diseases with statins: Gaps between updated clinical results and actual practice. Arch. Cardiovasc. Dis. **107**(3), 188–200 (2014)
2. Desai, C.S., Martin, S.S., Blumenthal, R.S.: Non-cardiovascular effects associated with statins. BMJ **349**, 37–43 (2014)
3. Cohen, J.D., Brinton, E.A., Ito, M.K., Jacobson, T.A.: Understanding statin use in america and gaps in patient education (usage): An internet-based survey of 10,138 current and former statin users. J. Clin. Lipidol. **6**(3), 208–215 (2012)
4. Hovingh, G.K., Gandra, S.R., Mckendrick, J., et al.: Identification and management of patients with statin-associated symptoms in clinical practice: a clinician survey. Atherosclerosis **245**, 111–117 (2016)
5. Hoffman, K.B., Kraus, C., Dimbil, M., et al.: A survey of the FDA's AERS database regarding muscle and tendon adverse events linked to the statin drug class. PLoS ONE **7**(8), e42866 (2012)
6. Finegold, J.A., Manisty, C.H., Goldacre, B., Barron, A.J., Francis, D.P.: What proportion of symptomatic side effects in patients taking statins are genuinely caused by the drug? Systematic review of randomized placebo-controlled trials to aid individual patient choice. Eur. J. Prev. Cardiol. **21**(4), 464–474 (2014)
7. Auer, J., Sinzinger, H., Franklin, B., et al.: Muscle- and skeletal-related side-effects of statins: tip of the iceberg? Eur. J. Prev. Cardiol. **23**(1), 88–110 (2016)
8. Liang, R., Borgundvaag, B., Mcintyre, M., et al.: Evaluation of the reproducibility of the naranjo adverse drug reaction probability scale score in published case reports. Pharmacotherapy **34**(11), 1159–1166 (2014)
9. Rosenson, R.S., Miller, K., Bayliss, M., et al.: The statin-associated muscle symptom clinical index (sams-ci): revision for clinical use, content validation, and inter-rater reliability. Cardiovasc. Drugs Ther. **31**(2), 179–186 (2017)
10. Mancini, G.B., Baker, S., Bergeron, J., et al.: Diagnosis, prevention, and management of statin adverse effects and intolerance: canadian consensus working group update (2016). J. Cardiol. **32**(7), 35–65 (2016)
11. Baigent, C., et al.: Cholesterol treatment trialists' (CTT) collaborators. efficacy and safety of cholesterol-lowering treatment: prospective meta-analysis of data from 90,056 participants in 14 randomised trials of statins. Lancet **366**, 1267–1278 (2005)
12. Kashani, A., Phillips, C.O., Foody, J.M., et al.: Risks associated with statin therapy. a systematic overview of randomized clinical trials. Circulation **114**, 2788–2797 (2006)
13. Anekwe, L.: Most patients who stop taking a statin are not intolerant, suggests study. BMJ **346**(1), 22–36 (2013)
14. Hou, Q., Li, S., Li, L., et al.: Association between SLCO1B1 gene T521C polymorphism and statin-related myopathy risk: a meta-analysis of case-control studies. Medicine **94**(37), 12–68 (2015)
15. Lee, H.H., Ho, R.H.: Interindividual and interethnic variability in drug disposition: polymorphisms in organic anion transporting polypeptide 1B1 (OATP1B1; SLCO1B1). Br. J. Clin. Pharmacol. **83**(6), 1176–1184 (2017)
16. SEARCH Collaborative Group, Link, E., Parish, S., Armitage, J., et al.: SLCO1B1 variants and statin-induced myopathy—a genomewide study. N. Engl. J. Med. **359** (8), 789–799 (2008)

17. Schmidt, E.W., Bender, W.: Sensitivity of creatine kinase and creatine kinase MB in myocardial infarction. evaluation of a new optimized standard method. Med. Klin. **74**(11), 391–395 (1979)
18. Brush Jr., J.E., Brand, D., Acampora, D., Goldman, L., Cabin, H.S.: Relation of peak creatine kinase levels during acute myocardial infarction to presence or absence of previous manifestations of myocardial ischemia (angina pectoris or healed myocardial infarction). Am. J. Cardiol. **62**(9), 534–537 (1988)
19. Huang, Z., Hu, Q., ten Teije, A., van Harmelen, F.: Identifying evidence quality for updating evidence-based medical guidelines. In: Riaño, D., Lenz, R., Miksch, S., Peleg, M., Reichert, M., ten Teije, A. (eds.) KR4HC 2015. LNCS, vol. 9485, pp. 51–64. Springer, Cham (2015). doi:10.1007/978-3-319-26585-8_4
20. Joy, T.R., Zou, G.Y., Mahon, J.L.: N-of-1 (single-patient) trials for statin-related myalgia. Ann. Intern. Med. **161**(7), 531–532 (2014)

Workshop on Mesoscopic Brainformatics (MBAI 2017)

Mesoscopic Brainformatics

Dezhong Yao[✉]

Key Laboratory for Neuroinformation of Ministry of Education,
School of Life Science and Technology, University of Electronic Science
and Technology of China, Chengdu 610054, China
dyao@uestc.edu.cn

Abstract. Brain science is a well-recognized frontier science with extensive and profound contents. For the past hundreds years, biological experimental studies are the main ways to understand the brain, but now, neuro-data and computing have grown up to be almost an equivalent important tool for uncovering the brain and brain disorders. It means that a fuse of "Brain" and "Informatics", Brainformatics, is becoming an area of science. In this paper, we take the brain research driven by information science as a new discipline – BraInFormatics – with "brain information acquisition", "Brain information decoding" and "brain information applications" as the main contents, and meso-scale problems are the main space waiting for this discipline. This paper explains the concept, scope and challenges with some examples, such as EEG zero-reference technique, brainwave music etc. developed in our lab.

Keywords: Brainformatics · Information acquisition · Information decoding · Information application · Meso-scale brain science

1 Introduction

Brain science covers problems of various spatial scales, from microscopic level (gene, biomolecule, neuron), mesoscopic level (neural mass) and macroscopic system level (Fig. 1). Each scale works specifically and acts differently, and requiring different research tools to look in. For the microscopic scale, gene sequencing, biochemistry, optical microscopy, electron microscopy and patch clamp etc. are the main research tools. For the macroscopic system level, electroencephalogram (EEG), magnetoencephalogram (MEG), magnetic resonance imaging (MRI), functional near - infrared spectroscopy (fNIRS), psychological and behavioral analysis are the main research technologies. Usually, microcosmic analysis typically processes off-body samples which are invasive and non in-situ analysis, but it can dip into the gene, protein molecule, synapse, ion channel, membrane potential etc. Macroscopic techniques may face the human brain directly as it is generally non-invasive and in-situ analysis, thus of more clinic values and cognitive science application.

For micro-scale brain science, constrained by biology, complete measurement of all activity and microstructure remains a wishful thinking. And even if such measurements were made possible, one could as easily get lost in the detail of a forest as in a large data set, just like one blind man feeling an elephant [1]. At the macro-scale, leading by

Y. Zeng et al. (Eds.): BI 2017, LNAI 10654, pp. 315–324, 2017.
https://doi.org/10.1007/978-3-319-70772-3_30

psychology (Statistics), many fundamental progresses are also made without having to account for costly physiological or anatomical measurements (Fig. 1). However, the limited information available outside the brain also remains much secret under the observed. Another prominent question is whether there is a meso-scale area which do not just fills up the physical gap between the micro- and macro-scales, but also opens a new world to fly our dream? Unfortunately, there is no effective tool for the meso-scale brain information acquisition, mathematical model and computation simulation are therefore the main tools in current practice. Thus meso-scale brain science is mainly an information science driven brain science, that's the main body of BraInformatics (Fig. 1).

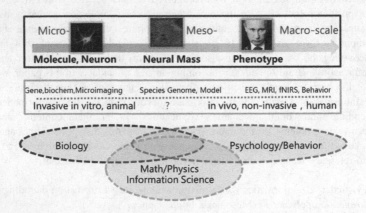

Fig. 1. Brain Science at different scales.

2   Meso-Scale Brainformatics

The meso-scale knowledge of brain is of specific significance not only because it is the critical bridge from the micro-scale details to the macro-scale phenomena, but also due to the fact that it may lead us to see the trees as well as the wood, based on its tight links with the micro- and Macro scales. As noted above, the three scale-dependent approaches are constrained respectively by genome/biology/biochemistry, mathematics/information Science, and Psychology/behavior (statistics) for the micro-scale, meso-scale and macro-scale efforts (Fig. 1), some scale-dependent sub-disciplines of brain science have been setup, such as cellular neurobiology, computational neuroscience and cognitive neuroscience in the past decades (Fig. 2). And some more specific sub-disciplines are emerging.

Actually, "when a workman wishes to get his work well done, he must have his tools sharpened". For brain science, the tool you are having would determine what kingdom you may be involved. In fact, if we look again at the brain science from the perspective of the main tools involved (Fig. 1), we will have a new view of the sub-disciplines: "Computational Neuroscience or Neurocomputing Science" driven by mathematics, "Electroencephalogram and Magnetoencephalogram" based on electromagnetism,

"Neurobiology" on chemistry and biology, and "Cognitive Neuroscience" on psychology and neuroimaging. In this way, the brain science study induced by information science would be titled as "Brainformatics", a new discipline founded on the fusion of information and brain science, thus we may take Brainformatics as an information driven brain science discipline [2], the main body being the knowledge of the meso-scale brain.

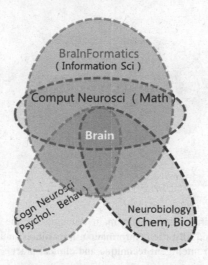

Fig. 2. Various sub-disciplines of Brain Science. Sub disciplines and the main tools adopted in parentheses. Brainformatics is the one driven by information science as the main tools.

3   The Scope of BraInformatics

Information science, which is based on information theory, system theory and cybernetics, covers the theory, method and system for information acquisition, transmission and processing, and various applications. Information science covers information theory founded by Shannon but not limited to it. Therefore, brainformatics is not limited to entropy theory, but to include brain information acquisition, decoding (the information mechanism of brain function and brain disorder) [3] and application (brain-like technology, brain-inspired technology, brain-computer interface technology and translational medicine) (Fig. 3). What should the brainformatics include in general? It can be listed according to the information detecting technologies we have. The existing technologies cover the behavioristics, mechanics (ultrasound), thermotics (infrared), electricity, magnetism, optics, biochemistry, metabolism and gene, such a scope is far beyond the well-known "neuron" or "neural network", thus the brainformatics is not the neuroinformatics. After obtaining the brain information, the coming challenge is the explanation or decoding of the data. It involves all aspects of the brain function and brain disease, and the knowledge will contribute to our understanding and protection of our brain. In the past decades, new experimental techniques are generating a wealth of information about the brain at different scales – from the levels of genome, single cells to brain circuits to behavior – but neuroscience still lacks effective tools for managing these massive data sets. We need to find new ways to organize, analyze, and extract meaning from neurodata, thus open a new big data-driven neuroscience and in so doing accelerate the pace of discovery [4]. Thirdly, how could we verify the decoding and understanding of the brain information? It depends on the generalization of the obtained model and assumption in facing various new aspects of brain function and new intervention of brain disorder.

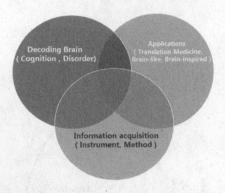

Fig. 3. The scope of Brainformatics. The main domains are the three: decoding brain to understanding cognition and disorder; information acquisition and the instrument, method behind; applications in both intelligent technique and clinical practice.

4   Brain Information Acquisition and EEG Zero-Reference Technique

Brain information acquisition involves two aspects: instrument and method. In recent years, driven by USA BRAIN Project [5] etc., a few breakthrough technologies emerged for accessing brain information, they are CLARITY technology, optogenetics technology, high field MRI brain imaging technology, flexible electrode technology, gene editing technology (such as CRISPR-9, Clustered Regularly Interspaced Short Palindromic Repeats). In this aspect, the Chinese advances include high resolution optical sectioning imaging technology [6], EEG-MRI information fusion technology [7], EEG zero reference technology [8] and so forth.

As an example, let's get into a little more details of EEG zero reference techniques [8]. EEG is a scalp potential recordings, as potential can only be measured as the potential difference between two points, we need to have one point as base, the reference electrode, and the other as the active electrode, and only when the reference is zero or constant, we can get the true activities at the active electrode (Fig. 4). However, there is no point on the body surface including the brain scalp where the potential is constant, thus no point on the body surface can be used as the gold reference. Figure 4 shows that for the same neural activities inside, the waveforms may be quite different when different references are adopted.

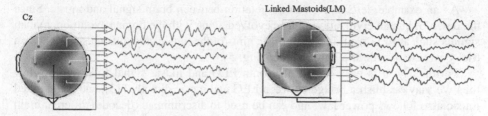

Fig. 4. EEG recordings with different reference. Left: Cz as reference, Right: Linked mastoids as reference. With the same sources underneath, EEG recordings may be quite different when different references are adopted.

EEG non-zero reference problem is a historical and fundamental problem since the human EEG discovered in 1924 by Hans Berger. In 2001, we took the lead to adopt the information science knowledge, the brain electric theory, that the scalp potential recordings, no matter what reference adopted, are generated by sources inside the brain, thus the different recordings with different references can be bridged together by the sources inside, then we developed a software method (reference electrode standardization technique–REST: www.neuro.uestc.edu.cn/rest) [8, 9] which was later repeatedly confirmed to be able to transfer an actual non-zero reference recordings to approximate zero reference based recordings [10], and now it was directly adopted more and more to get the true activities on the scalp surface to improve the later brain function or disorder discovery in many labs around the world.

5 Brain Information Decoding and Brainwave Music

Brain information decoding is in fact the understanding of brain information in a sense of brain cognition and brain disorder. The existing neural coding and decoding theory is a typical example. The current coding theory, including frequency coding, time coding, mode coding, collaborative coding, still cannot work in consent with the large amount of data accumulated in neuro-electro-physiological domain. In other words, the development of decoding theory is still at a relatively delayed situation than the data accumulation. Another issue is the brain network theory. It has now been recognized that the brain function is networked [11], which follows "small world" and "rich club" rules, and the networks are hierarchical, but the whole story of brain network is still blurred. The third problem is the brain's "Schrodinger equation" or "Einstein's equation" problem. Are there a few simple equations or principles which control the human brain's function? In past few decades, Dr. Karl Friston proposed "Bayesian brain" and "the principle of minimum free energy" [12]. Could they be the simple equation? All these problems are core problems in brainformatics in the future. In recent years, there are a lot of works conducted in understanding the information mechanism of brain function and brain diseases in China, such as Brainwave music [13], bi-directional regulation mechanism of absence epilepsy [14], effects of game training on insula plasticity [15] as well as the EEG brain network theory [16].

As an example, let's show you the relation between brain signal and music. Since the dawn of human civilization, music evolved along with the human evolution, human signals such as EEG would have intrinsic hints of the music, and thus music might be a tricky way to decode the human brain (Fig. 5).

Fortunately, it is easy to find that both EEG and music signal follow power-law thus we may establish a bridge between EEG and music directly [13], and the resulted music also follows power law, and can be used to discriminate (decode) different brain state [17].

Fig. 5. Brainwave music. Left: brainwave, right: music. With the intrinsic relation(both EEG and music following power-law), the brainwave (left) is translated to music (right).

6 Exploratory Application of Brainformatics and Apparatus-Brain Conversation

For the exploratory application of brainformatics, the EU human brain project (HBP) is a good example [18]. One task of HBP is to explore the internal cognitive mechanism of the brain through simulating normal brain function, for example, to understand the causality mechanism of cognitive behavior by comparison of various aspects between 'dummy human or SimAnimals' and 'real man or real animal'. The second goal is to simulate the brain mechanism of brain disease. It tries to find biomarker of brain disease, determine new targets, and explore the curative effect of new drugs or side effects by combining pharmacokinetics and systematic simulation study from healthy brain to morbid brain. The third point focuses on neural morphology, which aims to develop human-like control technology, such as automatic driving technology.

In general, any application of brainformatics is about the interaction between brain science and the environment, the most well-known concept is the so-called brain-computer interface (BCI). However, interface cannot characterize the whole story, we propose to use apparatus-brain conversation (ABC) (Fig. 6), where apparatus is not just computer or machine, it also represents the biological organs of a living system, and conversation represents both one-way and bi-directional communication in nature. In this way, ABC may have three categories, such as ABC-1 for "activating brain", here we assume the brain at good state, the problem is to have normal channel to let it activate, thus to get the neural signal pass through the channel to interact/control something outside, such as the brain-signal P300, SSVEP etc. based armchair control, and various artificial limb, visual or auditory prosthesis; ABC-2 for "modulating brain", such as animal robot, various neurofeedback; ABC-3 for "brain enhancement", such as the previous brainwave music, and action game where a brain decision becomes action through the biological organ, the hand, and the results feedback to the brain through ears and eyes, the repeating practice would enhance the related brain function and even change our brain [15].

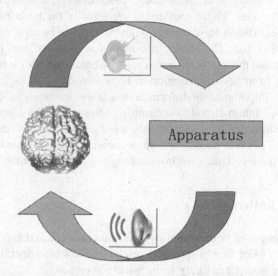

Fig. 6. Apparatus-brain conversation (ABC). Left: brain functionalized with perception/execution, making decision, memory and consciousness etc. Right: apparatus including both biological organs within normal health person and machine like computer outside the body. Conversation covers both one-way and bi-direction communication.

7 The Opportunities

Why should Brainformatics be assigned as an independent sub-discipline? Actually, "substance, energy, information" have now been recognized as the three basic elements of the world though the initiation of information science is later than that of physics, mathematics, chemistry and biology. It means that information science is now as important as the other main basic disciplines. In fact, our life and cell, including our brain and neuron, is actually a unity of "substance, energy and information". In this sense, publicizing Brainformatics is more than necessary, it would be an important routine to drive both the brain science and information science, making them help each other and fastening each other. Essentially, the journal "Brain Informatics" setup a few years ago already partly play the role to drive computational technologies related to human brain and cognitive [19].

In the wave of artificial intelligence, the brain information mechanism is the thinking spring of the artificial intelligence technology, Brainformatics will play an important role in such an age. The European Human Brain Project (HBP) initiated in 2013 took the brain simulation as the core, it actually reflects the European scientists' judgment of the future science and its economic and social values. Currently, with the development of society and economy, health, especially brain health, becomes the sign of the level of social development. It is estimated that brain disease may cost about 25% of the total medical burden in years 2020. Therefore, research on brain disease mechanism is already a timely crucial topic. In the past decades, various international enterprises have put in a lot of resources to identify the molecular target, so as to design a medicine for the therapy of brain diseases. Unfortunately, almost all of these efforts failed in the past decades. These challenges are actually the main background of the USA BRAIN project (Brain Research through Advancing Innovative Neurotechnologies) initiated in 2013, too. The "brain information acquisition" was put at the core of the BRAIN project and the aim is to search for new breakthrough for both brain disease and brain function from new information to be obtained.

In general, the importance of information science for brain study is already the consensus of the international community. Systematically pushing the new sub-discipline, Brainformatics, can not only satisfy the scientists' curiosity, but also meet the giant need of brain health, and such researches would receive much attention and support, and make a significant breakthrough in the near future.

8   Potential Challenges

What are the challenges of brainformatics - the brain research and application driven by information science? For such a question, each one may have specific opinion, however, the following issues are likely to be widely accepted.

(1) Brain aging: due to the desynchronized aging between physiology and brain function, brain aging disease is becoming a big challenge problem for the whole world. What are the determining factors of the brain aging, and are there any intervention techniques which can be adopted to stop or delay the brain aging process? The drug intervention is more emphasized in current medical system, but more and more people realized that it may not be the most effective method. How about to have some information-based technologies and methods to intervene brain aging?

(2) Major neuropsychiatric disorders enlarged by current rapid social informationization: We have to face more and more patient population with neuropsychiatric disorders and their medical burden getting heavier when we enjoy convenience brought by informationization. For these diseases caused by informationization, should we focus on using information approach to remit them?

(3) How to further improve imaging to dynamic, noninvasive, natural and multimodal integration. For example, the spatial resolution of the current functional magnetic resonance image is about 3 mm, how can we reach the desired 0.1 mm for functional column of brain in the coming ten years?

(4) The big data in brain science: Brain information covers multi-level, large-scale heterogeneous datasets ranging from gene to behavior, thus brainformatics is a typical issue of big data [20]. What can artificial intelligence technology do in this respect? What new development does artificial intelligence technology need to explore?

(5) Reverse and emulation brain: 'Reverse brain engineering' is one of the big challenges in the 21st century [21]. For brainformatics researchers, no matter 'brain-like', 'brain emulation' or 'brain-inspired' technologies are worth to develop, each of them has its own special target, and they all try to feedback the society to promote social progress and make human life better.

9 Conclusions

In conclusion, brainformatics, founded by the brain research driven by information science, on one side owes its growth to progress of information science in information-acquisition technology and computational capacity. On the other side, it also forces innovation of information science theory and method. Thus, it will facilitate the understanding of the brain information mechanism and develop engineering applications and clinical translation. Obviously, the development of the brainformatics discipline needs researchers with multiple discipline backgrounds and spirits of innovation to devote themselves to this work. In the brewing Chinese brain project, brainformatics will play an important role as the information mechanism of the brain is one of the main contents. This fact also means that Chinese scientists also attach great agreement on the importance of this crucial sub-discipline for the future.

Acknowledgements. This work is supported by was funded by the National Natural Science Foundation of China (Grant No. 81330032), the 111 project B12027.

References

1. Mitra, P.P.: The circuit architecture of whole brains at the mesoscopic scale. Neuron **83**, 1273–1283 (2014)
2. Yao, D.: Brainformatics: concept, scope and challenge. Chin. J. Biomed. Eng. **35**(2), 129–132 (2016). (In chinese)
3. Wu, T., Dufford, A.J., Egan, L.J., Mackie, M.-A., Chen, C., Yuan, C., Chen, C., Li, X., Liu, X., Hof, P.R., Fan, J.: Hick-Hyman Law is mediated by the cognitive control network in the brain. Cereb. Cortex. 1–16 (2017)
4. Kavli NDI. http://www.kavlifoundation.org/johns-hopkins-university
5. National Institutes of Health, BRAIN. http://www.nih.gov/science/brain/. Accessed 30 Dec 2013
6. Li, A., Gong, H., Zhang, B., Wang, Q., Yan, C., Wu, J., Liu, Q., Zeng, S., Luo, Q.: Micro-optical sectioning tomography to obtain a high-resolution atlas of the mouse brain. Science **330**, 1404–1408 (2010)
7. Lei, X., Yao, D.: Principles and Techniques of Simultaneous EEG-fMRI. Science Press, Beijing (2014). (in Chinese)
8. Yao, D.: A method to standardize a reference of scalp EEG recordings to a point at infinity. Physiol. Meas. **22**, 693–711 (2001)
9. Yao, D.: Is the surface potential integral of a dipole in a volume conductor always zero? A cloud over the average reference of EEG and ERP. Brain Topogr. **30**, 1–11 (2017)

10. Chella, F., Pizzella, V., Zappasodi, F., Marzetti, L.: Impact of the reference choice on scalp EEG connectivity estimation. J. Neural Eng. **13**, 036016 (2016)

11. Jiang, T.: Brainnetome: a new -ome to understand the brain and its disorders. NeuroImage **80**, 263–272 (2013)

12. Friston, K.: Active inference and free energy. Behav. Brain Sci. **36**, 212–213 (2013)

13. Wu, D., Li, C., Yao, D.: Scale-free brain quartet: artistic filtering of multi-channel brainwave music. PLoS ONE **8**, e64046 (2013)

14. Chen, M., Guo, D., Wang, T., Jing, W., Xia, Y., Xu, P., Luo, C., Valdes-Sosa, P.A., Yao, D.: Bidirectional control of absence seizures by the Basal Ganglia: a computational evidence. PLoS Comput. Biol. **10**, e1003495 (2014)

15. Gong, D., He, H., Liu, D., Ma, W., Dong, L., Luo, C., Yao, D.: Enhanced functional connectivity and increased gray matter volume of insula related to action video game playing. Sci Rep. **5**, 9763 (2015)

16. Li, F., Tian, Y., Zhang, Y., Qiu, K., Tian, C., Jing, W., Liu, T., Xia, Y., Guo, D., Yao, D., Xu, P.: The enhanced information flow from visual cortex to frontal area facilitates SSVEP response: evidence from model-driven and data-driven causality analysis. Sci Rep. **5**, 14765 (2015)

17. Lu, J., Wu, D., Yang, H., Luo, C., Li, C., Yao, D.: Scale-free brain-wave music from simultaneously EEG and fMRI recordings. PLoS ONE **7**, e49773 (2012)

18. European Working Group on brain planning, Human brain plan Roadmap. https://www.humanbrainproject.eu/roadmap. Accessed 30 Dec 2015

19. Zhong, N., Peng, H. (Editors-in-chief): Description of "Brain Informatics" http://link.springer.com/journal/40708. Accessed 30 Dec 2015

20. Yao, D., Li, J., Zhang, Y., Yuan, Q., Guo, D., Liu, T.: Neuroinformatics and coordinating facilities: concept, current status and development. China Basic Sci. **6**, 11–15 (2013). (in Chinese)

21. Grand challenges for engineering committee. Grand challenges for engineering. National Academy of Sciences (2008). (www.engineeringchallenges.org)

Special Session on Brain Informatics in Neurogenetics (BIN 2017)

The Development and Application of Biochemical Analysis of Saliva in the Assessment of the Activity of Nervous System

Chen Li[1], Pan Gu[2], Kangwei Shen[1], and Xuejun Kang[1(✉)]

[1] Key Laboratory of Child Development and Learning Science,
School of Biological Sciences and Medical Engineering,
Southeast University, Nanjing, Jiangsu, China
chenlee1203@163.com, xjkang64@163.com,
13505163516@139.com

[2] British Columbia Academy, Nanjing Foreign Language School,
Nanjing 210018, Jiangsu, China
derylGu@hotmail.com

Abstract. Biochemical molecules are substantial bases for the function of nervous system. The concentrations of specific molecules are closely linked to the activity and status of nervous system. This paper summarized our previous work on the relationship be-tween some active biochemical molecules and specific activities in nervous system. The results indicated that the levels of some biochemical substances may reflect the function and regulation of the nervous system, which derived easy, fast and cheap methods for assessing the activity and status in nervous system.

Keywords: Biochemical analysis · Nervous system · Stress · Biomarker

1 Introduction

Nervous system is a sophisticated part of human body. The electro-physiological signal is transmitted thought the synapses between neurons, these signal pathways constitute the bases for the function of nervous system. Many active molecules are playing pivotal roles in the transmission of signals, such as neurotransmitters and hormones. The involvements of active molecules in the actions of nervous system offer the measur-ability of the status and function of the nervous system. The alteration of the con-centrations of these molecules possibly reflect the activity, regulation, function or anomaly of the nervous system. In this study, we summarized the work involving the investigation of the concentrations of these special molecules in biological samples, including the development of analytical techniques and the application in different situations, as well as the significance of these studies.

Y. Zeng et al. (Eds.): BI 2017, LNAI 10654, pp. 327–333, 2017.
https://doi.org/10.1007/978-3-319-70772-3_31

2   The Development of Flow Injection Analysis of Salivary-α-Amylase and Its Application in the Evaluation of Stress and Anxiety

Salivary α-amylase (SAA) is one of the main protein secretions in saliva. The most-realized role of SAA is its enzyme function in digesting carbohydrate. Recently, many studies emphasize the role of SAA in stress reaction. The secretion of SAA is controlled by the sympathetic nervous system (SNS). When the SNS is stimulated, SAA is released from seconds to minutes and shows a dramatic increase. Therefore, SAA is proposed as a biomarker of stress response of SNS. Our group developed an analytical method for SAA based on flow injection analysis and applied this method into the determination of SAA during a running exercise task. Subjects were a group of university students who were to take a 100m dash running. Therefore, the exercise in relatively intensive which is considered as an acute stressor. Sub-jects offered three saliva samples at pre-running, post-running and 10 min post-running, respectively.

The analytical method for SAA is developed by combining flow injection sampling with UV-Vis spectrophotometry. The range of linearity of the method is 0–50 µg/mL; the correlation coefficient of the calibration curve is 0.9990; the recovery of the method is 97.8–105.9% (n-3). A single run of determination could be finished in 1 min. The method is fast, sensitive, precise and cheap.

Under dash running experiment, subjects' SAA levels increased dramatically after running, and decreased to the initial level after rest for several minutes (Fig. 1). The alteration of SAA concentration reflected the activation and recover of sympathetic nervous system [1, 2]

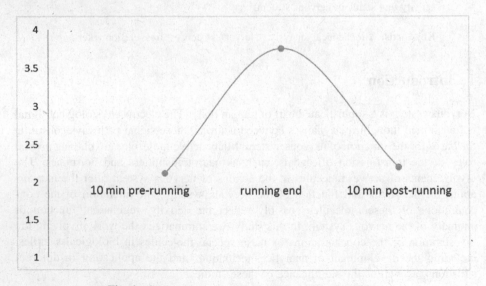

Fig. 1. SAA activity alteration during physical exercise

Apart from the assessment of the activity of SNS, SAA is associated with psychological states such as anxiety. In another study, we recruited some university students who were to take an examination. Prior to examination, they took the State-Trait Anxiety scale. They were grouped as low-anxiety group and high-anxiety group according to their scores on the scale. Their saliva samples were collected before examination, examination end and 10 min after examination. The results showed the correlation between the anxiety level and SAA activity (Fig. 2) [3, 4].

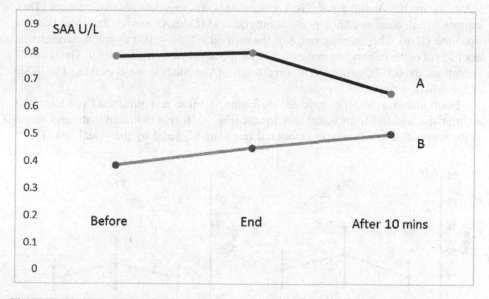

Fig. 2. SAA alteration during an examination in (A) low-anxiety group and (B) high-anxiety group

This study suggested the linkage between anxiety and SAA activity. In high-anxiety group, their SAA level was constantly growing during the whole procedure; while the low-anxiety group showed a rise-recover regulation. It seemed that the high-anxiety group had prolonged emotional reaction to stress, which was reflected by increased SAA even after the stress situation had ended.

3 The Development of HPLC Analysis of Salivary Amino Acids and Its Application in the Assessment of Emotional Reactions During Visual Task

Amino acids are molecules with a common structure that contains an amine group, a carboxylic acid group and a characteristic side-chain. They have critical functions to life with a direct involvement in metabolism by serving as the building blocks of proteins. Additionally, amino acids act as precursors for the synthesis of hormones and neurotransmitters. Some amino acids play the roles as neurotransmitters in nervous system. There-fore, the biochemical relevance of amino acids and their concentrations are of interest in the study of the activity of nervous system.

We designed a visual experiment which intended to trigger affections in subjects. Subjects were 9 university students aged 19–21 years. They were instructed to watch a three-dimensional cartoon which had strong emotion-al clips. Their saliva samples were collected before task, task end and 20 min post-task. Thirteen salivary amino acids were determined using a HPLC-UV method, i.e. i.e. glycine (Gly), arginine (Arg), ornithine (Orn), lysine (Lys), glutamine (Gln), glutamine (Glu), aspartate (Asp), threonine (Thr), taurine (Tau), alanine (Ala), γ- aminobutyric acid (GABA), serine (Ser), and hydroxyproline (Hyp). The linearity range of the method is 0.5 ∼ 100 μg/ml, the correlation coefficient of the calibration curves for 13 amino acids are 0.9934–0.9999. The limits of detection are 0.1–0.5 μg/ml. The precisions of the method were evaluated as RSD 0.3–8.8% (n = 6).

Four amino acids (i.e. glycine, threonine, alanine and ornithine) per-formed a regular rise and fall from visual task to rest (Fig. 3). It was indicated that these amino acids responded sensitively to emotional reactions triggered by the visual task. [5].

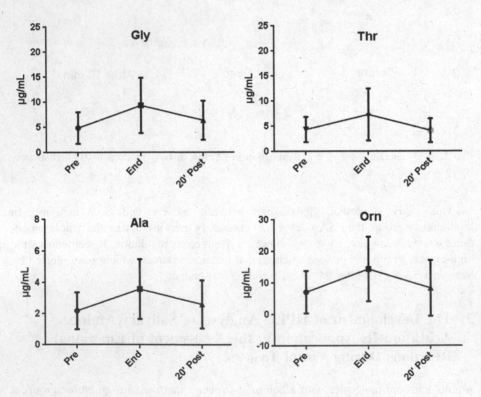

Fig. 3. Salivary amino acids alteration during visual task

During these 13 amino acids, Glu, Gln, Gly, Ala, and GABA are previously confirmed as excitatory or inhibitory neurotransmitters. However, only Gly and Ala showed significant alterations when participants were facing a visual stimulus, which indicated the selective activation of different neuro-transmitter systems. Additionally, this experiment discovered two amino acid which were potential indicator of the emotional actions.

4 The Development of Flow Injection Analysis of Salivary Histidine and Its Application in Assessing Stress

L-Histidine is an essential component of many proteins. With the active imidazole group, histidine residue offers the active site for many enzymes and directly participates in catalysis [6]. Apart from this, histidine is reported to act as a neurotransmitter or neuromodulator in central-nervous system.

Conventional analysis of histidine is carried out in HPLC which required cumbersome derivatization step. In our study, we developed a flow injection sampling tandem UV-vis spectrophotometry method in which histidine was determined in the dynamic balance of derivatization reaction in flow injection system. A single run of the analysis costed less than 4 min. The method had a wide linearity range of 5–200 μg/mL. The correlation coefficient of the calibration curve was $r2 = 0.9979$. The developed analytical method for salivary histidine was convenient, time saving and precise.

To analyze the relationship between salivary histidine and stress response, we conducted an experiment which induce a social stress to subjects. The subjects in treatment group were undergraduate students who were to give academic reports on their graduation oral examinations, while the control group did not give speeches. Salivary samples in treatment group were collected 20 min before speech, end of speech and 20 min after speech. The control group offered salivary samples on same time points. Then their salivary histidine concentrations were determined using the above mentioned method [7].

As it was shown it Fig. 4, participants who gave the academic reports showed significant fluctuations of salivary histidine. By contrast, the control group who had no speech tasks showed stable salivary histidine secretion. The results indicated the potential indicator role of salivary histidine for assessing social stress. The academic report task was a typical social stressor which involved cognitive function and executive function. This study sup-plied a novel approach for the assessment of the activity of these functions during social stress.

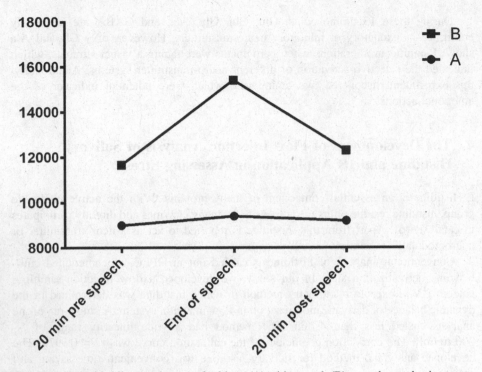

Fig. 4. Salivary histidine (peak area) of subjects (A) without task (B) gave the academic reports

5   Discussion and Conclusion

Biochemical molecules have important roles in the function of nervous system. Their roles supply the possibility of assessing the function and status of nervous system through their concentration and alteration. Our studies indicated that some biochemical substances in human saliva are useful to reflect some activities of nervous system, including stress response, anxiety, cognitive and executive function. This method is easy to handle and fast to obtain data, which is especially applicable in the high-throughput data analysis on the study on the nervous system. Additionally, the samples used in our work are non-invasive saliva, which are easier to collect and friendlier to participants.

Acknowledgement. This work was supported by the National Natural Found Committee of China (No. 81673230), the Social Development Research Program of Jiangsu Province Science and Technology department (No. BE2016741), the Open Project Program of Key Laboratory of Child Development and Learning Science of Ministry of Education, Southeast University (CDLS-2016-04), and the Fundamental Research Funds for Research and Innovation Projects to Postgraduate students in Central Universities and Colleges in Jiangsu Province (KYZZ_0070).

References

1. Chen, L.Q., Kang, X.J., Zhou, X.L., Gu, Z.Z., Lu, Z.H.: Flow injection spec-trophotometry determination of salivary alpha-amylase for stress evaluation. In: International Conference on Bioinformatics and Biomedical Engineering, Wuhan, China, pp. 1-3 (2011)
2. Chen, L.Q., Kang, X., Zhou, X., Wang, Y., Ma, L.: The establishment of flow injection analysis method for saliva α-amylase and its application in the psychological stress. J. Inf. Technol. 1(3), 119–123 (2012)
3. Gao, X., Zhou, R.L.: Inhibition function in selective attention of high trait anxiety undergraduates. Chin. J. Clin. Psycholo. 20(3), 288–291 (2012)
4. Huang, J.J., Chen, L.Q., Wang, Y., Shen, K.W., Li, C.: The development of on-line monitoring instrument for the determination of salivary α-amylase. Symposium on field test instrument and front technology, Beijing, China (2014)
5. Tang, W., Li, X., Wu, X.X., Wang, Y., Kang, X.J.: Salivary amino acids determi-nation and their changes in vision stress experiments. Open J. Soc. Sci. 01(6), 23–25 (2013)
6. Littauer, U.Z., Grunberg-Manago, M., Creighton, T.E.: The Encyclopedia of Molecular Biology, pp. 1911–1918. John Willy and Sons Inc., New York (1999)
7. Sun, J., Zheng, S.L., Wang, Y., Kang, X.J., Gu, Z.Z., Lu, Z.H.: The prelimi-nary investigation of salivary l-histidine under computer vision stress. In: The 7th International Conference on Cognitive Science, Beijing, China, 17-20 August 2010

Author Index